HS423   Hist/Cult St.   New ~~£3~~
£15.50

Engineered to S...

Engineered to Sell

European Emigrés and the Making of Consumer Capitalism

JAN L. LOGEMANN

The University of Chicago Press   Chicago and London

The University of Chicago Press, Chicago 60637
The University of Chicago Press, Ltd., London
© 2019 by The University of Chicago
All rights reserved. No part of this book may be used or reproduced in any manner whatsoever without written permission, except in the case of brief quotations in critical articles and reviews. For more information, contact the University of Chicago Press, 1427 E. 60th St., Chicago, IL 60637.
Published 2019
Printed in the United States of America

28 27 26 25 24 23 22 21 20 19   1 2 3 4 5

ISBN-13: 978-0-226-66001-1 (cloth)
ISBN-13: 978-0-226-66015-8 (paper)
ISBN-13: 978-0-226-66029-5 (e-book)
DOI: https://doi.org/10.7208/chicago/9780226660295.001.0001

Library of Congress Cataloging-in-Publication Data

Names: Logemann, Jan L., author.
Title: Engineered to sell : European emigrés and the making of consumer capitalism / Jan L. Logemann.
Description: Chicago ; London : The University of Chicago Press, 2019. | Includes bibliographical references and index.
Identifiers: LCCN 2019012198 | ISBN 9780226660011 (cloth : alk. paper) | ISBN 9780226660158 (pbk. : alk. paper) | ISBN 9780226660295 (e-book)
Subjects: LCSH: Marketing—United States—History. | Consumers—United States. | Immigrants—United States.
Classification: LCC HF5415.1.L64 2019 | DDC 381.089/09073—dc23
LC record available at https://lccn.loc.gov/2019012198

♾ This paper meets the requirements of ANSI/NISO Z39.48-1992 (Permanence of Paper).

Contents

Introduction: Consumer Engineers and the Transnational   1
Origins of Consumer Capitalism

Consumer Engineers as New Marketing Experts 7
Transatlantic Transfers and Transnational Dimensions
of Consumer Capitalism 13
Midcentury Marketing as Social Engineering 18

1   The Origins of "Consumer Engineering":   20
Interwar Consumer Capitalism in Transatlantic Perspective

The Emergence of Mass Marketing in the United States 22
American Perceptions of European Consumer Modernity 25
The Reciprocity of Transatlantic Consumer Transfers 29
Social Engineering between European Reform Movements and
1930s America 33

SECTION ONE   Transformations in Marketing and
Consumer Research

The Rise of Consumer Engineering: 41
American Marketing at Midcentury (1930s–1960s)

2   The Art of Asking Why: The "Vienna School"   44
of Market Research and Transfers in Consumer Psychology

Toward a Professionalization of Marketing Research in the
United States 47
Interwar Vienna and the Study of Modern Consumer Markets 50
Paul Lazarsfeld's Transatlantic Career in Market Research 55
The BASR and the "Vienna School" in Postwar American
Marketing Research 63
Social Scientists as Consumer Engineers 70

3   From Mass Persuasion to Engineered Consent: 73
    The Impact of "European" Psychology on the Cognitive
    Turn in Marketing Thought

    New Approaches to Survey Psychology and Consumer Motivations 75
    Wartime Research and New Perspectives on Mass Communication 78
    Kurt Lewin and the Impact of Experimental Psychology 83
    George Katona and the Advent of Behavioral Economics 89
    *Consumer Psychology and Social Engineering in Wartime
    and Cold War 95*

4   Hidden Persuaders? Market Researchers as "Knowledge   99
    Entrepreneurs" between Business and the Social Sciences

    *The Expansion of Market Research in American Industry,
    1930s–1950s 102*
    *The Drive for "Scientific" Marketing Research: Alfred Politz
    Research Inc. 108*
    Ernest Dichter's Institute for "Motivation Research" 114
    *Image and Brand: Market Research as Creative Consumer
    Engineering 121*
    Consumer Engineering and the Limits of Hidden Persuasion 126

    SECTION TWO   Designing for Sustained Demand

    "Tastemakers" or "Wastemakers"? Commercial Design at   131
    Midcentury (1930–1960)

5   The Designer as Marketing Expert: European Immigrants   133
    and the Professionalization of Industrial and Graphic Design
    in the United States

    Industrial Designers as Consumer Engineers 134
    European Immigrants and American Commercial Design 141
    Raymond Loewy, French-Born Star of "American" Industrial Design 145
    A "New Type of Artist" in Graphic and Advertising Arts 151
    *"Good Design" and the Aestheticization of American Consumer
    Capitalism 158*
    New Experts for America's Midcentury World of Goods 161

6   The Commercialization of Social Engineering?   163
    Adapting Radical Design Reform to American Mass Marketing

    *Ferdinand Kramer: From Standardizing Working Class Homes to
    Marketing Novelties 165*

Radical Modernism and Commercial Applications of Social Engineering 172
The American Bauhaus: Between Experiment in Totality and Design for Industry 176
Moholy-Nagy's Struggles with Corporate America 180
Business Ties of the Institute of Design 183
The American Legacy of European Design Reform 187

7   "Streamlining Everything": Design, Market Research, and the Postwar "American" World of Goods   193

Consumer Research at Raymond Loewy Associates 196
The Psychology of Packaging in the Supermarket Era: Walter Landor Associates 205
Brand Images and Corporate Identities 213

SECTION THREE   Transatlantic Return Voyages

Bridging Transatlantic Divides: Bringing Consumer Modernity "Back" to Europe   219

8   Corporate America and the International Style: The Transnational Network of Knoll Associates between Europe and the United States   222

Knoll Associates in the United States 224
The Use of Emigré Networks 229
Marketing Interior Design as Corporate PR 232
Exporting "American" Design as "International" Style 237

9   The "Return" to Europe: Emigrés as Cultural Translators and the Transformation of Postwar European Marketing   243

(R)emigrés as Transatlantic Mediators 246
Consumer Research in Postwar Europe 249
Ernest Dichter as Transatlantic Mediator 254
Commercial Design as a Transatlantic Transfer 260
"Good Design" as Cold War Cultural Policy 266

Consumer Engineering: Challenges and Legacies   271

Acknowledgments 281
Abbreviations for Archival Sources 285
Notes 287
Index 359

INTRODUCTION

Consumer Engineers and the Transnational Origins of Consumer Capitalism

During the second half of the twentieth century, few things appeared more quintessentially American than the suburban shopping mall. By the 1950s, enclosed shopping centers symbolized an affluent postwar society in which suburban homeownership went hand in hand with access to a dynamically expanding world of consumer goods. One of the premier architects behind these midcentury temples of consumption was Victor Gruen, whose design of several early centers such as Detroit's Northland Center and the Southdale Center earned him the nickname "Father of the Mall" among historians of retail architecture.[1] His shopping malls were a central feature of what historian Lizabeth Cohen has called America's postwar "consumers' republic," which promised economic growth and broad democratic access to material abundance.[2] This democratic promise was in many ways an illusion, to be sure, as spaces such as shopping malls catered primarily to a white, suburban middle class. In part, this was the result of increasingly segmented and targeted marketing approaches among American companies, retailers, and advertisers. Still, Gruen's vision for his new "shopping towns" was also inclusive and combined commercial and civic functions, shops and community features. Gruen aimed to design total shopping environments that corresponded to the elevated place that material consumption had attained in postwar America.[3]

Going beyond the mere introduction of new mass production and mass distribution technologies, midcentury consumer capitalism made commercial consumption an ever more encompassing experience for individuals and communities alike. This entailed the creation of a novel and expanding world of goods and of consumption spaces devised for specific consumers and market segments, a process that drew heavily on new insights from consumer research and commercial design. Gruen's shopping spaces speak to the way in which midcentury marketers had learned to create tailor-made shopping environments and to design consumer goods that appealed to psychological needs of consumers and targeted defined demographics. With the entire middle-class family—and especially women and children—in mind, Gruen argued that shoppers needed to be "surrounded by pleasurable experiences." Open spaces, artistic elements, and a coordinated graphic and visual design to accompany the merchandise were all prerequisites for the successful design of shopping centers. They would benefit shoppers, merchants, and communities alike.[4] While such attention to the aesthetics and psychology of consumption was not an entirely new feature of consumer capitalism, it became much more comprehensive, systematic, and dynamic at midcentury. Furthermore, consumer capitalism increasingly relied on a host of new experts such as architects, designers, and consumer psychologists to "engineer" and sell a new comprehensive system of mass consumption.

Victor Gruen, however, was an unlikely engineer of American consumer capitalism. He was not a native to the American cities he helped transform, nor had he been socialized in the country's consumer culture. Gruen was an immigrant from Austria, a Jewish refugee from the rise of National Socialism in Central Europe. Born Victor Grünbaum in Vienna in 1903, he was trained as an architect and briefly worked for famed industrial designer Peter Behrens before launching a career in modernist retail design for high-end Viennese stores. In interwar Vienna, Gruen had been enmeshed in a vibrant metropolitan culture of avant-garde art, he had sympathized with the Austrian Socialist movement, and for a while he was active in political theater and cabaret. After immigrating to the United States in 1938, Gruen built upon this European background even as he launched a new career within the context of American consumer capitalism. He worked for a New York industrial design studio, for example, helping prepare General Motors' Futurama display at the 1939 New York World's Fair. He also designed shop interiors for upscale retailers on Fifth Avenue before relocating to California where he started a small, independent architectural firm in 1941. As he began to draft plans for suburban shopping centers, however, Gruen still had the ideal of traditional European inner cities in mind; he envisioned them as community centers for a new suburban world.[5] His Old World vision became adapted to but also helped to shape a new type of engineered mass consumption in the United States.

The making of consumer capitalism, Gruen's story suggests, was informed by transnational transfers and transatlantic exchanges in ways that have been largely overlooked by historians. Instead, familiar narratives assert the expansive global reach of America's "irresistible empire" of consumer goods in the middle of the twentieth century and focus on the notion of "Americanization" abroad to which innovations in retailing and marketing were central.[6] Indeed, Victor Gruen did return to Europe frequently after the war in his professional capacity as architect and urban planner to advise European cities on such "American" innovations as prefabricated homes and retail centers. By the 1960s, however, he cautioned European cities not to simply imitate the shopping centers he had helped to pioneer in the United States, but rather to preserve the hearts of their cities by creating pedestrian shopping streets in downtown districts.[7] In this, Gruen's career points to the role returning European emigrés played in adapting and translating rather than transplanting an American consumer culture that they themselves had significantly influenced to postwar Europe. Postwar mass consumption evolved differently in Europe and the United States, as I have shown elsewhere.[8] Here, I argue that transatlantic exchanges in consumer marketing remained multidirectional even as U.S.-style mass consumption appeared to be the dominant global model. Indeed, European emigré experts significantly shaped the transformation of consumer capitalism at midcentury on both sides of the Atlantic.

Victor Gruen was just one of many outside experts in the expanding field of American marketing. Much like Gruen, a surprising number of these new consumer experts were European immigrants and emigré refugees. They excelled in areas such as market research and advertising psychology as well as in industrial and graphic design. The group included leading consumer researchers such as Hungarian-born George Katona and the Viennese sociologist Paul Lazarsfeld. The work of emigré psychologists including Ernest Dichter informed the study of consumer motivations in advertising and selling. Commercial artists such as Herbert Bayer from Berlin and, most prominently, French American Raymond Loewy pioneered new trends in graphic and industrial design during the 1930s and 1940s. Sought after by U.S. institutions and corporations, their methodological innovations and designs helped to promote the seeming ubiquity of the American "way of life" in the postwar decades. Their transatlantic careers and those of many other immigrants and emigrés in consumer marketing coming to the United States between 1919 and 1939 from Germany, Austria, France, Italy, or Scandinavia will provide the focus of this book's actor-centered approach to consumer history.

These consumer experts belonged to a larger cohort of well over a hundred immigrant experts in consumer design and marketing with influential careers peaking around the middle of the twentieth century. While they were a heterogeneous group, their transnational lives still shared many similarities with the experiences of Victor Gruen. They were born and socialized in Europe and received their first professional training in the metropolitan centers of the continent. Often critical of interwar capitalism, many were affiliated with or at least inspired by interwar European social reform movements. Examples include artists from Germany's famed Bauhaus school who, in exile, would consult for American consumer goods companies. Social scientists affiliated with the left-leaning Frankfurt Institute for Social Research similarly became part of radio research studies financed by commercial broadcasters in the United States. While most of the immigrants had already been engaged in commercial art and research work in Europe, moving to the United States brought them more fully into contact with the corporate world as they began to work for advertising agencies or as independent consultants. Like Gruen, more than a few of them also found themselves involved with the 1939 World's Fair, which would become an important prism for visions of midcentury consumer modernity in the United States. In different capacities, these experts helped to engineer the midcentury consumer's republic by developing new marketing tools that fundamentally informed the dynamic expansion of American consumer capitalism. As corporate and government advisors, furthermore, they acted as transatlantic mediators after the war, facilitating postwar transfers of marketing knowledge and commercial practices back to Europe.

Immigrants, of course, had long contributed to the emergence of American consumer society in manifold ways. They constituted important groups of consumers with distinct needs and tastes who helped shape the diverse American domestic market.[9] As immigrant entrepreneurs, they both catered to ethnic niche markets and introduced novel goods into the national mass market. We also find immigrants in the role of experts influencing the consumer economy, as economists and artists, as traders and intellectuals.[10] Besides architects such as Gruen, for example, one could point to the management consultant Peter Drucker, who helped promote the marketing orientation of companies, or to the numerous emigrés in Hollywood who helped transmit American commercial culture to domestic and international audiences starting in the 1920s.[11] This book, however, will focus on a narrower group of immigrant experts in consumer research and commercial design who became key protagonists of midcentury "consumer engineering." A new concept of expert-driven marketing, consumer engineering sought to systematically create a dynamic and ever-expanding world of goods. It was not simply a scheme to increase sales to individual companies but promised "prosperity" for society at large, defined in terms of a widespread wealth in new consumer goods. As such, consumer engineering became a driving force behind midcentury transformations in consumer capitalism. Requiring new forms of scientific and aesthetic knowledge, this transformation offered immigrant experts a path into prominent positions in U.S. marketing.

In analyzing the transatlantic careers of consumer engineers, this study pursues two broad and interrelated questions. First, it seeks to explain the rise of the dynamic world of goods that characterized midcentury consumer capitalism in the United States. Historians have long emphasized new modes of production and a Fordist economy of serialized mass production, which first came to full fruition during the early decades of the twentieth century.[12] By the middle of the twentieth century, this mass production economy aligned both with a Keynesian policy consensus around consumption-driven growth and with a consumer culture that accentuated the social importance and cultural symbolism of commercially produced goods.[13] The equally important marketing side of this phenomenon, however, remains less explored, aside from a substantial historiography on the development of advertising. To answer how consumer goods producers and retailers adapted their approach to consumers to match consumer capitalism's new emphasis on consumption-driven growth, I will pay special attention to the emergence of market research, consumer psychology, and commercial design as central aspects of modern marketing.

Marketing, striving to combine scientific predictability with innovative creativity, transformed consumer capitalism in ways that went well beyond the stratagems of advertising. Companies increasingly attempted to predict market developments, to understand consumer motivations, and to design new products and brands to satisfy consumer needs. Beginning in the 1920s, marketers debated ways to systematically shape those needs, stimulate new desires, and engineer environments that channeled consumer behavior in calculable ways. The careers of the European-born consumer engineers elucidate the professionalization of market research, consumer psychology, and commercial design at a time when the marketing field in general became more organized. Significantly, competing attempts to define the marketing profession as either "scientific" or "creative" in the decades between the 1930s and the 1960s opened up opportunities for very different types of outsiders. Imbued with "outsider knowledge" as Europeans with backgrounds in arts and academia, the emigré consumer engineers provided a crucial innovative impulse to American consumer capitalism.

The second leading question therefore is how and why consumer engineering was shaped by transatlantic exchanges. In its emphasis on the systematic study of markets, the exploration of the consumer's psyche, and the continuous novelty of design forms, consumer engineering appeared as quintessentially American to many contemporary observers. Yet these marketing innovations in the United States actually drew on methodological concepts and an aesthetic symbolism that developed in a transatlantic setting. Thus, the history of consumer goods marketing is not a story of American exceptionalism. Instead, the careers of immigrants point to the limits of the "Americanization" paradigm. Their stories illustrate the importance of transnational exchange processes particularly among such metropolitan nodal points as Vienna and New York, Paris and Chicago, and Berlin and San Francisco. The prevalent view that corporations alone drove marketing innovations similarly needs to be reassessed. The simultaneous crossing of national and disciplinary boundaries—between arts and academia as well as between governments, corporate actors and social reform movements—places consumer engineering within wider exchanges that constitute technocratic social engineering. Here, the transnational perspective helps us to ask about the place of marketing developments, which in interwar Europe were frequently tied to public actors, within a broader history of expert attempts to shape individual and social behavior in realms ranging from urban development and social policy to economic planning.

The midcentury timeframe coincides with the high point of technocratic "high modernity" in the Atlantic world.[14] Despite the widespread experience of economic adversity, the era was characterized by tremendous optimism regarding the ability of technology and of the social sciences to shape the world. The large wave of professionals, academics, and intellectuals fleeing Europe during the 1930s arrived in the United States at a time when economic hardship and New Deal policies made the country particularly receptive to new impulses from abroad. Bookended by the Great Depression and the global crisis of the early

1970s, the careers of the immigrants and emigrés thus unfolded during an era shaped by technocratic and political responses to the challenges of economic crisis, of war and Cold War. World War II in particular did not simply represent a break in the development of American consumer capitalism but rather acted as a great catalyst for consumer engineering efforts by facilitating the interplay of academics, government institutions, and private corporations. The 1930s and 1940s are now being recognized as the decades with the most dramatic productivity burst in the twentieth century, and they also produced marketing innovations that underpin the subsequent decades of affluence.[15]

The midcentury careers of our protagonists, finally, span a period that historians occasionally discuss as the "American Century" (in reference to Henry Luce's famous phrase) to emphasize the unprecedented and global influence of American political, corporate, and academic elites and their ideas and norms at the time.[16] Most of the European immigrants and emigrés discussed here began their professional activity during the 1920s, and they ended their careers over the course of the 1960s, now often as part of "American" business or academic elites. Tracing the transnational lives of consumer engineers across these decades reveals the internal dynamics shaping postwar consumer capitalism while simultaneously probing the larger social and political ramifications of marketing. The fact that these Europeans became part of the "American Century" should not obscure their European origins but rather help us reframe that era as a "Transatlantic Century" of vibrant transnational exchanges.[17]

Consumer Engineers as New Marketing Experts

Concerned about the continuing dynamism of American capitalism and its growing corporate bureaucracies, emigré economist Joseph Schumpeter observed in 1947: "In the large-scale corporation of today, the question that is never quite absent arises with a vengeance, namely, who should be considered as the entrepreneur."[18] Entrepreneurial innovation was a core characteristic of capitalist development for Schumpeter and essential for continued success. Such innovation could take many forms: new products and technological production processes, "new combinations" in corporate organization, and the opening of new markets. Entrepreneurial innovation for Schumpeter did not have to be a path-breaking new engine or Henry Ford's assembly line. Instead, he wrote, "it can be the Deerfoot sausage. To see the phenomenon even in the humblest levels of the business world is quite essential."[19] Marketing innovations were very much part of Schumpeter's concept of entrepreneurial innovation. "Capitalist achievement," he believed, "does not typically consist of providing more silk stockings for queens but in bringing them within the reach of factory girls in return for steadily decreasing amounts of efforts . . . The capitalist process not by coincidence but by virtue of its mechanisms, progressively raises the standard of living of the masses."[20] Schumpeter knew that by midcentury, American industry had achieved tremendous advances in systematic mass production in part through scientific management and economies of scale. Yet the slump of the Depression had demonstrated the vulnerabilities of the age of mass production and its dependence on mass demand enough for theorists such as Schumpeter to worry whether the large, corporate bureaucracies of the Fordist era were able to mount a creative response to the challenge of economic crises and to an increasingly competitive marketplace. Consumer capitalism needed innovations that went beyond efficiencies in production.

We can read Schumpeter's concerns as part of a larger midcentury debate on business responses to crisis. Already in 1932, the advertising professionals Roy Sheldon and Egmont Arens had suggested in a book called *Consumer Engineering* that consumer engineering, a new approach to systematic marketing, would be crucial for companies to respond to the challenges of crisis and competition.[21] Writing at the height of the Great Depression, the authors defined "consumer engineering" as "shaping a product to fit more exactly consumers' needs or tastes . . . in its widest sense it includes any plan to stimulate the consumption of goods."[22] Marketing would become the new driving force of American industry, they predicted, by adopting scientific standards to make markets more predictable while allowing companies to be innovative and to forge ahead of the competition. This meant that companies would embrace "obsolescence"—the process by which a product or an idea becomes outmoded or rapidly obsolete—as "a positive force."[23] Obsolescence, they believed, was becoming a defining feature of U.S. commercial culture and of "the modern American tempo." Much like Schumpeter's process of creative destruction through entrepreneurial innovation, obsolescence was a dynamic force that opened opportunity to the skillful marketer:

[O]bsolescence to the producer presents a threat and an opportunity. If he ignores its changes either within his factory or in the outside world of the consumers, he himself will be cast into the discard. If he keeps up with changing methods and markets his returns will be normal. And if he is able to outguess and outknow, to think ahead of the moment and keep his advance on obsolescence, both fashion and change will be his quickest servant and his harvest will be a rich one.[24]

Sheldon and Arens argued that marketing innovations were key to the perpetual prosperity of both companies and the economy as a whole.

Consumer Engineering thus contained a dual message for companies engaged in the competitive midcentury marketplace: be informed and be creative. Companies, the book argued, needed marketing professionals to stay abreast of market developments and "outknow" their consumers and "outguess" their changing tastes. Market research and new advances in consumer psychology (what the authors referred to as "humaneering") would be the primary tools to improve business knowledge about dynamic markets and consumer motivations. At the same time, companies needed to creatively adapt this knowledge by producing new goods that took advantage of new fashions, appealed to consumer desires, and even defined new trends and brought about new needs. Here, Sheldon and Arens identified graphic and industrial design and the proper aesthetic "styling" of goods as an important strategy for modern consumer marketing.

Both the collection and analysis of market knowledge and the production of novel, aesthetically exciting goods required the help of new experts as outside consultants or as members of new departments within the company. These new "consumer engineers," Sheldon and Arens noted, had "the double job . . . to fit the product and the promotion to the existing market [and] to create new needs and stimulate consumption by every possible means."[25] When I write of "consumer engineers" in the following chapters I will generally refer to professionals engaged in systematic attempts to understand and shape consumer behavior. More specifically, I will look at a new group of experts in market research, consumer psychology, and commercial design who became part of or affiliated with the emerging profession of marketing. As employees in newly created marketing departments and as independent commercial, academic, and artistic consultants, they proposed to fulfill the role of entrepreneur in the large-scale corporations of midcentury America.

Histories of marketing and advertising traditionally stress the decades around 1900 as the formative period or look at the "creative revolutions" advertising more individualistic lifestyles following the 1960s.[26] This book will make the case that there is a missing link in the history of marketing in the middle decades of the twentieth century, connecting the presumed era of the "mass market" and the later reign of "segmented marketing." This middle period consists of two parallel trends. First, from the late 1920s to the early 1960s we see the professionalization of marketing. This involved the growth of a formal body of academic marketing knowledge, which refined the implicit knowledge and intuitive research practices of modern marketing's early phase.[27] A new cohort of consumer engineers touted the systematic potential of marketing as a "social technique" by means of which they could shape and control mass markets.[28] With new broadcast media at their disposal, they anticipated the rise of a new age of mass persuasion in which professionally made commercial propaganda based on the latest scientific insights could influence the behavior of the consuming masses.[29]

At the same time, a second, more innovative trend emerged: midcentury marketers became increasingly aware of the dynamism and diversity of consumer markets and of the problems inherent in a reliance on mass-market strategies for large corporations. Much as Sheldon and Arens had proposed, motivation researchers like Ernest Dichter and industrial designers like Walter Landor began to emphasize targeted aesthetic appeals and creative approaches to segmented marketing, which paid attention to the transformative dynamics and growing diversity within the consumer marketplace. Marketers began to recognize the agency of consumers in defining the symbolic meanings of goods and consumption practices and they learned to identify opinion and taste leaders within consumer communication. The psychologically informed brand image strategies and the creative revolutions in consumer goods marketing of the 1960s and 1970s were a result of midcentury consumer engineering efforts to anticipate changing consumer needs.

Midcentury marketing was marked by the growth of a veritable "knowledge industry" of consultants that went far beyond the usual focus on advertising agencies. Emigrés such as Raymond Loewy and Alfred Politz counted among the pioneers of a growing field of design and research consultancies.[30] In general, consulting firms had become a central feature of the American economy during the middle decades of the twentieth century.[31] Following World War II, management consultancies became influential actors in the transatlantic transmission of "American" management concepts and practices.[32] Along with advertising agencies, these consulting firms were especially important in promoting new marketing approaches. Producing more than mere billboards and advertising copy, large agencies began to consult along a full line of marketing services from public relations and early customer research to packaging design during the interwar years. In addition, specialized consulting firms emerged in consumer research and in commercial design. Professional designers who presented themselves as "creative" marketing experts sometimes even saw themselves as competitors with specialized market researchers. The two groups bore many resemblances to each other during the midcentury decades. Both designers and market researchers positioned themselves as experts who could help companies "make" new markets for innovative consumer products and bring consumers to recognize needs and desires they might not have been aware of themselves. Thus, they fulfilled a function that Andrew Godley and Mark Casson have recently discussed as "diagnostic entrepreneurship" to underscore the growing importance of experts in market making.[33] Whether as independent consultants or as specialists within new corporate departments, design experts and market research experts claimed to bring innovative knowledge into corporate processes.

Business historians have recently begun to pay greater attention to the role of design in modern capitalism, focusing, for example, on the fashion industry and related sectors of the economy where the conscious marketing of fashion cycles became increasingly important in the first half of the twentieth century.[34] In addition to fashion industries, scholars have stressed the importance of implicit, artisanal knowledge of design and changing tastes for many sectors of the consumer goods industry since the nineteenth century.[35] By midcentury, however, design knowledge became more explicit as designers strove toward professional organization. Design historian Jeffrey Meikle has chronicled the rise of prominent industrial design studios from the later 1920s and early 1930s, which began to serve as consultants to the largest American corporations.[36] By midcentury, the "styling" of goods had become a vital aspect of the automobile industry, and well-known professional designers such as Loewy or Herbert Bayer began to influence the product and image strategies of many Fortune 500 companies.[37] Advertising agencies and advertising departments similarly saw an influx of formally trained graphic artists at a time when aesthetic concerns and a discourse about the need for a "modern" look influenced a consumer goods industry seeking to recover from the Great Depression.[38] From streamlined automobiles and refrigerators to iconic, recognizable logos, this midcentury interplay of commerce and culture opened opportunities for well-known avant-garde artists.[39] Historian Regina Blaszczyk has cautioned against overestimating the influence of elite designers and of the prominent "tastemakers" of the era on the development of consumer goods and popular taste, and indeed the colorful and overly exuberant "populuxe" style of many postwar consumer goods clashed with the strict functionalism of the era's leading modernists, including the Bauhaus emigrés.[40] Nevertheless, I argue, these "tastemakers" were part of a transatlantic elite of designers who played a significant role as pioneers and as teachers in shaping the aesthetic of midcentury consumer capitalism. Studying their work from the perspective of their commercial consulting activities and their business practices reveals the growing interrelationship of design and marketing in midcentury consumer capitalism. Design consultancies furthermore set out to turn aesthetic change into meticulously planned marketing strategy.

Like the history of product design, the story of market research has become a major theme of business historians only in recent years. Sporadic efforts by producers, retailers, and advertisers to trace sales and survey customers in order to improve sales and distribution date back well into the nineteenth century.[41] By the 1960s and early 1970s, however, market research had become an integral part of marketing management, with a refined set of qualitative and quantitative methods to allow for a fine-grained segmentation of consumer markets and a sophisticated exploration of consumer motivations and attitudes. The growth of market research and its increasing importance to corporate strategy making, historian Ingo Köhler has argued, were in part a result of crises and of increased competition in consumer markets. Especially at times when companies faced uncertainty and buyers' markets, market research promised the comfort of increased predictability and the potential for new customer segments.[42] This was true for the midcentury decades in the United States—and with some time lag in Western Europe—as market research underwent a substantial professionalization process.[43] Marketing scholars have traced the roots of particular market research methods such as motivation research, and they have highlighted the work of a few prominent individuals such as Ernest Dichter and Paul Lazarsfeld.[44] Historians have noted the growing influence of psychological thought on advertising research as well as the important interactions between early consumer studies and emerging empirical social science research.[45] This book will weave these stories together to make a broader argument about a search for predictability within American corporations. This quest led not only to the professionalization of market research, but also to a transnational exchange between European emigrés in psychology, sociology, and other behavioral sciences and consumer goods corporations in the United States.

Consumer engineering and the rise of commercial design and market research between the 1930s and the 1950s illustrate what sociologist Andreas Reckwitz has described as the "paradox of innovation" in mid-twentieth-century capitalism.[46] On the one hand, Reckwitz argues, this stage of economic development was characterized by a Weberian emphasis on formal rules and predictable standards. The importance of bureaucratic and technical rationality was most clearly exemplified by the rise of scientific management as an ideal and by the model of the large, Fordist corporation that succeeded through standardized production and economies of scale. The rise of market research easily fits such a Weberian paradigm of an administrative business organization. On the other hand, Reckwitz continues, these economies were also marked by a countervailing, Schumpeterian impulse that emphasized the "creative destruction" inherent in modern capitalism. New technologies, dynamic markets, and the logic of continuous capital accumulation forced companies into constant change and innovation. This led not only to the establishment of R&D units for technological innovation, but to design units specialized in aesthetic innovation as well. In Reckwitz's reading, this conflict between the need for standardization and predictability and the need for continuous change gave rise to organizational attempts to make innovation permanent and opened the door for aestheticization of economic processes. By the 1960s and 1970s, as Fordist economies of scale increasingly gave way to segmented "economies of speed," Reckwitz argues, a series of "creative revolutions" in advertising and in other fields infused a dynamic of innovative creativity into marketing in particular and into the economy in general.

The careers of consumer engineers in design and market research provide an empirical window into this paradox of innovation at midcentury. Beginning in the interwar years, this paradox applied in particular to those consumer goods industries in which standardized mass production had most thoroughly taken hold. Collecting information and creating designs in response to the challenge of obsolescence, new marketing professionals strove for ways to bring permanent innovation to this sector of the economy. Designers advertised themselves as systematic engineers of creativity who created appealing goods and brand ensembles to fit changing consumer tastes. Consumer researchers, in the meanwhile, claimed to offer "creative research" and a more "scientific" understanding of "the consumer" that would allow for prediction and planning as well as for entrepreneurial innovation. Here, the emigré consumer experts laid the foundation for marketing transformations, which drew heavily on transatlantic knowledge exchanges.

Transatlantic Transfers and Transnational Dimensions of Consumer Capitalism

Marketing appeared as a peculiarly "American" phenomenon to mid-century observers both in the United States and in Europe. Especially the postwar years witnessed the construction of a "modern" West under the leadership of U.S. experts and institutions. European societies frequently perceived postwar reconstruction as an "Americanization" of their politics, culture, and economies. The Marshall Plan, productivity missions, as well as American products and popular culture all added up to what appeared to be a massive flow of transatlantic transfers from the United States to Western Europe, expanding the reach of America's "irresistible empire."[47]

Historians, however, have recently qualified the "Americanization" argument with regard to marketing and consumption in important ways. First, they have stressed the importance of knowledge adaptations in marketing transfers. While Europeans looked to the United States, marketing practices were not imported in any straightforward manner but instead selectively adapted to fit local needs.[48] Second, Europe had long developed a sophisticated consumer economy of its own, and patterns of consumption, distribution, and marketing did not converge toward an American model, but transatlantic differences in areas such as credit and retailing remained palpable throughout the interwar and postwar decades.[49] Third, studies in transnational history have underscored the reciprocity of transatlantic exchanges especially for the early part of the twentieth century when American observers paid close attention to European social and economic developments.[50] Even at the height of American power, the Atlantic was not a one-way street, but rather transnational influences continued to shape U.S. society as well.[51] Theorists of globalization have thus abandoned the notion of cultural homogenization through consumption inherent in notions such as "Americanization" (or even "McDonaldization"). They have instead championed concepts of cultural hybridization, stressing the emergence of a mélange of interrelated and locally specific consumption practices to which migration and transnational knowledge transfers contributed significantly.[52]

Exploring the transatlantic careers of emigré consumer engineers offers a transnational reading of American history and contributes to its globalization. A few marketing professionals have already acknowledged the transatlantic mélange out of which fields such as consumer research and commercial designs emerged.[53] Earlier studies have treated the impact of a small number of prominent individuals, and they have discussed the influence of emigrés on select aspects of advertising and commercial design.[54] This book breaks new ground by providing the first comprehensive account of a broad array of transatlantic careers and transnational exchange in marketing. To this end, it draws on a rich and growing literature on elite migration, cultural translation, and emigré intellectual transfers.

With some exceptions, the majority of the immigrant consumer engineers featured in this book were part of the forced migration of Europeans to the United States following the rise of fascism and Nazism, because of either their political leanings, such as the designer Ferdinand Kramer, or their Jewish descent like the market researcher Ernest Dichter and many others.[55] The tightening of U.S. immigration restrictions following World War I provided severe administrative obstacles to most European refugees except for a relatively small number of professionals, academics, and intellectuals who frequently benefited from existing ties or aid networks. Still, the influx of more than 130,000 Jewish and political refugees had a tremendous impact on numerous professions and academic fields in the United States from physics and art history to public administration.[56] Historians have detailed the numerous challenges faced by refugees adapting to professional cultures and academic systems in the United States as well as the range of their migration experiences, depending on professional background, status, ethnicity, and especially gender.[57] Above all, they have emphasized the contributions these individuals made to transatlantic knowledge transfers in specific areas, including disciplines relevant to the study of midcentury marketing. Scholars have pointed out the importance of emigrés to U.S. developments in fields ranging from economics to art and design, individual and experimental psychology, and the empirical social sciences.[58] In an older, more traditional reading, this "intellectual migration" amounted to an almost unprecedented "brain drain" by means of which European knowledge enriched American academia and society.

Recent transnational studies, however, have dispelled the notion of a singular midcentury brain drain through forced emigration in several ways. The knowledge transfers facilitated by emigrés fleeing persecution in Europe must be contextualized within a longer history of transatlantic transfers that date back through centuries of American immigration history. The phenomenon of immigrant entrepreneurship, exemplified in this book by market researcher Alfred Politz or furniture designer Hans Knoll among many others, with its cross-border transfers of technology and business models, has been a constitutive feature of the American economy from its early beginnings.[59] Emigrés in the field of business were part of a longer history of migration (and return migration), which influenced economic development on both sides of the

Atlantic.[60] Especially since the Progressive Era, as Daniel Rodgers and others have shown, a myriad of "Atlantic crossings" has left a profound imprint on numerous professions and various fields of American society and politics.[61] Such transnational discourses in areas ranging from housing and municipal development to labor conditions, social services, and economic planning were institutionalized through transnational organizations during the interwar years.[62] The Rockefeller Foundation, the Carnegie Endowment, and other American foundations were particularly prominent in fostering a systematic circulation of knowledge through cultural and exchange programs for scholars and other elites. Furthermore, these transnational organizations and networks played an instrumental role in enabling the later emigration of academic refugees. Although World War II and the manifold crises of the 1930s and 1940s posed a challenge to this process, they did not constitute a complete break.[63] Thus, midcentury transfers associated with the Europeans fleeing National Socialism should be understood not as a singular event but rather as a distinct chapter in a much broader history of elite migration and mobility. The story of consumer engineering thus interweaves different forms of transatlantic exchanges and knowledge circulation spanning the interwar and postwar years.[64]

This book traces metropolitan professional networks in marketing research and commercial design that spanned the Atlantic world—a multidirectional flow of people, publications, and ideas with little regard for national borders. Indeed, the kind of transfers relevant for this story often occurred not so much across national borders or transatlantic cultural divides as between different professional settings. Knowledge transfers between academics and corporate managers or between advertising executives and avant-garde artists also meant bridging divides that were often as deep as if not deeper than those owing to language or national background. Such interdisciplinary crossings and fractured careers were a frequent result of forced migration. Exile in particular required flexibility from emigrés, and the economic necessity of finding employment in a new country demanded that they look beyond their familiar professional environments. When universities and art schools had few positions to offer, a corporate contract might pay the bills.

Biographical approaches and the study of individual careers are ideally suited to reveal this complex interplay between the various dimensions of transnationality.[65] Until recently, biography has not been at the center of transnational history, but the perspective of individual professionals such as, for example, social researcher Paul Lazarsfeld, who moved back and forth between metropolitan centers like Vienna and New York, allows us to analyze constitutive processes of cross-border adaptation and transfer on the micro level and to connect the at times "disembodied" transnational sphere of global discourses and organizations to concrete local developments. Taking individual migrants as the starting point for investigations of cross-border exchange processes connects both with a newfound emphasis on elite migration among migration scholars and with a growing interest in "transnational lives" among historians of globalization.[66] In various capacities, migrants and exiles have been recognized as important "cultural brokers" who facilitated cross-border exchanges and the adaptation of knowledge across cultural or institutional divides.[67] Instead of a passive diffusion of ideas or practices, the active and creative role of migrants in transfer processes gets highlighted through the concept of "cultural translation," which presupposes the transformation of cultural content.[68] In particular, individual biographies can provide insights into the agency of migrants as translators of culture, which proved particularly crucial in the area of marketing and mass consumption.

Ultimately, this is not just a story of individual careers or of the impact of emigration on marketing as a professional field. Instead, the cultural translations of emigré consumer engineers provide a concrete case study for a process that historian Anselm Doering-Manteuffel has termed reciprocal "Westernization."[69] Coined originally with an eye toward political culture and in opposition to overly simplistic notions of postwar "Americanization," Westernization emphasizes the growing importance of mutual interrelations and influences between Western European and North American societies at the middle of the twentieth century. Political, intellectual, and professional elites were central actors in this process, serving as "transatlantic intermediaries" and cultural brokers.[70] In part, "Westernization" stresses the political and cultural influence of the United States in Europe in the context of a cultural and intellectual Cold War.[71] Recent studies, however, have provided a more reciprocal understanding by emphasizing European voices in transatlantic exchanges. Emigrés from Weimar Germany, for example, helped shape Cold War concepts of muscular democracy in the United States, revealing a transnational dimension to the postwar American national security state.[72] American consumer capitalism, I suggest, was similarly affected by cross-border exchanges and mutual Westernization, which pertained not only to specialized disciplines but to core aspects of the consumer economy from its aesthetic forms to its dynamic engagement with consumers.

Midcentury Marketing as Social Engineering

The transnational perspective suggests that midcentury marketing experts were more than simply profit-minded "captains of consciousness" who manipulated consumers for commercial gain.[73] Instead, their efforts to reshape the material environment of consumers fit within a larger movement of "social engineering" that spanned the Atlantic world.[74] While social engineering typically refers to government planning efforts to mold social behavior or to bold reform visions of social scientists, administrators, or urban planners, corporations, too, were engaged in efforts aimed at social planning and control. Not coincidentally, many of the expert social scientists and architect designers at the center of this study shifted almost seamlessly between academia, government contracts, involvement with social reform movements, and corporate consultancy.

Consumer engineering was akin to other efforts to reshape societies at the height of an era of planning euphoria that some historians have labeled "high modernity." Culminating at midcentury, "high modernity" was characterized by an enthusiastic belief in the powers of science and social sciences and of technocratic solutions to social and economic problems. It was widely shared across national boundaries and political ideologies.[75] In the context of New Deal America, for example, consumer engineering's emphasis on planning and predictability was part of a transnational movement. The rise of Keynesian policies aimed at creating aggregate demand and shaping consumer behavior dovetailed with marketing's promise to create continuous prosperity through engineered obsolescence. Whether at the level of government or of the corporation, the 1930s were an "engineering decade," and professionals of all stripes tried to emulate the practices and the mindset of engineers.[76] An all-pervasive spirit of efficiency and rationalization informed even the world of art and design as artists of the so-called "machine age" embraced utilitarian functionalism as the stylistic language of modernity.[77] Even the organic curves of "streamlining" and aerodynamic styling, Christina Cogdell has argued, can be read as an expression of this era, which combined a fascination for technology with progressive notions of "uplifting" the tastes of consumers and modernizing their life-worlds.[78] Despite their social reform backgrounds, emigré consumer engineers were frequently less concerned with the continuing inequalities of consumer capitalism and its reproduction of stereotypical gender roles. Still, much like their counterparts in other disciplines, consumer researchers and commercial designers eagerly sought to "improve" society and make consumers fit for the "modern" era.

The technocratic focus on "modernizing" society connects the interwar period seamlessly to the post-1945 era with its now global discourse on modernization.[79] Marketing practice at the middle of the twentieth century, finally, cannot be properly understood without reference to the Cold War at a time when access to goods and the material standard of living became part of the systems competition between East and West.[80] Those consumer researchers interested in demographic analysis and attitudes of consumers were part of what has been discussed as the "Cold War social sciences." Through their links with government, corporations, and institutions such as the Ford Foundation, the social scientists were enmeshed in Cold War policies of global impact.[81] As much as consumer capitalism was a driving force of social change at midcentury, it was itself constrained by larger cultural and political forces of technocratic modernism and a cultural Cold War. In this context, the emigré experts at the center of this story played crucial roles as transatlantic cultural intermediaries, translating American-style modernity to consumers and cultural elites in Cold War Western Europe and beyond.

ONE

The Origins of "Consumer Engineering": Interwar Consumer Capitalism in Transatlantic Perspective

In January 1930, German graphic designer Lucian Bernhard contributed a lengthy article on the importance of "beauty" for American industry to a brand-new section of the trade journal *Advertising and Selling*. Manufacturers, Bernhard claimed, had long focused exclusively on efficiency in engineering and on the mechanics of production; outside of advertising art, they were reluctant to let artists and other outsiders meddle in their business. Yet, he noted, even if the consumer desire for good design was still "quite unconscious," an increasing demand for beauty in commercial goods could be expected. Advertisers, he advised, should turn to style as a "new angle of distinction" now that the technical performance of a given product could be taken for granted. Bernhard called on industry to embrace a "new school of artists" who were willing and equipped to bring their art to merchandise.[1] In a buyer's market, new experts were needed to help companies market more desirable goods.

In essence, Bernhard made the case for bringing a new group of "consumer engineers" into American marketing.[2] Also in 1930, advertising executive Earnest Elmo Calkins wrote about the need for "consumption engineering" as a new "business science" that would combine market research with consumer-oriented design innovations to create new demand by means of "artificial obsolescence."[3] Advertising executives and sales managers, to be sure, had long seen themselves as expert professionals analyzing consumer markets. The psychology of buyers and the aesthetics of commercial goods and advertising had been prominent topics in professional circles during the early decades of the twentieth century. The collapse of demand during the Great Depression, however, lent new urgency to marketing efforts geared at appealing to consumers. Advocates of consumer engineering wanted to make existing marketing practice more systematic. At the same time, they became receptive to outside experts, looking to the worlds of academia and modern art.

American business, Lucian Bernhard furthermore argued, also needed to look to Europe for inspiration. Some U.S. advertisers, he believed, had already caught on: "For years advertising agencies have been watching with keen interest the development of this 'art in industry' movement, especially in Germany, France and Austria . . . European efforts to create a new beauty in machine products have pointed the way to new life in industrial merchandising."[4] Bernhard's own transatlantic career was a testament to the vibrant interwar exchanges in the commercial arts. He had been a prominent poster artist in Germany before and during World War I and one of the pioneers of modernist "object posters" for brand goods and war loans alike. By the mid-1920s, Bernhard had relocated to New York City where he eventually became a partner in Contempora Studio, producing graphic arts and interior designs.[5] The history of interwar marketing is usually written in terms of the "Americanization" of European advertising or of the European reception of Fordist mass consumption. Analyzing the debates surrounding consumer engineering, by contrast, demonstrates the reciprocity of Atlantic crossings in interwar marketing with regard to elements of design and of the psychology of consumption.

This chapter thus pursues a twofold aim. First, it contextualizes the notion of "consumer engineering" within broader developments in interwar marketing and mass consumption. The impact of marketing innovations in styling and consumer psychology can only be understood before the backdrop of an American consumer society, which already had developed sophisticated mechanisms of mass production and distribution. Retailers and advertisers in particular had accumulated a specialized body of sales expertise on which a new generation of marketing professionals could build. As the Depression hit, a range of further factors contributed to the spread of consumer engineering efforts. The systematic creation of demand now became of interest to industry and the state alike. In addition, prevailing technocratic approaches to government and management favored the rise of "scientific" methods as well as of experts of all stripes. Interwar preoccupation with efficiency and rationalization of social and economic processes, finally, dovetailed with a circumscribed but growing fascination with modernist forms and functionalist designs, part of a broader aestheticization of American commercial culture.

Much more than often assumed, secondly, this was a transatlantic story. Interwar Americans still looked to Europe, especially when it came to innovations in style and design or to the psychological appeal of consumer goods and luxury products. While American goods and advertisers indeed spread across the globe during the 1920s, European metropolitan centers in many ways held their own with regard to retail facilities and modern methods of sales and advertising. Focusing on Germany and Central Europe in particular, the chapter traces European marketing developments during the interwar years. What set European marketing apart early on was the prominent role of public actors and state institutions with regard to consumption as well as an emphasis on social reform among various new experts engaged in commercial design and consumer research. When some of these European experts in design and social research were forced out of Europe by the rise of National Socialism, they brought new marketing knowledge to the United States. As consumer engineers, the emigrés intensified an already ongoing "cross-cultural fertilization," to use emigré intellectual Paul Tillich's term..[6] Their knowledge of design and consumer psychology promised to be adaptable to American midcentury efforts in social and consumer engineering by state and industry alike.

The "Horn of Plenty": The Emergence of Mass Marketing in the United States

"What is wrong with the horn of plenty?" Sheldon and Arens asked their readers in 1932. Like a generous cornucopia, American industry had been providing food and furniture, clothing and automobiles in never before seen quantities, leading straight into the Great Depression.[7] While industrial mass production had overcome the perennial problem of scarcity, the authors now identified "underconsumption" as the new challenge for business, citing Henry Ford and other "consumption-minded" business leaders as their key witnesses. Marketing practice had already improved greatly, they acknowledged. Retailers had sped up the distribution chains, making shopping "quick and easy" with modern chain stores, convenient packaging, and advertising, which enticed consumers directly, creating demand for specific goods. Striving for predictability and innovation at the same time, consumer engineering promised a recipe for more scientific marketing and the systematic creation of new consumer demand.

Indeed, mass marketing was not an innovation of the Depression era in the United States. Product innovation and new packaging forms, historians Gary Cross and Robert Proctor have recently argued, shaped consumer behavior beginning in the late 1800s; "packaged pleasures" such as cigarettes, candy bars, or phonograph records were engineered to entice new consumption habits.[8] American industry, furthermore, was never as focused on standardized mass production as traditional narratives of Fordism have suggested. Instead, business historians have demonstrated that the "Second Industrial Revolution" saw many producers and entire branches of the consumer goods industries oriented toward production in smaller batches, which produced an endless stream of novel designs. Craftsmen and designers in ceramics, furniture, and other household goods were intimately attuned to shifting popular tastes, and experienced experts in the employ of manufacturers and retailers "imagined consumers" and their desires.[9] The advocates of consumer engineering hoped to put such practices on a more scientific footing and to apply them across the economy more generally.

With the turn of the twentieth century, mass marketing found new forms and professional expressions in sales, manufacturing, and retailing.[10] Consumer goods producers increasingly relied on recognizable brands and sophisticated distribution networks, which established a more direct link between manufacturer and consumer. Catalogues such as Sears made the new wealth of goods available to a truly national audience. Retailing also became more elaborate and differentiated in its appeals to consumers. Large department stores served as "palaces of consumption" with lavish displays and elaborate shop windows that sparked the consumer imagination.[11] The world of goods they presented aimed to entice desire among an expanding buying public. Retail historians have recently emphasized the role of independent shops and of the more mundane grocery trade in "modernizing" consumer goods distribution in the United States.[12] By the 1920s, American industry could rely on an efficient distribution machinery to market its wares.

Retailers and manufacturers increasingly harnessed mass media magazines such as the *Lady's Home Journal* and the *Saturday Evening Post* to shape consumer expectations and new social standards of consumption.[13] While it was still out of reach for many, including most African

CHAPTER ONE

Americans and those living in rural poverty, American households aspired to a middle-class "standard of living" which was reflected in home furnishings and proper dinnerware as much as in up-to-date equipment for kitchens and bathrooms.[14] Professional advertising agencies were at the forefront of selling this American consumer modernity. Advertisements had long provided much more than product information, but instead sold notions of progress and offered consumers cues to proper behavior in a quickly changing world.[15] Around the turn of the century, some advertising experts even developed early systematic forms of "hard sell" and "soft sell" approaches, relying on market research and information or on aesthetic and psychological appeals respectively.[16]

Advertisers pursued their business in an increasingly strategic and professional manner as agencies became nationwide and, particularly after World War I, even global entities.[17] Advertising trade journals such as *Printer's Ink* led the way in a movement toward more professional forms of selling, heeding the call for new forms of "scientific marketing" to complement Frederick Taylor's "scientific management" in production for the realm of distribution. From retailing and door-to-door salesmanship to the marketing of national brands such as Ford and Coca-Cola, the first decades of the century witnessed the emergence of professional salesmen and marketing experts and the early development of "scientific selling."[18] Even prior to the Great Depression, then, most major corporations in the United States had recognized the importance of mass marketing. Instead of a single-minded focus on production, they began to develop brands and considered their corporate image as it related to consumers and the broader public.[19] The crisis of the Depression made consumer-oriented marketing ever more critical in competitive industries. Both manufacturers and retailers invested even more than before in marketing strategies aimed at anticipating and creating consumer demand.[20]

What, then, was new about the call for consumer engineering at the beginning of the 1930s? The boom of the 1920s had laid a solid foundation for their ideas, Sheldon and Arens acknowledged, yet they planned to go further: "If the depression put a temporary stop to the consumer gold rush, it ushered in the more systematic mining of the engineer. Today wise executives are laying their shafts for the richer ores that are not so obviously on the surface."[21] On one level, the marketing program the authors prescribed simply made explicit those trends in marketing practice which had been gathering steam over the course of the previous decades. This would entail the creation of specialized marketing departments and of a more fully developed marketing profession over the coming decades. A growing number of companies began to move toward a strategy of "marketing management" that employed new forms of "merchandising," i.e., the conscious planning of product innovations for the consumer market.[22]

On a second, more fundamental level, the engineering mindset and the emphasis on scientific methodology characterized a new generation of consumer experts who presented themselves as having a more thorough grasp of consumer markets and their psychology and a new, proactive attitude. New experts in "humaneering," Sheldon and Arens admonished in 1932, needed to tackle "the much deeper and subtler problems [of] the sociologist and the psychologist."[23] Such an approach required detailed analysis, but it also raised the possibility of conscious manipulation. Consumer engineering, the authors noted, was the "science of finding customers, and it involves the making of customers when the findings are slim."[24] In either case, specialized market research would be crucial for this endeavor as it allowed for forecasting market developments and consumer tastes. Of the new experts, this required a degree of scientific understanding combined with a great deal of creative intuition. The modern marketer or ideal "consumer engineer," according to Sheldon and Arens, had to be "mechanically minded and inventive, but above all he must be an artist with a lively imagination."[25] This opened the door to both academics and artists to play a much greater role within professional marketing over the coming decades.

Embedded within this combination of the technocrat and the artist, furthermore, we find a tension between the marketer as a sober analyst and as a creative visionary. Efforts to make marketing more systematic and scientific were accompanied by consumer engineering's other central tenet, its emphasis on aesthetics and style. The more competitive the market, the greater the need for distinguishing products through design. Recurring style innovations fostered consumer demand and added a broader sense of cultural dynamism and modernity to American products. In order to keep up with the times, *Consumer Engineering* suggested, U.S. companies would do well to embrace the modernist forms of avant-garde art of the interwar era. In this, the authors notably turned to Europe for inspiration.

Beauty in Industry: American Perceptions of European Consumer Modernity

When it came to design (especially with regard to high-priced luxury goods), American marketing experts during the early 1930s felt as though they could still learn from Europe. Art promotes business, Sheldon and Arens proclaimed, and, in their view, the inspiration for the true artist remained in Europe. Here, they felt, both modernity in design and the concept of fashion change appeared to be most developed, as for example in the market for women's clothing, which was still squarely dominated by Paris, where Sheldon himself had lived for a while.[26] While Sheldon and Arens scorned the elitism of European industrial design as incompatible with the demands of the machine age, they still saw enormous potential in European modernism for American industry. Interested in design for machine and mass production, they highlighted German immigrant Kem Weber's innovative bent-lock furniture designs for mass production as one example of industrial styling from a "new type of artist" who combined modernist aesthetics with the demands of mass marketing.[27]

The use of design and artwork in marketing was, of course, not entirely new at the beginning of the 1930s. In fact, Earnest Elmo Calkins had been among the first to emphasize the importance of product design for sales as early as the 1910s, and his Chicago-based advertising agency Calkins & Holden had pioneered the use of modernist imagery and graphics in advertising. By World War I, graphic design was widely recognized as an effective sales tool.[28] During the late 1920s, however, aesthetics attracted additional attention in U.S. business culture amidst a growing sentiment (especially among East Coast elites) that American product design needed to catch up with modernist trends across the Atlantic.[29] The 1925 Paris Exposition of industrial goods was widely seen as a wake-up call to American design experts.[30] *New York Times* commentator Miriam Beard observed that European countries not only led the United States in style and design but also translated this into an advantage in trade and industry. Americans by contrast were "giants of efficiency" but lacked taste, ideas, and proper industrial arts schools. European industry and governments, contemporary observers warned, had realized that style and design were "an economic resource."[31]

Whereas interwar Europeans were fascinated with American mass production and marketing methods, in the realm of commercial design the transatlantic flow of ideas was very much a two-way street.[32] A 1928 exhibit of international design at Macy's department store in New York City highlighted the modernity of German art exemplified by the designs of Bruno Paul along with the Austrian designs of Josef Hoffmann and modern Italian shop interiors by Giovanni Ponti of Milan.[33] A similar exhibit of glassware at the Metropolitan Museum of Art was praised for its European contributions while American manufacturers were chided for spending too little on arts departments and creative design.[34] The American business community increasingly shared in the notion of American "backwardness" when it came to design and fashion goods. In a speech dramatically titled "America at the Knees of Europe," Ira Hirschmann, vice president of upscale department store Saks Fifth Avenue, decried "the fact that American department stores still must look to European designers for leadership." Before an audience of sales executives, Hirschmann echoed a call for U.S. industry to embrace styling and design as valuable in themselves.[35]

Interwar American observers frequently regarded modern European art as the ideal fit for the "machine age" of mass production. While the radical functionalism of the Bauhaus school did not find a wider U.S. audience until later in the 1930s, different variants of design modernism found their way across the Atlantic already during the 1920s.[36] American art critic Edward Alden Jewell called on American industry in 1928 to bring a more "pure" modern design and simple forms to factories and machines. From textiles and furniture to automobiles and locomotives, industrial artists should develop the aesthetic side of practical and commercial products following European examples.[37] Not by coincidence did Jewell present Calkins as one of the champions of a new drive for commercial modernism in the United States; it was in this atmosphere of widespread clamor for modern design that the advertising journal *Advertising and Selling* issued its first special issue called *Advertising Arts* in January of 1930.

During the early 1930s, *Advertising Arts* became a prominent outlet for American coverage of European commercial modernism. In his contribution to the first issue, Earnest Calkins argued for the importance of modern art in advertising particularly for those iconic goods, such as radios and motor cars, which embodied the dynamic "age we live in."[38] Much like the authors of *Consumer Engineering*, contributors to *Advertising Arts* firmly believed in the sales potential of design. In 1931, Abbott Kimball developed the business case for good design, which, he believed, provided a sales stimulus and a new "basis for competition." Beauty in design could help to build upscale distribution networks, entice consumers to "trade up," or create a sense of obsolescence. Good design opened up new markets by reaching new customers and, finally, added to the identity of a brand: "coordinated beauty gives identity to a line, and makes one product sell another."[39]

Using art as a profit maker required new experts to join American corporations. An advertisement by the Industrial Art Council acknowledged that many U.S. businessmen were still "a little frightened at the idea of calling in a long-haired art expert who may be good enough at his own game but a complete wash-out in horse sense and practicality." To allay such fears of impractical artists, the council offered the services of their professionals with extensive "experience in turning art to business purpose."[40] According to designer Joseph Sinel, the new artists mediating between art and commerce were reform-minded educators of the consuming mass, solving the problem of shoddy products and tasteless form.[41] They needed to be both "artists with a lively imagination" and at the same time marketers endowed with an "uncanny foreknowledge of changes in public taste," new "packaging engineers" who enabled the successful promotion of American consumer goods.[42] Reflected in such debates we find the nascent industrial design profession, which would come into its own over the following decades.

In almost every one of its early issues the editors of *Advertising Arts* took a close look at European developments in graphic and industrial design. New York advertising consultant Nathan Horwitt, for example, paid tribute to the marriage of arts and industry and to "that handful of artists in Europe who endeavored, without patronage, to experiment in new forms of expression, in particular the expression of the nascent age in which we are living."[43] Specifically, Horwitt singled out the poster art of France and Germany, modernist graphic arts magazines such as *Das Plakat* and *Gebrauchsgraphik*, and modern designers and architects such as Le Corbusier, Josef Hoffmann, and Lucian Bernhard. The art director of publisher Condé Nast discussed magazine layout innovations of the Bauhaus and of the Werkbund publication *Die Form*, while other articles surveyed European retail interiors or reported on the ad posters of French designer Jean Carlu and innovative German typefaces.[44] Modernism, designer Walter Dorwin Teague noted in 1931, "got off on the wrong foot" in America while it was already a full-blown movement in metropolitan Europe.[45]

By the beginning of the 1930s, a growing number of Americans in arts and business alike were eager to overcome their country's perceived deficits in design and the commercial arts. The New York business community saw a need for a new school of applied arts with an integrated department of marketing.[46] Not surprisingly, European immigrant artists such as Lucian Bernhard were on the forefront of commercial America's turn to modernism. A 1932 Philadelphia exhibition called "Design for the Machine" set out to illustrate the expanding reach of contemporary design across American industry. Among the artists and objects featured most prominently were Russian-born graphic designer Alexey Brodovitch, a kitchen design from Danish-born Gustav Jensen, furniture by German Americans Kem Weber and Marianna von Allesch, and brass and copper ware by Walter von Nessen, as well as fabric designs of the Contempora group.⁴⁷ As modern design became an increasingly influential factor in midcentury consumer marketing, the number of immigrants and emigrés active in this field would grow even further, as later chapters will demonstrate. Significantly, these transatlantic mediators of modern commercial design facilitated transfers from a metropolitan consumer culture in Europe that had witnessed its own innovative developments in consumer marketing during the interwar years.

Metropolitan Modernity in Europe: The Reciprocity of Transatlantic Consumer Transfers

Historians of consumption in interwar Europe typically stress the importance of the "American model" for contemporary Europeans with its Fordist emphasis on affordable mass consumption.⁴⁸ Subsidiaries of American finance companies helped bring credit marketing to the European automobile market during the 1920s, while Hollywood movies and popular culture helped to spur on the consumer imagination of consumers across the continent. Large U.S. advertising agencies became active in European markets, and they were particularly important in spreading American notions of professional marketing. From swing music and chewing gum to market analyses, "Americanisms" featured prominently in interwar European commercial culture.⁴⁹

Modern marketing, however, was hardly an American invention, and there was more reciprocity to transatlantic transfers than the story of "Americanisms" implies. A metropolitan commercial culture had emerged in Europe as well, which became increasingly differentiated and professionalized during the interwar years. In cities such as London, Paris, Vienna, or Berlin, sophisticated marketing practices influenced the lives of elites and the urban masses alike.⁵⁰ In interwar Berlin, for example, observers noted the increasing prominence of advertising as a feature of metropolitan life. Besides shop windows, magazines ads, and billboards, the spread of neon light fixtures, skywriting, cinema, and (later) the radio made advertisements nearly ubiquitous, especially during the relatively prosperous mid-1920s.⁵¹ Across Europe, a veritable "consumer revolution" occurred around the turn of the century with increasingly professional approaches to advertising, commercial design, and consumer research.⁵² Against the notion of a European-American dichotomy, historians have recently emphasized the transatlantic coevolution of marketing methods and the importance of transnational transfers in multiple directions during this process.⁵³

In Europe, too, the rise of advertising constituted a central element in the development toward professional marketing. After 1900, both graphic design and psychological considerations took on a bigger role in the advertising industry, and commercial posters and print ads in new high-quality magazines became more artistic, aiming to lure consumers toward new consumption experiences. At the same time, advertisers pushed to improve their credentials by creating professional organizations and dedicated training programs, while trade publications such as *Seidels Reklame* (Vienna and Berlin) set new professional standards.[54] By 1929, there were as many as fifteen professional associations in the field of advertising and marketing in Germany alone, and nearly 50,000 full-time professionals worked either independently, for advertising agencies, or for corporate advertising departments.[55]

Again, the retail sector was particularly influential for the creation of a modern infrastructure of distribution.[56] The decades around 1900 are known as the "golden age" of European shop design as avant-garde artists such as Adolf Loos in Vienna and Bruno Taut in Berlin created modern shop interiors. French Art Deco influenced the interior design of many large stores by the 1920s, while Dutch modernists of the De Stijl movement left their imprint on Amsterdam's premier department store De Bijenkorf.[57] They created shopping environments that would entice customers to spend and elicit desires for "modern" goods. Department store windows and displays were dressed by professionals who were especially trained and organized in trade associations.[58] German window dressers, who hosted an international congress in 1928 in Leipzig, increasingly made use of psychological considerations in creating displays meant to catch the eye of the consumer.[59] Not just department stores but also many small retailers underwent processes of rationalization and modernization, which laid the true foundation for the mass marketing of commercial goods and for new brand-name products.[60]

As in the United States, European companies became more marketing-oriented during the early twentieth century, if perhaps at a somewhat slower pace. Newly established advertising departments promoted national brands such as Odol mouthwash, Continental bike tires, Kupferberg sparkling wine, Dr. Oetker baking powder, Tobler chocolates, or Pelikan pens.[61] Companies used an array of new marketing strategies that included pricing (rebates), packaging, and sales organization as well as widespread advertising and innovative marketing through sales agents and early automated vending machines.[62] Producers of branded goods had a particular interest in consumer research; over 150 firms in Germany alone had departments devoted to forms of psychotechnics and applied psychology as early as 1922 directed at employees and customers alike.[63]

Such interest in targeted consumer appeals dovetailed with the use of modern art and design in corporate marketing.[64] One especially prominent example for the integration of design was the work of Peter Behrens at the electric company AEG in the years prior to World War I. Behrens was involved in the design of the company's logos and products, and he shaped AEG's advertising as well as the internal communication down to the design of offices and shop-floor interiors.[65] Behrens's work was representative of the ideals of the Werkbund, an influential organization of designers and industrialists founded in 1907. Following the ideas of Hermann Muthesius, the Werkbund dedicated itself to a mission of bringing "modern form" to industry and to adapting the ideals of craftsmanship to an age of machine production.[66] At its founding congress, Werkbund members bemoaned the decline of artisanal skill and artistic form in industrial production and called for the "organized stimulation of consumers," whose aesthetic senses and understanding of product quality needed to be trained.[67] Far from simply backward-looking, however, the Werkbund soon assembled leading modernist artists and architects such as Josef Hoffmann and Bruno Paul and cooperated with firms such as Siemens and Bahlsen. Over the coming decades the group organized a series of influential exhibitions, which, along with the Werkbund journal *Die Form* (published between 1924 and 1934), informed the Central European debate about modern form in art and industry.[68]

Artistic Modernism found its way into commercial advertising as well. Turn-of-the-century French poster art and German advertising posters by graphic designers including Ludwig Hohlwein in Munich or Lucian Bernhard in Berlin straddled the line between art and commerce. Particularly the clean and bold look of the *Sachplakat* (or object poster) managed to capture viewers' attention and signaled a sense of modernity that was welcomed by many consumer goods producers. European advertising posters of the era developed a distinctive modern look, which presented a sharp contrast to the increasingly copy-heavy advertisements in American mass circulation magazines.[69] Graphic design magazines such as *Das Plakat* helped sustain the modernist turn in advertising and established German graphic artists among the leaders of an international avant-garde. Beginning in the 1920s, the magazine *Gebrauchsgraphik* not only covered American developments for a German public but, featuring bilingual contributions in English and German, also promoted the reputation of German graphic artists internationally.[70]

Across the metropolitan centers of Europe, design advocates strove to bring a social and aesthetic reform impulse into the consumer goods sector. Initially, the emphasis was on small-scale luxury production such as at Vienna's Wiener Werkstätte.[71] After World War I, German designers became increasingly interested in producing standardized, low-cost objects, which would be affordable to middle-class consumers as well. Werkbund designers in particular pursued the ideal of a standardized, "typified" product that would combine the goals of rationalized mass production and a shared social standard of material consumption with the aesthetic demands of a good, functional form.[72] While the ideal of standardized goods was at odds with a commercial consumer culture driven by fast-paced cycles of fashion, even the most radical proponents of functionalist design at the Bauhaus school for design (located in Weimar and later in Dessau) directly engaged with consumer goods producers. Indeed, Bauhaus artists and their students left a significant mark on product design for many German companies during the 1920s and 1930s, as exemplified by the cooperation of designer Wilhelm Wagenfeld with Schott glassworks in Jena.[73] Their modernist design forms became characteristic particularly for upscale consumer goods in interwar Germany and reflected a professional concern among designers to produce "good design" for a commercial mass market.

New approaches to consumer research similarly informed interwar European marketing. Students at the Bauhaus, for example, not only took courses on the basic principles of business and marketing but, at least in part, also received schooling in consumer psychology and the psychological ramifications of design.[74] Individuals in different disciplines such as economists Rudolf Seyffert in Cologne and Fritz Giese in Stuttgart or psychologist Edmund Lysinski helped to establish early research in marketing psychology that sought to understand consumers in their sociodemographic context.[75] Vienna counted among the first centers for systematic market research as the Wirtschaftspsychologische Forschungsstelle began to conduct sophisticated market surveys for Austrian, Swiss, and German companies by the early 1930s. In Nuremberg Wilhelm Vershofen's Institut für Wirtschaftsbeobachtung der Fertigware was established as a precursor to the later Gesellschaft für Konsumforschung (GfK), forming another leading center for consumer research.[76] Vershofen became an important broker between the disciplines of psychology and economics in interwar Germany, and starting in 1934 the GfK began to conduct nationwide consumer surveys for marketing purposes.[77]

If only slowly and sporadically, professional marketing research began to make its way into the business practices of German retailers and consumer goods producers during the interwar years. One prominent venue was the trade journal *Verkaufspraxis*, which began publication in 1925. Its editor, Victor Vogt, was particularly interested in bringing American innovations in marketing to a German-speaking audience.[78] Besides topics such as brand goods or chain store distribution, contributions to the journal analyzed the motivations of consumers, discussed the psychology of sales pitches, and looked at the ways in which advertising affected sensory perception.[79] The most prominent practitioner to advocate a systematic approach to marketing in interwar Germany was certainly Hans Domizlaff, who worked as a consultant to the cigarette manufacturer Reemtsma, among other companies. Domizlaff's *Markentechnik* (brand technics) strategy involved elements of segmented marketing with a carefully designed mix of luxury and mass market brands, but also entailed the use of rebates, trading cards, and elaborate package designs.[80] While they were exceptional, one can consider individuals such as Vogt or Domizlaff as early European protagonists of a move toward consumer engineering much like Earnest Elmo Calkins or Egmont Arens in the United States. In fields from window dressing and advertising arts to product design and consumer research, Europe, too, witnessed the rise of new professionals with marketing expertise.

Social Engineering between European Reform Movements and 1930s America

When European professionals, scholars, and artists fled to the United States in growing numbers by the 1930s, they came from metropolitan commercial cultures that hardly lagged behind their U.S. counterparts in sophistication. One important European peculiarity, however, was the prominent role of state and civil society actors from consumer cooperatives and social reform movements to local and national governments. While Europeans lacked the broad-based access to mass-produced goods available to Americans of various social strata, state and municipal governments along with numerous other organizations from labor unions to the Werkbund were increasingly involved in "engineering" access to mass consumption. Before the backdrop both of the emerging New Deal state and of a business community in need of renewed legitimation following the Great Depression, American consumer engineers were interested in this social aspect of European marketing as well.

The boundaries between the economic and political dimensions of marketing were blurred in interwar Europe. Hans Domizlaff, for example, embedded his ideas on brand technics within a broader conception of "social marketing" that emphasized the importance of advertising in winning public trust for business and government alike.[81] This, of course, brought commercial advertising into uneasy proximity to political propaganda, which in the German case became painfully clear with the cooptation of advertising and marketing by the National Socialist regime.[82] Even before, the interwar German state sought to regulate consumption by developing the notions of a minimal standard of living and by engaging in efforts to ensure the provisioning of basic needs. While government regulators were skeptical of advertising and "wasteful" spending, Domizlaff and other German brand technicians argued that modern marketing could actually serve as a crucial social technology in ensuring future prosperity not just for individual companies but for the national economy at large.[83]

After 1933, the National Socialist state systematically employed marketing techniques for political ends despite an often contradictory stance toward commercial consumption.[84] This included attempts to enforce "truth in advertising" and efforts to "Germanize" the language and appearance of newspaper advertising.[85] Along with Jewish-owned businesses, some American ad agencies were forced out of the German market or had to change ownership.[86] While many consumer researchers and modern commercial artists as well as retailers and marketers with Jewish background or overtly leftist leanings were forced to flee Central Europe over the course of the 1930s, however, most marketing practitioners stayed and found ways to accommodate themselves with the new regime. Indeed, the NS state recognized the popular appeal of mass consumption, and promises of radios, automobiles, and other "people's products" for average German households soon became part of the state's propaganda repertoire.[87] When it came to marketing during the Nazi era, historian Jonathan Wiesen has observed, "racist utopia and commercial utopia went hand in hand."[88] In its efforts to shape public culture and social behavior and to exploit material consumption for political gain, the Nazi state brought aspects of social engineering immanent in modern marketing practice most directly to the fore.

The political dimension of marketing, however, was by no means a distinctly German issue, but instead played out in various forms across interwar Europe. In many countries, World War I had already served to heighten the link between notions of citizenship and nascent consumer identities.[89] In interwar Britain various government and quasi-government organizations engaged in propaganda campaigns to promote the consumption of public services or of British-made goods, utilizing market research data and modern advertising techniques.[90] Smaller nations such as Austria and Switzerland had long been interested in fostering "patriotic" consumption habits among their "consumer citizens."[91] Besides such efforts geared at strengthening national economies, social reform programs especially on the municipal level also utilized the emerging toolset of marketing. The social democratic government of "Red Vienna," for example, not only planned for the public provision of housing and mass entertainment, but drew on early studies in social and consumer research for these efforts.[92] Reformers in Sweden strove to make the country's emerging welfare state compatible with the promise of modern consumerism as experts envisioned their "people's home" (*folkenhemmet*) as a publicly engineered variant of a modern consumer society.[93] Across Europe, interwar social engineers were in various ways engaged in understanding and shaping consumption practices to advance the public interest.

In the midst of the crisis of the early 1930s, American advocates of consumer engineering similarly couched their call for professionalized marketing within a broader social agenda. The marketing methods they proposed, Sheldon and Arens believed, would be a "new technique for prosperity," not just for individual companies but for society as a whole. While the principal actors of their consumerist vision were large private enterprises such as General Electric and DuPont rather than the state, the administrative structures of these corporations similarly bet on the power of science and engineering to shape social and economic relations. Their technocratic experts with backgrounds in psychology, sociology, statistics, or economics would be tasked to reconfigure America's consumer culture to promote economic growth driven by expanding demand. More than simply a sales agenda, they envisioned broader social changes if science began to "dominate the business world." Consumer engineering, the authors promised, would lead to a reconsideration of social problems that "but a few years ago were only agitated by the Bolsheviks," and much akin to fellow technocrats across interwar Europe, they exuded faith in "human technology to solve the human problem."[94]

The notion that scientific marketing was relevant for American social development and political economy could draw on several aspects of interwar commercial culture. Early proponents of public relations such as Edward Bernays, for one, had consciously blurred the lines between political and commercial "propaganda." By the 1930s, many corporations relied on PR not only for sales and marketing but also for political purposes in an attempt to regain public trust in private enterprise. At the same time, social reformers, companies, and public entities alike engaged in efforts to shape the behavior of the consuming public. Reform-minded experts in "home economics" classes were active in educating homemakers to become "rational consumers" in the spirit of Progressive reform movements.[95] Educators and advertisers similarly targeted children in efforts "to raise consumers" and familiarize future generations not only with specific products but with the practices and the proper mindset of "modern" consumption.[96]

In a growing body of popular advice literature, consumer advocates such as Stuart Chase and Frederick Schlink popularized a "science of purchasing" by the late 1920s. They aimed to shape consumers into discriminating buyers able to exercise their influence on the consumer marketplace.[97] Many of these consumer activists and organizations of the nascent consumer movement such as Consumers' Research (founded 1928) saw themselves as fundamentally opposed to perceived manipulative practices on the part of professional advertisers and marketers. Much like the later consumer engineers, however, they too were imbued with the technocratic spirit of social engineering and the belief that consumer behavior could and should be shaped and directed by expert knowledge.[98] Different visions of "the consumer" thus stood in competition as reformers and commercial interests each advanced their own agenda for a modern consumer society. In fact, Sheldon and Arens understood *Consumer Engineering* as a direct response by business to what they called the changing mindset of the "American housewife" by the 1930s. She had become better educated as a consumer, they claimed, and now acted more discriminating in her choices in what was increasingly a "buyer's market."[99] Rather than outright manipulation, they argued, professional marketers consequently had to engage in more subtle efforts to modify social consumption while acknowledging the growing agency of consumers.

In their desire to understand and influence consumer markets, corporations and social reformers were increasingly joined by the American state during the era of the New Deal. Scores of new government agencies got involved in improving standards of living for a population hard hit by the Depression. The distribution of goods and the development of prices came into the purview of state administrators along with efforts to modernize housing and to expand the infrastructure for new electric appliances.[100] New Deal liberalism, as Lizabeth Cohen and others have argued, recognized the importance of mass consumption within the American economy and emphasized the active role of consumers as "citizens."[101] This consumerist turn in American economic and social policy would entail a growing demand for consumer expertise on the part of government agencies for years to come. Midcentury marketing would ultimately be shaped by close interactions with government agencies interested in recent advances in market analysis and consumer research.

What defined the midcentury transformation of consumer capitalism was, first and foremost, a more systematic use of modern marketing techniques, which had developed over previous decades on both sides of the Atlantic. Its claim to modernity entailed the inclusion of new scientific and artistic elements into marketing practice. Second, the program of consumer engineering needs to be understood within the broader social and political context in which professional marketing evolved during the interwar era. This is crucial for the story of transnational transfers which will unfold over the following chapters, because not only did actors and ideas move from Europe to the United States and vice versa, but we also find a great deal of exchange between public administrations, reform movements, arts, and academia on the one hand and commercial corporations on the other. The protagonists of consumer engineering were representative of a diverse group of technocratic social engineers engaged in retooling and "modernizing" society in the wake of the Great Depression and a perceived crisis of liberal market economies.

The third defining feature of the consumer engineering program was its emphasis on creative competition. Despite its technocratic outlook, consumer engineering underscored the importance of dynamic change. Much like Joseph Schumpeter, Sheldon and Arens emphasized the disruptive character of innovation and the constant need for competition through better services and better products that produced the speed of the "modern American tempo." The continuous demand for novelty and the seemingly irrational and rapid turnover in fashion and styles was ultimately a "force for good," they believed, even if it impeded their own efforts in systematic forecasting and planning. Ultimately, these were expressions of a process of creative destruction, which put corporations continuously to the test.[102] A new cohort of midcentury designers and consumer experts promised to ensure the continuous dynamism of consumer capitalism without paying much attention to its social costs and environmental externalities. They would add to its creative and innovative power but also amplify its indulgent, materialistic, and wasteful aspects. Incidentally, this expansion of consumer research and commercial design opened up new opportunities for outside experts and transatlantic transfers. The careers of the European immigrants and emigrés at the center of the subsequent chapters can thus serve as a lens to bring changes in marketing practice into focus.

SECTION ONE

Transformations in Marketing and Consumer Research

The Rise of Consumer Engineering: American Marketing at Midcentury (1930s–1960s)

The midcentury decades saw the emergence of what Lizabeth Cohen has called America's "consumer's republic." Advancing a democratic promise of widely shared material abundance, mass consumption became a defining feature of American capitalism.[1] Consumption came to occupy a central position in the worldview of American liberalism, and it increasingly shaped the everyday dynamics of U.S. politics. What made this newfound material abundance possible was in part a profound productivity gain among American corporations in response to crisis and war. U.S. industry slowly recovered from the shock of the Depression, and after the end of World War II, large corporations enjoyed an improved public perception as harbingers of free enterprise and consumer modernity based on scientific principles and efficient organization.

Transformations in marketing were equally responsible for creating the abundant world of American goods at midcentury. Most obvious to consumers was the spread of new retailing formats such as large self-service supermarkets where enticingly packaged brand goods seemingly sold themselves or the rise of engineered consumption environments, including new suburban shopping centers. Between the 1930s and the 1960s the United States also witnessed a "golden age" of broadcast media, radio and subsequently television, which created new national audiences and new opportunities for advertisers. Madison Avenue's ad agencies enjoyed a period of almost unparalleled cultural and economic influence, but their account managers and creative directors represented merely the tip of the iceberg of a much more fundamental expansion of professional marketing. Large consumer goods producers now had specialized marketing departments, and newly minted marketing professionals turned ever more systematically to the study of markets and consumers. They sought to shape the postwar world of goods, from market segmentation and brand images to corporate identities and product personalities. Consumers, long thought of as an amorphous and irrational mass, became the object of careful sociological and psychological study.

The midcentury, then, was the moment of the consumer engineers and of a new group of professionals in market research and product design who occupied positions at the intersection of industry, government, and academia. With the rise of the New Deal state, the U.S. government increasingly demanded consumer expertise, and World War II would prove to be a catalyst rather than a break for the development of market and survey research. Nazism in Germany and the specter of the Cold War contributed to a broad-based interest in questions of propaganda and mass persuasion. Yet consumer engineers frequently acted as social engineers in their own right, engaged in efforts ranging from social planning on the community level to the promotion of liberal consumer capitalism. Many leading consumer researchers of the era came from academic and social reform traditions fundamentally different from the culture of corporate America. In some ways, they would remain outsiders to the world of advertising and business. In other ways, however, they adapted and provided creative input to American consumer marketing. As independent or academic consultants, they contributed to marketing's efforts to perpetuate the dynamic postwar economy.

The chapters in this section trace the influence of European immigrants and emigrés on the professionalization of American marketing in general and market research in particular between the 1930s and the 1960s. Chapter 2 follows the transatlantic careers of the "Vienna School" in market research and the influential institutional networks between industry and academia, which Paul Lazarsfeld and his colleagues helped to forge. As consumer research emerged as a new profession in the United States, emigré scientists provided academic legitimacy to the field. Transnational transfers in communication sociology, social psychology, and behavioral economics, as chapter 3 argues, helped to transform prevailing ideas about "the consumer" and the dynamics of marketing communication. A new focus on consumer motivations and the social fields in which consumption took place owed much to the work of emigrés such as Kurt Lewin and George Katona. They laid the groundwork for the postwar rise of a new generation of marketing consultants, which Vance Packard vilified as "Hidden Persuaders." With German emigré Alfred Politz and Austrian emigré Ernest Dichter as their most prominent faces, these independent research consultants represented the new overall importance of consumer research in the U.S. economy. Politz and Dichter serve as case studies for a postwar (self-)perception of consumer researchers as science-based engineers of modern mass consumption and, at the same time, for professional marketing's claim to be a creative force within consumer capitalism.

TWO

The Art of Asking Why: The "Vienna School" of Market Research and Transfers in Consumer Psychology

In the wake of the Great Depression, forecasting consumer demand became an issue of central importance not only to American corporations but also to the New Deal state. Corporate marketing specialists and government administrators alike were looking for new ways and better methodologies to study consumption and to analyze consumer behavior. In January of 1934, Alexis Sommaripa, a product and consumer research specialist at the DuPont chemical company, wrote to the director of the National Recovery Administration's Consumer Advisory Board about a shared interest in far-range forecasting of consumer needs. Sommaripa, acting in his capacity as committee secretary for the Business Advisory and Planning Council to the Department of Commerce (a body comprising executives of large corporations from General Motors and AT&T to Sears Roebuck), emphasized the importance of understanding "consumer satisfaction" and of differentiating between the consumer behaviors of "people of different income and occupational levels."[1] The United States, he believed, needed to develop more sophisticated approaches to "studying consumer buying motives." For this, Americans could benefit from looking abroad: "Although quantity measures for various goods have been developed to a certain extent in this country," Sommaripa observed, "their psychological approach has been very superficial. We understand that Vienna being the present center of the science of psychology has developed very important methods of analysis as applied to consumer needs."[2] Early market research efforts in the United States had provided some numerical sense of what and how much people bought, but the data said little about the "why" or about the considerations that structured demand and buying decisions.

The interest of Sommaripa and of New Deal administrators in Viennese consumer psychology was focused on one person in particular: Paul Lazarsfeld, a young social psychologist who had come to the United States the previous year to study American marketing and research methods. Soon, however, Lazarsfeld set out to teach his American colleagues the "art of asking why," advocating qualitative approaches to reveal consumer motivations.[3] His career not only illuminates the degree to which market and consumer research became more methodologically sophisticated over the course of the 1930s and 1940s but also highlights the impact of transatlantic knowledge transfers facilitated by a small group of European-born immigrants. Well into the 1950s and 1960s, Lazarsfeld and several of his fellow emigrés from Vienna and elsewhere would come to play influential roles in American market research. In her study of America's postwar consumer society, Lizabeth Cohen briefly credits a group of emigrés with laying the methodological groundwork for a shift toward motivational market research and segmented marketing with significant ramifications for postwar American consumer culture.[4] Contemporary marketing professionals shared this assessment. When the *Journal of Advertising Research* portrayed eight "founding fathers" of ad research in its June 1977 issue, their list included three emigrés, Alfred Politz, Hans Zeisel, and Ernest Dichter. Several of the others, including Frank Stanton, were intimately tied to Lazarsfeld and his group.[5]

A close connection between corporate marketing and the academic social sciences was crucial for this development. The professionalization of marketing research and efforts in "consumer engineering" drew on new insights in fields from social psychology to communication studies, which thrived at midcentury because of transatlantic knowledge circulation. This chapter follows the exemplary transatlantic careers of members of the "Vienna School" of market research. The group emerged from the Wirtschaftspsychologische Forschungsstelle, a social research institute associated with the University of Vienna during the early 1930s. Besides Paul Lazarsfeld, the group most prominently included the sociologist Hans Zeisel as well as the motivation research specialists Herta Herzog and Ernest Dichter. Their careers suggest a more transnational understanding of midcentury American consumer capitalism, with European—in this case particularly Viennese—influences shaping marketing practices, which consumer historians still often regard as a quintessentially "American" phenomenon of psychological consumer manipulation.[6]

The story of the "Vienna School" reveals the growing importance of "scientific" marketing and an increased exchange between business and academia at midcentury. Transfers took place on several levels, and the subsequent chapters will analyze the role of individual emigré scholars, of the professional networks they formed, and of the research concepts and methodologies they developed between Europe and the United States. Lazarsfeld and his fellow emigrés helped to introduce new organizational forms ranging from research institutes to independent consulting firms, which connected academic research with commercial marketing. Speaking before the World Association for Public Opinion Research in 1967, Hans Zeisel listed several key aspects of what he regarded as the legacy of the "Vienna School" for American marketing. The group had facilitated a close connection between academic theory and commercial practice, and they had established the psychology of purchase acts as a field of academic inquiry. They also developed precise and psychologically informed interview methods, Zeisel argued, which emphasized the limits of rational consumer decisions and the role of subconscious motivations for consumer behavior. More than introducing Freudian psychology, however, the group combined qualitative methods with empirical and mathematical methods. This interplay of qualitative and quantitative approaches to consumer research was partly rooted in research conducted in interwar Vienna. It was then further developed by the emigrés and their American collaborators over the course of their careers in the United States to become a central aspect of consumer engineering. The Vienna School, Zeisel boldly claimed in retrospect, had put the "first sharp tools [into] the toolkit of empirical social research."[7]

The significance of the group around Lazarsfeld for the development of transatlantic empirical social research has been widely acknowledged.[8] Here, the focus will be on the reception of their methods within American marketing. At a moment when American marketers were searching for innovative approaches to consumer psychology that went beyond the long dominant behaviorist stimulus-response, these European academics were widely welcomed in the United States. Yet many emigrés, often Jewish and politically left-leaning, came from academic backgrounds that were steeped in social reform traditions; this made them attractive to New Deal social engineers but complicated their integration into corporate marketing contexts. Not only did academic and commercial goals come into conflict, but individuals such as Paul Lazarsfeld, whose socialization and original research agenda were heavily intertwined with the socialist reform movements of "Red Vienna," had more than just a transnational divide to bridge.

Toward "Scientific Marketing": Calls for Professionalization of Marketing Research in the United States

The success of transfers of any kind depends on the willingness of "receiving partners" to openly engage with new ideas and concepts. In the present case, the field of American marketing research was actively searching for new methodological input from the academic world as claims to "scientific methodology" promised to bolster the standing of marketing experts as members of a new "profession." Calls for "scientific marketing" in analogy to "scientific management" had been reverberating within the field at least since the 1910s, building on work by Walter Dill Scott and Arch W. Shaw.[9] Advertising agencies were among the first to apply new academic concepts to practical marketing research. Already during the 1890s, N. W. Ayer had undertaken early systematic efforts in "campaign planning," which by the 1920s entailed research and consulting on market developments, sales strategies and organization, and even package design.[10]

Following World War I, J. Walter Thompson (JWT) emerged as the prototypical full-service agency. Under the leadership of Stanley Resor, JWT strongly emphasized its "scientific" approach to advertising: "Advertising must be scientifically prepared. Nothing must be taken for granted."[11] The agency employed the psychologist John Watson, arguably the most prominent proponent of behaviorist stimulus-response research during the 1920s, to systematically research consumers' behavior and their responses to advertising. Following Pavlov's experiments with dogs, such behaviorist research explored possibilities of conditioning consumers to react to advertising stimuli.[12] For full-service ad agencies market and consumer research became essential to their "scientific" self-understanding.

Specialized market research firms began to offer their services during the early decades of the twentieth century, but their real growth period came during the 1930s and 1940s. The A. C. Nielsen Company (est. 1923) offered market analyses and consumer demand projections based on household studies that asked consumer panels about what they stocked in their pantries. Studies in media use for marketing purposes had been pioneered by the Curtis Publishing Company (*Saturday Evening Post, Ladies Home Journal*), which set early standards for consumer reception research.[13] Institutions specialized in public opinion surveys such as the firms of Archibald Crossley (est. 1926), Elmo Roper (est. 1937), and the American Institute of Public Opinion, founded in 1935 by George Gallup in cooperation with advertising executive David Ogilvy, also conducted commercial consumer research. By the late 1930s, the Market Research Corporation of America (est. 1934) likely had the largest contingent of interviewers across the United States, conducting elaborate surveys and publishing the trade journal *Market Research*.[14] The Psychological Corporation, finally, had been established already in 1921 by James Cattell and connected academic psychologists with corporate clients.[15] Its "psychological sales barometer" drew on the expertise of sixty academic psychologists to survey changing customer preferences regarding various brands. The efforts of the Psychological Corporation count among the early attempts at systematically unearthing consumer motivations.[16]

In short, consumer research (ranging from market surveys to psychological analyses) had become part of the American marketing landscape at the beginning of the 1930s. Outside the largest consumer goods corporations, however, this type of research was neither common practice nor particularly sophisticated in its statistical and psychological methodology. This would change over the course of the following decades. Heeding the call for "consumer engineering," marketing specialists increasingly engaged in what they termed "merchandising": they planned products based on customer expectation and demand.[17] Merchandising entailed an orientation of the product styling and packaging toward sales potential and thus required companies to know their markets much more closely than ever before. Marketing experts envisioned a "new consumption era" in which the distribution of goods would be transformed by insights from the social sciences, and earlier, "spasmodic" efforts in salesmanship had to give way to "more definitively and scientifically planned campaigns for the consumption of goods."[18]

To match such confident rhetoric, market researchers sought a professional identity that would put them on a par with other scientists and engineers already established in the business world. While early marketing men had learned their trade in the practical business world, in sales or retailing and advertising, by the early twentieth century marketing experts came to the field with decidedly academic backgrounds. Leading proponents of professionalization including Hugh Agnew, Paul Converse, George Hotchkiss, Charles Parlin, Stanley Resor, Walter Dill Scott, and Arch Shaw had been trained in disciplines such as economics and psychology; and many of them would later in life take on university positions themselves or publish textbooks on aspects of scientific marketing.[19] Their work helped usher in a much broader professionalization process during the 1930s. Over two decades, the number of colleges and universities offering training in marketing nearly doubled from 299 in 1930 to 588 by 1950, growing at a much faster rate than higher education overall. The breadth of specializations also expanded, ranging from advertising and sales management to market research.[20]

Those employed in marketing positions at corporations and advertising agencies thus increasingly came with specialized training. Professional organizations gave further cohesion to the emerging field. The Professional Organization of Marketing Teachers formed as an outgrowth of the American Economic Association in 1915, and twenty years later the organization claimed 375 members. In 1930, the American Marketing Society was created with a membership composed more heavily of practitioners than academics. Both organizations ultimately merged to form the American Marketing Association, combining their journals, the *American Marketing Journal* and the *National Marketing Review*.[21] The number of publications on specific aspects of industrial marketing and marketing research grew in similar fashion over the 1930s and 1940s with over a hundred titles published between 1930 and 1949 alone, more than twice the number published in all years leading up to 1930.[22] The 1936 merger of the *American Marketing Journal* and the *National Marketing Review* into the *Journal of Marketing* as an academic periodical for the field, finally, gave perhaps the most significant impetus for the further development of scientific marketing thought.

Within the pages of that journal, the marketing field saw a vigorous debate over the question of how "scientific" marketing already was and whether it should strive to emulate the natural sciences.[23] Lyndon Brown's 1937 handbook, *Market Research and Analysis*, devoted an entire chapter to the general concept of the "scientific method" and its particular application in fields such as psychology, sociology, or statistics. Brown's aim was to familiarize businessmen with the "rapid growth of scientific method in marketing" and to emphasize the importance of scientific methodology.[24] Other surveys of the field also argued that recent progress in research meant that the profession had begun to develop a discrete body of knowledge.[25] Throughout the early years of its

CHAPTER TWO

publication, the *Journal of Marketing* closely tracked ongoing academic consumer research across universities in the United States, emphasizing the connection of commercial practitioners to academic investigation.[26] This very exchange between academic and commercial research provided career opportunities to European emigrés with expertise in social research, statistics, economics, or psychology and allowed them to help shape the nascent American marketing profession. Thus, when the American Marketing Society published its first handbook, *The Technique of Marketing Research*, in 1937, Viennese emigré Paul Lazarsfeld contributed several chapters.[27]

Interwar Vienna and the Study of Modern Consumer Markets: The Wirtschaftspychologische Forschungsstelle

Early twentieth-century Vienna provided a promising setting for the emergence of research on modern consumer culture.[28] The city possessed a vibrant intellectual climate that combined an interest in subjective individualism and aesthetics with an attention to the development of new methodologies in the empirical sciences.[29] Modernism in art and architecture flourished as did new approaches to design.[30] Viennese intellectual life did not confine itself to specialized academic circles but integrated into a broader milieu of coffeehouses and salons as well as into the city's large Jewish community.[31] Several schools of psychological thought influenced Viennese intellectual circles, ranging from the individual psychology of Sigmund Freud's psychoanalysis and the more politically engaged work of Alfred Adler to Karl and Charlotte Bühler's interest in developmental and social psychology.[32]

From an American perspective, Vienna's diverse population with its large proportion of immigrant and minority groups in some ways resembled the consumer markets of Chicago or New York at the time, in which goods and the visual language of advertisements and design helped to provide a shared frame of reference. Politically, by contrast, interwar Vienna differed from its American metropolitan counterparts. Social Democratic "Red Vienna" pioneered efforts in "municipal socialism" with housing projects, social work experiments, and an emphasis on public and adult education for its working class population. This interest in public development spawned research on consumer standards of the "masses." A prominent example was the work of reform economist Käthe Leichter who during the early 1930s studied household budgets of female household laborers and conducted surveys for the Vienna Chamber of Labor (*Arbeiterkammer*).[33]

The intellectual and political environment of "Red Vienna" had a decisive impact on the members of the Vienna School of market research. Hans Zeisel recalled Freud and Adler's work in depth psychology as direct influences on the research of the group when it came to hidden motivations and the "bounded rationality" of consumer action. The behavioral insights regarding instincts of Viennese zoologist Konrad Lorenz as well as the causality debates and the methodological rigor of the so-called Vienna Circle of philosophers of science, he claimed in retrospect, also informed the thinking of the group. Finally, Zeisel added, the city's socialist reform movement and its efforts to improve living conditions and educational and recreational possibilities for the working class spurred on their research agenda.[34] In a 1961 interview, Lazarsfeld similarly reminisced about the intellectual and political coffeehouse culture of Vienna, and his immersion in it as a youth through his mother, who had had significant standing within Social Democratic social circles.[35] Like his first wife and co-worker Marie Jahoda, Lazarsfeld was very active in socialist youth organizations, and this involvement informed their early research on youth development and other social phenomena during the 1920s and 1930s.[36] Jahoda later recalled an interest in the project of "Austromarxism" as a driving force of their early work. This, she claimed, had brought her, Lazarsfeld, and others to the social sciences in the first place.[37] The metropolitan culture of interwar Vienna thus shaped the beginnings of the emigrés' careers.

From the outset, however, the Viennese context had transnational dimensions as well. Most of the group began as assistants or students of Karl and Charlotte Bühler, who had established the University of Vienna's institute for psychology in 1922. They received sustained support from the American Rockefeller Foundation. Indeed, Charlotte Bühler traveled to the United States on a Laura Spelman Rockefeller Memorial Fellowship in 1924/25 and later taught as a visiting professor at Barnard College and Columbia University. Her husband also spent time in the United States before he was arrested for his political engagement following the 1938 Nazi takeover of Austria.[38] Thus, the Viennese group was exposed early to American advances in empirical sociology, such as the "Middletown" study by Robert and Helen Lynd, which had a lasting impact on Lazarsfeld and his colleagues. Similarly, exchanges of the Viennese scholars with other European cities were significant, and Berlin psychologists such as Wolfgang Koehler and Kurt Lewin, for example, frequently appear in the theoretical discussions of early market studies. Both Dichter and Lazarsfeld, furthermore, spent time in Paris, providing them international experiences prior to their emigration.[39]

The origins of the Wirtschaftspsychologische Forschungsstelle date back to the late 1920s. An association of sponsors was incorporated in 1931 with a membership list that included academics such as economists Ludwig von Mises and Oskar Morgenstern as well as prominent Austrian industrialists, businessmen, and representatives of labor and commerce institutions. The institute was conceived as an "Austrian Institute for Motivational Research [*Motivforschung*]" tasked explicitly to study the psychology of markets, understand consumer behavior, assess public social programs, and generally improve the conditions of economic life in Austria.[40] Paul Lazarsfeld, who had come to psychology by way of a degree in mathematics, became the driving force behind the institute and its research. His interest in psychological motivations was coupled with a strong desire to innovate and improve nascent empirical survey methodology.

The institute was engaged in a broad array of empirical social studies until it was shut down by the Austrian government for political reasons in 1936.[41] Much of their research had been derived from Vienna's municipal social policy programs including a community study of a new housing development in the suburb of Leopoldau. The Forschungsstelle also conducted listener surveys for Vienna's public radio broadcasting station RAVAG that count among the first audience studies for new mass media.[42] The most well known study of the institute was a comprehensive community study on unemployed workers in the industrial town of Marienthal. Following a recommendation by Austrian socialist leader Otto Bauer, the social scientists shifted their research focus from working class leisure culture to the social and cultural effects of unemployment following the Great Depression. The sociographic study documented the disruptions of the Depression and explored the coping mechanisms of laid-off workers in a way that cemented the academic reputation of Lazarsfeld and his collaborators at the time.[43]

These studies were partially funded by commercial market research, which paid the bills of the Forschungsstelle.[44] The inspiration to conduct market surveys, Lazarsfeld later claimed, had come from observing an American student interviewing consumers in Vienna about their use of soap for an American company. Getting European companies to pay for market surveys was initially a tough sell, however.[45] The academics thus directly approached business associations and corporations. Customer surveys, Lazarsfeld explained before the advertising committee of

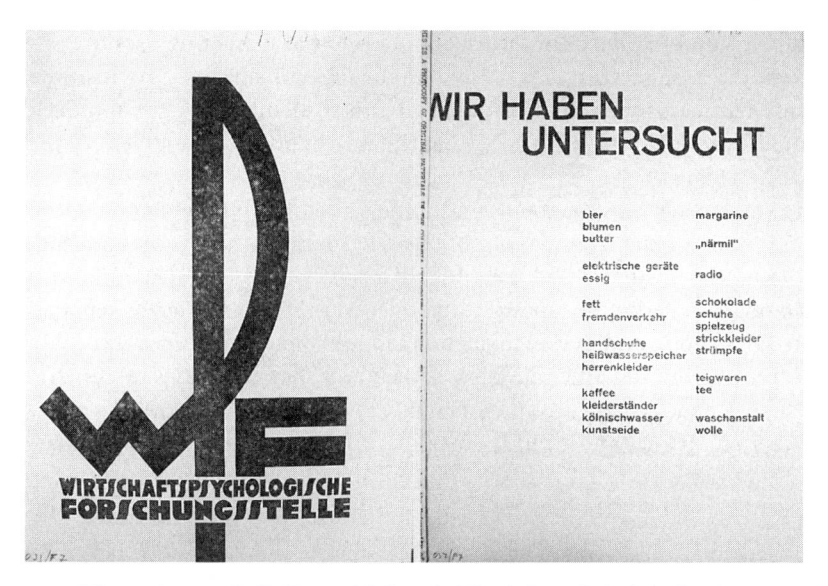

2.1 "What we have studied": PR materials from the Wirtschaftspsychologische Forschungsstelle, c. 1933. Courtesy of Paul F. Lazarsfeld Archives (University of Vienna) and with permission of the Rare Book & Manuscript Library, Columbia University.

the Berlin chamber of commerce, could reveal individual needs and particular motives. Why do customers base their buying decisions on store displays when it comes to shoes, while personal recommendations come into play more prominently when they are shopping for clothes? How do consumers differentiate between various brands? The key to proper market research, according to Lazarsfeld, was to tease out of respondents all that they actually knew about their own consumption preferences and shopping habits. Psychological factors such as assumptions about quality and especially suspicions about flaws informed consumer action, and businesses, he emphasized, had to learn to see their products and markets through the eyes of the consumer.[46]

The market studies were more than a source of revenue. Lazarsfeld, Herta Herzog, and other researchers at the institute used them as practical fieldwork to develop general theories about "true" consumer motivations, "consumer biographies," or the impact of advertising on decision making.[47] The studies helped them to refine interview methodologies designed to uncover subconscious factors in decision-making processes,[48] and researchers published articles and dissertations based on the commercial research.[49] Still, the reports produced by the Forschungsstelle were careful to give clear advice about potential marketing uses to their customers, suggesting advertising strategies and slogans. The institute very consciously wedded commercial and academic work, an approach that Lazarsfeld would apply several times over after he emigrated to the United States.

Lazarsfeld's move across the Atlantic served the budding commercial ambitions of the Vienna Forschungsstelle. Between the early and mid-1930s, the institute had established a small European network with branches in Berlin, Budapest, Stuttgart, and Zurich. Herta Herzog reported in 1933 about studies on Prague tourism, milk and electronics consumption in Berlin, the Stuttgart beer market, and shoe purchases in Zurich.[50] The Forschungsstelle thus increasingly looked beyond the confines of the Viennese market and conducted studies for companies in various industries ranging from Bally shoes and Närmil malt products to producers of electronic goods. The institute offered a "sales and consumption barometer" to which companies could subscribe and which provided quarterly updates on regional sales developments and the buying behavior of particular social groups based on surveys of at least ten thousand respondents annually. More comprehensive analyses for individual firms offered detailed assessments of markets and leading competitors. Clients offered enthusiastic feedback, crediting sales increases and reduced advertising costs to the "scientific work" and the "practical advice" of the Forschungsstelle.[51] The overall intention was to shift consumer research from intuition to observable facts.[52]

Nearly all of the institute's surveys paid close attention to differences in social class, and many open-ended questions aimed at the psychological factors involved in buying goods. One pilot study on the marketing of prebrewed coffee, for example, saw the potential of this product largely with working class consumers who thought of coffee as an affordable sustenance good rather than as an article of refined enjoyment. Still, the study suggested the use of the slogan "my coffee" (*mein Kaffee*) to advertise the ready-made coffee to petit bourgeois women who were psychologically invested in preparing their own coffee. The slogan was intended to elicit a feeling of pride and achievement among consumers of the commercially made product that merely required to be reheated.[53] The institute's research strove to unearth the factors that motivated purchases for different groups of consumers. A study for a men's clothing retailer found that consumers said that "quality" above all was the deciding factor in their purchasing decisions. Further probing uncovered, however, that most men had only a very vague notion of what quality actually entailed and that they simply chose clothing stores their friends had recommended or where the sales personnel had "an air of reliability."[54] While the institute remained a small operation plagued by recurring financial woes, Lazarsfeld and his collaborators became early providers of psychologically based marketing advice in 1930s Europe.

Most of the academics involved with the Forschungsstelle did not have a problem with commercial research. Marie Jahoda reflected back in an interview on the "dilemma" of doing capitalist market research in the context of "Red Vienna." To the group, she recalled, all of their research, including the market studies, served the larger purpose of improving the living situation of the "proletarian consumer."[55] As their careers crossed the Atlantic during the 1930s, the socialist background of the "Vienna School" eventually became more and more of a faint memory. Initially, however, the very social reform interests that informed their work helped some of them to gain the attention of New Deal reformers interested in class and consumption. Their attention to social market segmentation at a time when "averages" and the "mass" still dominated much market research and their use of psychological methodology beyond traditional behaviorist models were of interest to corporate America as well.[56]

"Doctor in America": Paul Lazarsfeld's Transatlantic Career in Market Research

Paul Lazarsfeld came to the United States in 1933 on a grant from the Rockefeller Foundation. He had applied for the fellowship with the express purpose of studying American market research and survey techniques. His objective, as he explained in the grant application, was the "study of American organizations in which psychology is applied to *economic* problems." "I should thus like to learn . . . those methods used in America in connection with market analysis, salesmen training, advertisement technic etc." His first stop was to be Columbia University in New York, where not only Robert Lynd was teaching, but research in the psychology of advertising was being conducted as well. Subsequent places on his proposed itinerary included the Department of Labor and New York University's Bureau of Business Research, along with private advertising agencies and market research firms in New York City. He also planned a short stop in Boston to study market analysis at the Harvard Business School, "of which one hears so much in Europe."[57]

In part, Lazarsfeld's trip was motivated by an attempt to interest American firms in retaining the services of the Wirtschaftspsychologische

DR. LAZARSFELD
Analytically it all begins with deliberation

Consequently their lack of knowledge drives them to retailers their friends recommend or where the salesmen seem to inspire confidence.

CONCLUSION

And on the basis of that conclusion, retailers were urged to inspire confidence of customers by store & window displays that showed how textiles were woven, how strong the fabric is, etc., etc. Both analysis and suggestion, as subsequently tried out, proved "greatly successful."

Other fields in which successful work has been carried on: laundry, beverage, shoes, vinegar, patent medicines.

COMMENT

Of his American hosts Dr. Lazarsfeld last month said that he was "appreciative of their great achievements," approved the fact that they don't put too much premium on academic face.

2.2 Coverage of Paul Lazarsfeld (1901–76) in the U.S. trade press. "Doctor in America," *Tide: The Newsmagazine for Advertising and Marketing*, Nov. 1934. Courtesy of Paul F. Lazarsfeld Archives (University of Vienna) and with permission of the Rare Book & Manuscript Library, Columbia University.

Forschungsstelle to conduct market surveys for their products in Europe. Soon after his arrival, he reported back to Vienna on plans to bring the so-called "Dr. Gallup Test" to Europe. Gallup had developed an ad recall test in which interviewers and respondents assessed together if and to what degree subjects were able to recall magazine ads to which they had been exposed. Lazarsfeld developed ties to the advertising firm Young & Rubicam and was hopeful that their London office might commission the Forschungsstelle to do market research studies. The introduction of the "sales barometer" methodology in Vienna, finally, rested in part on Lazarsfeld's association with the Psychological Corporation. Soon after his arrival, they hired him as an advisor, providing income to send back as remittances to subsidize the Vienna institute.[58] The Wirtschaftspsychologische Forschungsstelle expected both to profit and to learn from the transatlantic exchange, as Lazarsfeld was repeatedly reminded in subsequent letters from his colleagues back in Vienna.[59]

However, Lazarsfeld was soon convinced that he also had much to teach the Americans. In a 1934 letter he lamented the dire state of affairs of psychology in the United States. The few serious academics such as John Watson who were involved in American market research, he observed, merely lent their name for financial gain. Trained European experts, once they were recognized as such, he explained, were therefore "welcomed with great enthusiasm" in the field of market analysis.[60] He quickly published articles based largely on Austrian research in academic journals such as the *Harvard Business Review* and *Market Research* in which he detailed the institute's Viennese attempts to tease out consumer motivations and explained their approach to survey methodology. In addition, he prepared a manuscript on the use of psychology in marketing to be published with Harper Brothers.[61]

The advertising trade publication *Tide* portrayed Lazarsfeld's work and appearance on the American scene in 1934 under the headline "Doctor in America." The "Viennese psychologist" was introduced to readers as the "portly, bespectacled Dr. Paul F. Lazarsfeld," who was chiefly concerned with the study of "motives" as they enter into the purchase of goods. Making an argument for why the study of psychology was useful and should be employed commercially by marketers, the article related examples from Lazarsfeld's "European stock of experience" such as studies on ready-made clothes retailing and attitudes regarding sales personnel. Where the "psychologist finds his special role," the "doctor" from Vienna suggested, was in assessing and interpreting the factors (product attributes, outside influences, impulses, etc.) that make up the deliberation process of consumers.[62]

Like many other successful European emigrés, Lazarsfeld crafted a persona for himself—of an innovative and creative European academic with a rigorous intellect—which would help him advance his career in the United States. Of course, he recognized that interwar American corporations had been more open to psychological work than their European counterparts and that much commercial research work was already being done. However, he later recalled this to have been "on a terribly unsophisticated level. So that . . . if a European really specialized in the work, he'd just run rings around them."[63] In a letter from the early 1930s, Lazarsfeld claimed that European university professors, much like commercial advertisements, had a powerful impact on what he perceived as an impressionable American public.[64] More than anything else, such statements speak to Lazarsfeld's confident personality, which greatly facilitated the transatlantic transition in his career. Despite the prevalence of anti-Semitism in U.S. academia as well, he persuaded the Rockefeller Foundation to extend his stay in the States at a time when the political situation in Austria was growing increasingly troublesome for a leftist scholar.[65]

One key to Lazarsfeld's success in emigration was his ability to build professional networks. From the start, he made numerous connections among academic and commercial consumer researchers. As a Rockefeller fellow he provided a link between the foundation and the Psychological Corporation where he cooperated with Rensis Likert, who at the time was working on the impact of advertising at New York University's department of psychology.[66] Robert Lynd, meanwhile, had gotten him in touch with the Market Research Council in New York, which invited him to speak on his research in March of 1934.[67] Advertising executives from J. Walter Thompson and other agencies were interested in his views, and Lazarsfeld was in discussions with General Foods and its advertising firm Benton & Bowles about a large-scale study on attitudes and consumer behavior regarding food consumption.[68] Also in 1934, he met with General Electric marketing managers in Schenectady, New York, about survey projects they were planning in cooperation with Cornell psychologist John Jenkins, while Eastman Kodak corresponded with him about advertising research.[69]

Lazarsfeld thus quickly established himself as an active participant in the nascent field of American market research. Especially close was his association with David Craig at the University of Pittsburgh. Craig headed the Research Bureau for Retail Training and was similarly connected to the Psychological Corporation. He collaborated with Lazarsfeld on a survey on consumer acceptance of new rayon-fiber hosiery that was sponsored by Kaufmann's Department Stores in Pittsburgh. In early 1935, Lazarsfeld proposed to establish Craig's bureau as a "clearing house for methods of market analysis" in the United States, much as the Wirtschaftspsychologische Forschungsstelle had worked for its European network. Writing to Craig (who immediately relayed the letter to his corporate collaborator Edgar J. Kaufmann), he urged that the Pittsburgh bureau should offer both industry-wide market studies at regular intervals for all area retailers and confidential studies for individual clients. Lazarsfeld insisted that the demand for psychological surveys was "steadily increasing": "During the last 15 months [following his arrival in the United States], I have been consulted again and again by different institutions as to the best way of formulating questionnaires, setting up experiments, conducting interviews, and so forth. American market research is now at a state where it needs theoretical foundations and guidance for its activities." Thus, he recommended academics to act swiftly before less "conscientious" private entrepreneurs could swoop in and "damage the reputation of business research."[70]

During the mid-1930s, Lazarsfeld established himself as a transatlantic knowledge broker with one foot firmly planted in Europe and the other in the United States. All the while he kept up contact with his home base in Vienna, exchanging cables, for example, with Hans Zeisel about approaches to retail sales training based on the latest research in Pittsburgh.[71] This continuing connection to Europe also bolstered his American career insofar as he could leverage his European expertise with both corporate and state contacts in the United States. His early affiliation with the Psychological Corporation, the New York–based private market research company whose officers included the advertising psychologists Walter Dill Scott, Henry Link, and Rensis Likert, followed that premise. In 1935, Paul Achilles, managing director of the Psychological Corporation, wrote to Gallup and other corporate clients, recommending Lazarsfeld as a contact in Europe that summer: "His Bureau for Psychological Field Work will represent us on the continent, and the London office of the National Institute of Industrial Psychology, which is active in market research, will represent us in England."[72] American market researchers saw him as a promising boost of local expertise in supporting their attempts to make inroads into European markets.

As it had been in "Red Vienna," consumer research was of interest to state agencies in New Deal America as well. Lazarsfeld was in contact with the Bureau of Foreign and Domestic Commerce about the possibility of market studies for American foreign trade. Here, too, he stressed the importance of local knowledge, pointing out that, compared with America, "European markets are much narrower, competition is stronger, so much so that they need more detailed information to characterize the market situation."[73] On domestic demand he advised the Consumer Advisory Board, as mentioned at the outset of this chapter. Based in part on his Viennese experiences, Lazarsfeld developed plans for wide-scale household consumption studies that would allow government agencies to discern the patterns and motives of household spending, focusing on processes of household decision making as they influenced macroeconomic developments.[74] The "consumer" rather than the "worker," Lazarsfeld reported back to Austria, had become the model of "economic man."[75] To mid-century social scientists like Lazarsfeld, the study of consumer behavior thus offered the key for future conceptions of economic and social democracy.

Despite his numerous connections to American corporations and to the New Deal state, academia rather than politics or commercial research would remain Paul Lazarsfeld's primary field of activity in the United States. Following the Vienna model, he established his academic position by building survey research institutes.[76] The first such institute was established at Newark University in 1936 and grew out of research funded by another New Deal agency, the National Youth Administration. Much like the Wirtschaftspsychologische Forschungsstelle in Vienna, the Newark center conducted empirical and psychological social research while getting much of its financing through commercial market studies.[77] Social research and consumer research again went hand in hand at Newark for Lazarsfeld and his team of newly hired American researchers. They conducted a study on milk consumption in the youth market for the Milk Research Council of New York, and media use became a significant focus with a study on magazine reading in American cities for the trade journal *Advertising and Selling* and a survey on the use of home movies for Eastman Kodak.[78] Lazarsfeld continued to draw on his networks in market research, collaborating with the Psychological Corporation on a study of the perception of textiles made from artificial materials such as rayon. The study was sponsored by DuPont Chemicals and supervised by the same product and marketing executive, Alexis Sommaripa, who had expressed his interest in Vienna market research as a member of the Business Advisory Council a few years earlier.[79] Lazarsfeld's "Viennese fiscal principle" that academic research should draw on public and private sponsorship for scholarly surveys was ideally suited to applied social research closely aligned with (but formally separate from) commercial and government interests.

In 1937, the Newark-based research center was absorbed into the newly established Office of Radio Research (ORR).[80] The office, well-known for its prominent role in early communication studies, was initially tied to Princeton University and supported by a Rockefeller grant for research into commercial radio's potential for educational programming (while explicitly barring research overtly critical of commercial radio itself). Princeton psychologist Hadley Cantril, who had ties to the Gallup organization and was a founding editor of the journal *Public Opinion Quarterly*, was named director of the new research office along with Frank Stanton, the research director and later president of the Columbia Broadcasting System (CBS). In search of a managing director, they turned to Lazarsfeld on the recommendations of Robert Lynd and Cantril's mentor, Harvard psychologist Gordon Allport. Lazarsfeld's practical experience with research on radio listening for the RAVAG in Vienna had attracted the attention of U.S. scholars, and the real center of the "Princeton" Office of Radio Research would soon be located in Newark.[81]

In Europe as in the United States, radio was still a new and politically controversial medium during the 1930s. Its potential for mass communication appeared as both promise and menace, and in the United States at least, the political "radio wars" saw a commercial broadcasting framework succeed over state and public broadcasting models with the Communications Act of 1934.[82] Focused primarily on entertainment programming sponsored by commercial advertising, the large broadcasting networks developed an interest in tracking their audiences. The broadcasters as well as advertising agencies regarded listeners as potential consumers, and firms such as Crossley conducted telephone surveys to ascertain who listened to what and when. This commercial research, however, had problems in methodology (phone surveys, for example, reached only the wealthier strata of society) and was beholden to the interests of its clients and thus frequently biased. Listener attitudes and the more fundamental effects that radio had on consumers were issues beyond the scope of much of the existing research. George Gallup of the American Institute of Public Opinion noted in 1937: "We know virtually nothing about the influence of radio on listeners, or the reasons why persons listen to certain programs and not to others."[83] The Office of Radio Research promised to provide such independent research on the psychology and motives of radio audiences and on the effects that radio had on its listeners. Still, this academic effort was closely watched by and remained heavily intertwined with commercial interests.[84]

Research at the Office of Radio Research affected marketing and advertising in the United States more broadly. The significance of the radio project lay in its application of new survey methodologies in audience and communication research. Lazarsfeld and his collaborators worked with media use data (provided by CBS and advertisers) on a new scale and introduced new qualitative techniques such as the Stanton-Lazarsfeld program analyzer and panel studies. In 1939, the *Journal of Applied Psychology* published a special issue on radio research under Lazarsfeld's editorship that publicized the findings of the research group. Corporations and marketing professionals had been involved in this work since its very inception, as CBS funding and programming data provided a cornerstone of this research. As he had done elsewhere, Lazarsfeld drew on a host of corporate contracts and collaborated with commercial organizations. The Gallup organization, for example, occasionally supplied polls to be analyzed by project members, and, for a fee, the research departments of both CBS and the advertising agency McCann-Erickson used the new methodologies in their own work.[85]

What up to then had simply been Lazarsfeld in emigration now became a broader "Vienna School" of market research at the Office of Radio Research. As political conditions in Europe worsened by the late 1930s, more and more emigrés could be found in the orbit of Lazarsfeld's research institutions, many coming from the Vienna Forschungsstelle. In addition to a core group of American researchers including Hazel Gaudet and Marjorie Fiske, Lazarsfeld managed to bring several key members of the Vienna institute to the United States. Most important, he introduced Herta Herzog, who became his second wife in 1935, to American academia as a specialist in audience research. Herzog's dissertation, "Voice and Personality," was an early mass survey experiment that analyzed the physiognomics of the human voice as it came over radio.[86] As an employee of the Radio Project, she co-authored an analysis of the famous "Invasion from Mars" panic among listeners, which had followed Orson Welles's 1938 radio broadcast of H. G. Wells's *War of the Worlds*.[87] Herzog participated in the development of qualitative and quantitative research methodology in Newark and New York, studying daytime serial broadcasts among other topics. Hans Zeisel, who had managed the Vienna Forschungsstelle during Lazarsfeld's absence, now also joined the new American research project.

Other emigrés, too, became involved with the Radio Project. Most well-known is probably Theodor Adorno, who wrote on radio and music for the project until his theoretical style clashed with both Lazarsfeld's more empirical focus and the expectations of those funding the research. The emphasis on commercial broadcasts, furthermore, was at odds with Adorno's critical social science agenda.[88] Still, the fellowship at the office offered him a first opportunity to come to the United States, following Max Horkheimer and other members of the Frankfurt Institute for Social Research, which had relocated to New York City in 1935. Both Lazarsfeld and Adorno continued to share an interest in modern mass consumption and communication as their careers progressed.[89] Whereas Adorno and his Frankfurt colleagues never quite felt at home in the United States personally or academically, Lazarsfeld and his Viennese group transitioned across the Atlantic with much greater ease and had less hesitation in crossing the divide between academia and commercial research.

As the relative importance of radio research within the group grew, the project moved from Newark to a new Manhattan location in the fall of 1938. One year later, after tensions between Lazarsfeld and Cantril, the Rockefeller funds were officially transferred from Princeton to Columbia University, where Lazarsfeld took up a lecturer position in sociology.[90] By the early 1940s, yet another institutional structure took shape that continued the basic model of the Viennese Forschungsstelle in American exile, the Bureau of Applied Social Research (BASR). At Columbia, even more Viennese emigrés could be found, coordinating and conducting research in various capacities, including Hedi Ullmann and Ilse Zeisel who joined the group as Rockefeller fellows in 1940. Ernest Dichter, who had been a student of psychology in Vienna and worked for the Forschungsstelle after Lazarsfeld left and would later become a motivation research expert, now also became part of the wider BASR group as he got his start in the United States. For the field of American academic marketing and consumer research, this link between early 1930s Vienna and early 1940s New York would prove quite consequential.

The BASR and the "Vienna School" in Postwar American Marketing Research

The Bureau of Applied Social Research represents most fully Lazarsfeld's model of an independent research institution attached to a university with strong ties to both academia and commercial organizations. The new institute at Columbia relied heavily on corporate funding especially during its early years, and, in return, its research and members had a profound influence on postwar market research in the United States. In 1986, Rena Bartos, senior vice president at J. Walter Thompson Company, reflected on the origins of qualitative marketing research for an Advertising Research Foundation symposium. She discussed early experiments with group discussions and focus groups and the development of hypothesis design and questionnaire structures, as well as early attempts to integrate qualitative and quantitative methodology. Repeatedly, Bartos's account returned to the pioneering work of the group around Lazarsfeld, Herzog, and Zeisel, which she herself had first encountered as a student at Columbia.[91]

Over the course of the 1940s and 1950s, the BASR emerged as an important center for sociological survey research with studies of subjects ranging from media use and wartime public opinion to the psychology of philanthropy, literacy, and urban industrialism, mobility expectations, and communication flows.[92] Starting in 1942, the sociologist Robert Merton acted as an associate director of the institute together with Lazarsfeld, who eventually received a regular professorship at Columbia's department of sociology. Bureau members were employees of Columbia University, and the training of graduate students was an important aspect of its work. Beyond the university, the Bureau saw itself as a service provider to public service organizations without the funds to hire professional research consultants. They conducted contractual research for civil society groups such as the American Jewish Committee as well as for numerous government organizations including the War Department and the Office of War Information. In a memo marking the tenth anniversary of the research program, the Bureau listed among its contributions to social research methodology the "panel technique" of repeated interviews, the program analyzer and content analysis in media studies, and the "focused interview."

Research at the Bureau fell into several broad categories. Besides mass communication research, BASR scholars employed community studies to investigate race relations and other "problems which grow out of the life of modern communities." A "research services" division bundled the Bureau's marketing research operations, which studied consumer reactions and extended "beyond mere polling practices to the investigation of problems involving the factors that determine the opinions of various groups."[93] Much as in Vienna, the research of the Bureau thus combined community studies and economic sociology with consumer research in ways that made their work both academically and commercially applicable.

During its early years, the Bureau depended on commercial contracts for a large share of its operating budget.[94] The BASR had numerous corporate clients especially from the broadcasting and publishing industries, including the networks NBC, CBS, and ABC, publishers such as McGraw-Hill and McFadden, and magazines like *Tide*, *Time*, and *Life*, as well as advertising agencies. Besides American-born colleagues such as Merton and Marjorie Fiske, the Bureau's wider network of BASR collaborators included a whole group of emigré scholars ranging from the media psychologist Rudolf Arnheim (1904–2007) and the sociologist Paul Neurath (1911–2001) to the Frankfurt School sociologists Leo Löwenthal (1900–1993) and Theodor Adorno, who all supported themselves financially in part through Bureau studies. The Vienna group, however, contributed a particularly large share of early BASR studies: Ernest Dichter studied psychological radio programming, Herta Herzog surveyed daytime serial listeners and studied the effects of Kolynos Tooth Powder and Bisodol commercials, while Paul Lazarsfeld wrote on soap operas and produced numerous other studies for CBS. Hans Zeisel and Marie Jahoda similarly produced studies for the Bureau.[95] Jahoda had come to the United States only in 1945, following years in exile in London where she had been connected to the British diaspora of the Frankfurt Institute for Social Research, but had also been involved with market studies for a furniture company.[96] Building in part on work done in Vienna over a decade earlier, the BASR prided itself on establishing the type of psychologically informed qualitative market research characteristic of the postwar decades.[97]

While Lazarsfeld himself began to shift attention away from market research during the 1950s and 1960s, the BASR never abandoned commercial studies.[98] The Bureau, for example, was involved in the Ford Motor Company's launch of the infamous "Edsel" brand, for which the BASR supplied a series of studies on car buying, "brand image," and "personality."[99] Ford had been developing plans to introduce a new mid-price make to rival models in General Motor's Buick and Oldsmobile divisions since the early 1950s. In the context of the postwar boom years, the "ideal personality" of this new brand was supposed to appeal to "the younger executive or professional family on its way up." With this goal in mind, Ford market research turned to the BASR to track consumer perceptions of the Edsel in several waves of a panel study before and after its release in 1957.[100] Bureau studies inquired into what "social and product images" were attached to the car by different segments of the market; they asked what influenced such images and under which conditions attitudes and perceptions actually led to purchase actions.[101]

The notion of "brand images" was still relatively new to 1950s marketing, but the analysis of consumer perceptions and decisions clearly connected to traditions of motivation research within Lazarsfeld's institute.[102] Ford utilized several commercial market research providers for the launch of the Edsel, but Lazarsfeld and others touted the cooperation between the company and the BASR as a model for a joint corporate-academic undertaking. Ultimately, however, this academic-commercial relationship encountered severe tensions. The Edsel project turned out to be a failure, much like the car itself, which suffered from poor design choices, an economic recession, and, despite the tremendous research efforts, from a lack of a clearly defined personality in the minds of many consumers.[103] The project thus hints at the limits of consumer engineering efforts with their promise of adding scientific predictability to corporate planning.

Close cooperation with corporate partners also raised ethical questions about academic integrity and the boundaries of commercial involvement. Lazarsfeld, to be sure, was less concerned about this than some of his colleagues. After he had left his position as BASR director in 1949 to become chairman of the sociology department, he still continued to push for the Bureau's involvement in commercial research.[104] Even in retirement Lazarsfeld remained involved as an advisor to corporate America. For AT&T, for example, he helped foster research in telecommunication, and he was hired by the tobacco industry, giving insight into the motivations of smokers for Philip Morris in return for a personal fee of $10,000.[105] When the American Association of Advertising Agencies approached him about a possible study of "the social role of advertising," Lazarsfeld was eager to cooperate. In collaboration with Columbia marketing professor John Howard, Lazarsfeld was to study both the effects advertising had on individual consumers and its unintended consequences for society at large. Asking for the study to be placed within the framework of the BASR, Lazarsfeld wrote to Howard: "I have throughout my academic career always been interested in the study of advertising and consumer behavior from the social science point of view. I would want, therefore, to give this project all my personal attention."[106] The study was a clear attempt by advertising interests to counter mounting criticism of advertising practices from social scientists and consumer activists by searching for empirical evidence that their social effects "might not be as bad as the critics say."[107] As with his work for the tobacco industry, however, such service to corporate interests seemingly did not much concern Lazarsfeld, who had come a long way from his early academic work in "Red Vienna."

Others of the Vienna group left academia entirely to pursue influential commercial careers, including most prominently Herta Herzog, Hans Zeisel, and Ernest Dichter. Dichter established himself as an independent marketing consultant soon after his emigration, as discussed in more detail in chapter 4.[108] Following his arrival in the United States in 1937, Dichter had initially made a career for himself by conducting marketing research not just for the BASR but also for advertising agencies and corporations such as Proctor & Gamble. In 1947, he established the independent Institute for Motivational Research, which produced research utilizing depth psychology and Freudian concepts to undergird the development of ad campaigns. Along with Pierre Martineau, Dichter is considered to be one of the pioneers of commercial motivation research, which became a crucial tool for Madison Avenue advertisers during the 1950s. Rather than relying on empirical market research and polling, Dichter used in-depth interviews with smaller panels to uncover the hidden motivations of consumers. Still, his methodology clearly built on work he had done with the Forschungsstelle and on concepts of consumer motivations developed there.

While less publicly visible than Dichter, Herta Herzog was arguably one of the most influential women in advertising during the postwar decades. Herzog and several other women including Hazel Gaudet, Rose Goldsen, and Ilse Zeisel formed the core group of researchers at the Office of Radio Research and the Bureau. Female researchers typically developed questionnaires and coded interview results, and Thelma Ehrlich Anderson trained interviewers for field and survey work, including the influential 1945 Decatur study, which provided the empirical basis for Lazarsfeld's conception of "opinion leaders" and the "two-step flow" of communication. Between 1937 and 1948, historians Allison Rowland and Peter Simonson have observed, not only did women perform much of the legwork involved in empirical survey research, but female lead authors also appeared on half of all ORR/Bureau reports in those years. In studies of female soap opera listeners, radio quizzes, or children as radio listeners, Herzog paid close attention to the psychology of listeners as it influenced their use of the medium. She eventually became associate director for consulting studies and was a vital force behind new efforts in focus group research. The public credit for her academic achievements, however, often went to male colleagues including Hadley Cantril or, in the case of focus group research, to the sociologist Robert Merton.[109]

Herzog transitioned from the academic to the commercial world in 1943, when she was hired by Marion Harper to do qualitative studies for the research department of McCann-Erickson. Like many advertising agencies at the time, McCann-Erickson was heavily involved in the programming of commercial broadcasts and had collaborated with Herzog already at the BASR. To accompany Herzog's research, the agency obtained an exclusive license for the Stanton-Lazarsfeld program analyzer, which, by having a small group of sample audience members push "like" or "dislike" buttons, allowed for a detailed tracking of audience responses to various types of programming. Herzog had an interest in innovative research, experimenting with an "eye-camera" for marketing research, a device to track and record eye movements developed by German American psychologist Eckhard Hess at the University of Chicago. Projective personality tests, such as those developed by New York psychologist and Belarus immigrant Karen Machover, also found first implementation at McCann under her supervision. When the agency created Marplan, its own research subsidiary for market and PR research, Herzog became its chairperson. She was responsible for the training of

2.3 Herta Herzog (1910–2010) representing McCann-Erickson, c. 1951. Courtesy of McCann Worldgroup.

researchers both in the United States and abroad and in 1959 spent a year doing consulting work for the agency's West German office.[110]

From the 1940s until her retirement form commercial advertising in 1970, Herzog advanced the format of the focus group interview along with other small group approaches to unveiling consumer motivations. Despite her intellectual exchanges with Frankfurt School theorists, she was an early advocate of the "uses and gratification" perspective on media use. Herzog saw consumers not as passive objects of marketing efforts but rather as active participants in mass communication processes, deliberately using media to satisfy specific needs.[111] After her return from Germany, she became partner in a small think tank, Jack Tinker Partners, devised by McCann under the leadership of Marion Harper. At Jack Tinker, outside of the corporate structures of the agency, Herzog could experiment more freely with new psychological approaches. She was primarily interested in "the relationship we have to the products we buy," as journalist Malcolm Gladwell has observed, treating common household products such as Alka-Seltzer as the "psychological furniture" that made up the world of modern-day consumers.[112] She, too, counts among the early protagonists of psychological motivation research, chairing at one point the committee on motivation research for the Advertising Research Foundation. Building on her academic qualifications, Herzog played a crucial role in bringing qualitative market research to Madison Avenue advertising firms.

Hans Zeisel, by contrast built his career on improving empirical methods in commercial research, for which he was labeled the "forensic sociologist" among the field's "founding fathers" by the Journal of Advertising Research. When Marion Harper hired Herzog for her qualitative research, he asked Zeisel to join the team at the same time for his quantitative skills. Zeisel had come to the United States in 1938 with his wife, noted designer Eva Zeisel (née Striker). Besides his affiliation with the BASR, he had been employed by the Defense Department and by the advertising agency Benson and Bowels during the early 1940s. Zeisel worked at McCann-Erickson from 1943 to 1951 and later became a board member at Marplan, the research arm of the advertising conglomerate around McCann, which in 1961 changed its name to Interpublic Group. While at McCann, Zeisel published widely in marketing journals on forms of data presentation and on problems of questionnaire methodology and cross-tabulation, as well as on the need for extended interviews in media research.[113] To Zeisel, promoting methodological breakthroughs in survey and questionnaire design in the field of market research, improving cross-tabulation, and bringing a closer integration of psychological and statistical approaches were among the greatest achievements of Lazarsfeld and the Vienna group.[114]

Zeisel's 1947 *Say It with Figures* was a primer on quantitative methods in the social sciences that over the following two decades saw five new editions and translations into several languages.[115] The text engaged questions of processing survey data such as how to treat "Don't Knows" and "No Answers," which presented a perennial methodological challenge to market surveys. Zeisel discussed biases in panel study samples and a number of other challenges to quantitative survey methodology. Most important, he was interested in extrapolating causes and reasons from the data by means of careful analysis and interpretation. In this he followed his mentor Lazarsfeld quite closely, who wrote the foreword to several editions of the book. Looking back in 1968, Zeisel noted that "the book [had] its origin in the work of the Bureau of Applied Social Research at Columbia University and, in the more distant past, in the work of the old Vienna Institute for Psychological Market Research."[116] The legacy of the Forschungsstelle thus extended both to the quantitative and to the qualitative research methods of the postwar market research profession.

Social Scientists as Consumer Engineers: Transatlantic Exchanges and the Development of Market Research

The rise of consumer engineering entailed the expansion of systematic, methodologically refined and psychologically informed market research between the 1930s and the 1950s. Tracing the history of the "Vienna School" of market research helps to illuminate the growing interest in consumer psychology and in the exploration of consumer motives and attitudes among American corporations, media, and advertising organizations. Despite their common interests and prominent roles, however, the group around Lazarsfeld was not a unified "school" in any narrow sense. As the careers of Adorno, Herzog, Zeisel, Dichter, and others show, the emigrés were hardly a homogenous group transferring a coherent concept of "scientific" market or consumer research. Hans Zeisel, who in some ways became the historiographer of the group, was careful to note these differences when writing about the legacy of the Vienna research center. The empirically inclined Zeisel scoffed at Dichter's heavily qualitative methodology and at times liberal use of Freudian interpretations and sexual allusions. Introducing the notion of a "Vienna School," Zeisel (in consultation with Lazarsfeld) created a counternarrative to Dichter's attempts to position himself as the real innovator in research methodology in interviews and publications in which he reliably deemphasized the role that psychology and the quest for hidden motivations had played already in the work of the Wirtschaftspsychologische Forschungsstelle.[117] The search for motives and the "art of asking why," Zeisel noted, had always been part of the broader Viennese tradition.

The research centers Lazarsfeld directed in Vienna, Newark, and New York illustrate the increasingly close ties between academic social research and commercial market research at the middle of the twentieth century. Not only the combination of empirical survey methods and psychological motivation research was central to the Forschungsstelle and to its "successor" institutions in the United States, but also reliance on commissioned corporate research as a major source of funding. Lazarsfeld and some of his American colleagues regarded this type of research center, the university-affiliated, independent research institute, as an institutional "transfer from Europe."[118] The intellectual and methodological transfers that resulted from these exchanges during the 1930s to 1950s will be at the center of a subsequent chapter. For now, it is important to note that the academic institutions at the heart of this chapter also provided a training ground for a new and diverse generation of market and consumer researchers in the United States. Lazarsfeld's department of sociology at Columbia, for example, counted among its students both Rena Bartos, later an advertising executive at JWT, and David Caplovitz, whose studies on poverty and consumption during the 1960s would fuel the second wave of the American consumer movement.

Corporate marketing departments were interested in the work of academic social scientists. As suggested by the example of the Office of Radio Research, independent, academic research institutes offered the business world the kind of innovative, basic research that they could not always justify themselves. The relationship between business and academia was also fraught with conflicts, however. In a 1959 article for the *American Journal of Sociology* Lazarsfeld observed that business interest in the social sciences had steadily increased since the 1930s so that "in the 1950's a typical volume [of the *Harvard Business Review*] devoted about one-third of its content to problems of social science." Paradoxically, he went on, this interest was not reciprocated, as the majority of sociologists in American liberal arts colleges had an "ideological bias against business."[119] Here Lazarsfeld and his collaborators were different. As consumer engineers, they were unafraid to collaborate with commercial or government organizations and felt that commissioned market research could provide social scientists with valuable insights into broader problems pertaining to the relationship between individuals and their social environment.

The number of emigrés straddling the line between business and academia was in part a direct result of their emigration experience. Very few had the fortune of finding commensurate academic positions right away as they transitioned across the Atlantic. Commissioned studies and commercially funded work provided a way to pay the bills even for those such as Adorno who took a decidedly critical stance toward American consumer capitalism. Despite their origins in the intellectual and political climate of interwar "Red Vienna," however, most members of the Viennese group had few scruples in this regard. To the degree that they looked beyond the financial and methodological potential of their commercial work to its ideological context, they were inclined to see market research as part of a broader pursuit of social engineering. The market surveys, much like their community studies of unemployed workers, offered targeted insights into the life-worlds and behavior of consumers that, the social scientists hoped, would ultimately serve to better the material conditions and the standard of living of the population at large.

They were part of a transnational movement in social engineering that connected social scientists in interwar Vienna with American New Dealers and corporate consumer engineers. The emigré academics promised a more scientific approach to marketing demand management. While they initially lacked experience in American markets, they offered a form of systematic psychological and statistical knowledge that could be universally applied to consumers in Vienna as much as in New York. The emigrés also leveraged their European networks and experience to help expand American marketing back in Europe. They continued to be in involved in exchanges between the continents, making their careers truly transatlantic. When, for example, Hans Zeisel spoke in Vienna about his "Vienna School" in 1967 before the joint meeting of the World Association for Public Opinion Research (WAPOR) and the European Society for Opinion and Market Research (ESOMAR), the meeting (attended as well by Dichter and Herzog) was in many ways a homecoming for the group.[120] As we will see in the final section of this book, Lazarsfeld himself returned to Vienna frequently with the aim of promoting his brand of empirical social research in Europe. In this respect, too, the story of the Vienna market researchers serves as a prototypical case study for transatlantic exchanges in the field of midcentury consumer engineering.

THREE

From Mass Persuasion to Engineered Consent: The Impact of "European" Psychology on the Cognitive Turn in Marketing Thought

In 1935, at a time when both state and business institutions were challenged by crises, the Austrian American public relations specialist Edward Bernays diagnosed a dire need for more sophisticated interaction with the public.[1] To influence audiences effectively, public relations had to gain a better understanding of "the public," which, he argued, should be seen not as a unified mass but rather as divided by classifications such as geography, income, social ties, and political beliefs. Practitioners of mass communication had to determine how they could target group leaders and use their key role in moving public opinion and perceptions. The "propagandist," Bernays wrote in the parlance of the 1930s, furthermore had to identify the "great basic motivations" from hunger and self-preservation to pride and patriotism in order to win over any audience for commercial or political gain. Understanding cognitive and emotional factors influencing human action and studying the function and effects of various forms of modern media, Bernays believed, was central to "democratic propaganda" and should not be left to communist or fascist regimes across the globe.

In calling for a more nuanced conception of the consuming public, Bernays presaged central developments in consumer research of the coming decades. Echoing the demands of the emerging marketing profession, Bernays called for broad-based analyses of public attitudes based on a "scientific foundation" by "a new type of technician."[2] Building on and going beyond the individuals and networks discussed with regard to the Vienna School, this chapter demonstrates the broader relevance of transatlantic transfers for American marketing thought. Indeed, in all of the key areas identified by Bernays—in the study of opinion leaders and group dynamics, in the analysis of consumer motives and attitudes, in the investigation of the role of symbols and their perception, and in the field of media communication—European emigré scientists would play a central role in shaping the American discourse. The growing influence of Freudian psychoanalysis on U.S. marketing was but a small part of a broader current of exchange that also included the adaptation of interwar European social and Gestalt psychology.

At midcentury, academic conceptions of "the public" and "the consumer" underwent dramatic changes. Notions of a largely undifferentiated "consuming mass" prevalent among interwar researchers had coincided with the rise of survey research beginning in the late 1920s as pollsters tried to find the "average American."[3] At the same time, the public was conceived of as a fundamentally inchoate body and a passive target for mass media persuasion—an irrational "mass" to be managed by technocratic elites.[4] Such ideas about the public were closely linked to contemporary conceptions of "the consumer" among advertising practitioners. Whereas neoclassical economists since the late nineteenth century had imagined "the consumer" as a rational "homo oeconomicus" carefully weighing costs and benefits in the marketplace, a very different image prevailed among marketing and communications specialists. To advertisers, "the consumer" was much more of an irrational actor who could be swayed with various forms of "propaganda." Corresponding to the ideas of mass psychology, early consumer psychology thus tended to describe possibilities of manipulation in terms of relatively simple behaviorist stimulus-response (S-R) models as employed for example by John Watson for J. Walter Thompson.[5] With the right mass media stimulus, advertisers assumed they could condition and sway the "consuming mass" into desired behavior.

State and corporate actors alike were eager to employ consumer psychology in social engineering efforts. Over the course of the 1930s and

1940s, however, their understanding of consumer psychology, of the composition of the public, and of the function of mass media changed significantly.[6] The Great Depression, the New Deal, and especially World War II acted as catalysts on the field of consumer research as both state and industry tried to mold the consuming public. Experts like Bernays began to reject simplistic assumptions about the impact of propaganda on malleable "masses." Instead, consumers came to be seen as a diverse and socially contextualized group that was not as easily swayed by mass media messages. Consumer behavior and attitudes were increasingly understood as complex phenomena that required analysis drawing on a variety of approaches from individual and social psychology to the probing of cognitive and perception processes. Building especially on wartime research, marketers and pollsters began to pay attention to group dynamics and to the segmented nature of American consumer markets.

This transformation in consumer theory not only represents a significant chapter in the history of American marketing but was also part of a larger story of transatlantic exchanges in the social sciences.[7] Paul Lazarsfeld and his Vienna School elaborated conceptions of consumer motivations and offered new survey methodologies. They also contributed to the emergence of mass communication scholarship in the United States with a new understanding of communication flows to consumers that emphasized the "limited effects" of mass media messages. Other emigré social scientists such as Berlin-trained psychologists Kurt Lewin and George Katona were instrumental in transforming ideas about the social psychology of consumption and about the formation and impact of consumer attitudes respectively. Both Katona and Lewin came from within the larger orbit of the Berlin Institute for Experimental Psychology. Many of its members, who had been leading protagonists of Gestalt psychology in interwar Germany, fled to the United States during the 1930s where their work informed a growing interest in cognitive processes among marketing psychologists.[8] Transatlantic knowledge transfers contributed to a fundamental reorientation of marketing and consumer research into a behavioral science concerned with motivations, cognitive processes, and social dynamics.

Beyond Behaviorism: New Approaches to Survey Psychology and Consumer Motivations

Bernard Bernays was not the only marketing expert clamoring for a more sophisticated methodology in communication and consumer research by the mid-1930s, and commercial clients demanded increasingly finer gradations and segmentations of mass survey audiences.[9] Hadley Cantril, editor of *Public Opinion Quarterly*, in 1937 pitched what would become the Princeton Radio Project to the Rockefeller Foundation by emphasizing the shortcomings of current consumer research: "'Mass' has been the cry of the radio industry. They would lead us to believe that the tastes of the audience are almost the same from one economic group to the next."[10] This proposed research project, which Paul Lazarsfeld would eventually head, would by contrast differentiate between various groups of listeners (or consumers) classified, for example, by age or by vocational and educational background. The project, furthermore, would address the "fundamental questions" of "*why* people listen and what effect their listening has on their attitudes, conduct and information."[11] The radio project thus serves as an exemplary case for new empirical research challenging existing paradigms about mass communication during the 1930s.

Mass communication studies, public opinion polling, and commercial market research were heavily intertwined fields in interwar America. The three leading polling organizations at the time all got their start in market research. Archibald Crossley, who founded Crossley Incorporated in 1926, had first built up the research department of the J. H. Cross advertising agency. Elmo Roper had done survey work for J. Walter Thompson in the early 1930s and later counted Standard Oil, Ford, the mail-order firm Spiegel, and media businesses such as Time and NBC among the clients of his young firm. Before George Gallup founded the American Institute for Public Opinion in 1935, finally, he had been director of research for the advertising agency Young & Rubicam, and his Gallup method was already used by corporations such as Lever, General Foods, and the Hearst Sunday Papers. The A. C. Nielsen Company similarly grew through its connection with commercial audience research. Consumer surveys, pollsters such as Roper claimed, could "reestablish the older 'direct contacts' between businesses and their customers."[12]

Paul Lazarsfeld soon became a prominent contributor to methodological debates about the usefulness of polling in assessing consumer sentiment. Of course, the Vienna emigrés were neither the first nor the only scholars interested in the psychology of consumers. Henry Link of the Psychological Corporation had published a comprehensive handbook on the use of psychology in advertising in 1932, and psychologically informed ad campaigns, such as the infamous Listerine mouthwash advertisements preying on fears of social stigmatization due to "halitosis," had already become common practice in the industry.[13] Much like John

Watson, however, most American marketing psychologists adhered to the dominant behaviorist S-R models, which looked at external stimuli and measurable responses but treated consumer decision processes and motivations as a "black box." Link was dismissive of surveys that asked consumers to self-report on their motivations, which he rejected as unreliable and unscientific.[14] In contrast to a subsequent generation of consumer engineers, these behaviorist market psychologists did not believe it possible to systematically "look inside" the minds of consumers.

Lazarsfeld's short article "The Art of Asking WHY in Marketing Research" was one of his first and most successful interventions in the American debate on consumer motivation research. The article critiqued the failure of standard surveys to expose the motivational setups of consumers: it explicitly pushed beyond the behaviorist "black box" by asking why consumers made the choices they did. Lazarsfeld advocated a more sophisticated and psychologically informed line of questioning and the use of qualitative, detailed interviews, which allowed for open questions yet posed them in a way that remained systematic and amenable to statistical tabulation. Market research, Lazarsfeld argued based on his Viennese experience, needed to progress beyond the simple accounting of goods acquired and money spent by tackling more subtle questions of how people "feel" about commodities, what features were of special importance, and "why they buy the way they do." Consumer motives could be understood as the specific connections between individual needs, social influences, and the decisions that ultimately led to the act of purchase.[15] Yet only methodologically refined approaches to qualitative information would be useful, he emphasized elsewhere; the collected information had to be systematic, reliable, and open for quantitative analysis.[16] Such an empirical and still "scientific" approach to revealing motivations, he believed, would ultimately allow for typologies of consumers, and thus for a way of breaking down the "consuming mass" into discrete segments.

Applying his European experiences and theoretical considerations to a new context, Lazarsfeld offered American marketing scholars a new way to theorize consumer action. Beginning in the late 1930s, he and his colleagues at the Radio Project and later at the Bureau of Applied Social Research (BASR) conducted numerous small group interviews to complement existing consumer surveys. Herta Herzog and Robert Merton developed the "focused interview" as a standardized procedure in the context of wartime communications research and propaganda analysis.[17] Panel studies provided another tool to systematize and analyze consumer motivations as groups of interviewees were repeatedly interviewed over a longer time frame. Less costly than large-scale surveys, informal panels had been pioneered during the 1920s by the food industry and by media businesses. Lazarsfeld's main contribution to panel studies was their methodological refinement with regard to both the setup of the panel (dealing, e.g., with selection biases) and the process by which panels could be conducted "scientifically."[18] He saw a decisive advantage in the panel's ability to probe consumer attitudes more deeply through qualitative questions and in their long-term perspective, which captured trends and changes rather than temporary snapshots.

Lazarsfeld and his colleagues were part of a broader transformation in U.S. consumer research, which was increasingly informed by more complex perspectives on consumers as individuals within groups. Consumer motivations became a legitimate object of study, and an emphasis on systematic data collection helped correlate purchasing patterns with markers such as occupation and education. By the early 1940s, motivation studies and audience persuasion attained additional significance with the coming of the war. Besides corporations, the state and political parties became ever more involved in the development of consumer and media research.

The Limits of Propaganda: Wartime Research and New Perspectives on Mass Communication

Government funding during World War II was instrumental for the expansion of the social sciences in general and of media studies in particular.[19] Speaking before the American Marketing Association in December of 1952, Paul Lazarsfeld presented two central insights of recent research regarding the interrelation between mass media and personal influence. First, he noted, people were affected by mass media differently depending on the personal environment in which they found themselves. Second, many people received the content of mass media messages not directly, but rather secondhand through so-called "opinion brokers." Directors of advertising, he warned, were "not sufficiently aware of how greatly people are influenced by the groups in which they live."[20] To a significant degree, these basic lessons for commercial communication, which Lazarsfeld related to postwar marketing experts, were drawn from large-scale government research devised and conducted during the war.

Historians have long acknowledged the importance of World War II for the development of American consumer society. A wartime growth consensus centered on consumption emerged during the 1940s, which would prefigure postwar economic policy in the United States for decades to come.[21] The war saw an exceptional expansion of regulatory efforts, such as rationing, price controls through the Office of Price Administration, and restrictions on consumer credit, all with tremendous impact on the daily lives of American consumers.[22] War-related policies such as the GI Bill and especially its provisions for housing also helped to lay the foundation of the postwar consumer society. A militarized "warfare state," finally, mobilized large parts of American industry and labor, and it targeted consumers and their behavior more directly than ever before.[23] World War I had already seen concerted efforts in both social research and state propaganda; World War II efforts now required information about consumer markets and consumer behavior on an even greater scale.

Even if the American home front experience pales in comparison to that of other combatant nations, state efforts directed at the consuming public still count among the largest attempts in social engineering and mass persuasion in U.S. history.[24] Between 1942 and 1945, the government spent over $200 million on propaganda activities at home and abroad.[25] State agencies became involved in surveying consumer behavior and molding consumer opinions and expectations. The capacities for large-scale surveys expanded in Washington, while massive marketing campaigns for war bonds and information campaigns for "proper" wartime consumption were conducted. Shoring up support for the troops and calling on consumers to recycle or save crucial raw materials, wartime advertising frequently intertwined national interest and commercial intentions as corporations demonstrated their allegiance to the common cause. In a way, the war presented a unique public relations opportunity for American industry in general and for the embattled advertising profession in particular.[26] Even more consequential for the development of postwar mass consumption than these advertisements, however, was the research that went into creating such wartime communications.

Government-sponsored consumer research expanded massively as hundreds of social scientists moved to Washington. The number of researchers employed by the federal government nearly doubled during the first six months of the war, and nearly all leading communications scholars, including Harold Lasswell, Hadley Cantril, Samuel Stouffer, and many others, participated in some capacity. Commissioned and coordinated especially by the Office of War Information (OWI), survey research reached new qualitative and quantitative dimensions.[27] Like many of his colleagues, Paul Lazarsfeld became involved in this aspect of the war effort. He served as a consultant to the Research Branch of the U.S. Army's Division of Morale as well as to the OWI's Bureau of Intelligence (and to its predecessor, the Office of Facts and Figures [OFF]). As a group, the social researchers advised on survey studies and on understanding and manipulating civilian morale.[28]

Wartime surveys increasingly targeted subjective factors such as attitudes and opinions. Yet the introduction of qualitative elements into survey methodology was still the subject of controversy. As survey research projects expanded in size and significance during the war, a contentious methodological debate emerged over the use and validity of open-ended questions in surveys. The conflict was carried out in the pages of *Public Opinion Quarterly* and focused on the introduction of attitude surveys with open-ended questions at the Program Surveys Division of the USDA, under the leadership of Columbia-trained social psychologist Rensis Likert. Labeled a "depth interview," the approach was criticized by skeptics such as Henry Link for being of limited value in assessing true consumer motivations.[29] Demonstrating his standing in the field, proponents of both the qualitative and the quantitative side in this debate now invoked Paul Lazarsfeld's research on consumer motivations. He himself published a mediating intervention in a subsequent issue of the journal, but Lazarsfeld clearly approved of the open-interview methodology employed by Likert's USDA studies, noting their suitability for complex scale ratings.[30] The kind of qualitative yet quantifiable survey questions that Lazarsfeld had long pushed for in an effort to uncover consumer motivations now became widespread practice in the context of the war.

World War II was also "radio's first war," and the communication researchers at the Bureau of Applied Social Research (BASR) were involved in a wide variety of activities regarding audience research.[31] The Bureau took part in wartime debates on the effectiveness of propaganda and mass persuasion. It analyzed audience reactions to propaganda film series such as "Why We Fight."[32] Robert Merton's 1946 *Mass Persuasion* was a large-scale study of the appeals and effectiveness of war bond drives that Lazarsfeld had suggested.[33] Similar studies on the home front included one of magazine reading before and after Pearl Harbor for the OWI (1942) and weekly reports on shortwave radio broadcasts to Germany and Italy. An audience test of the OWI pamphlet "Negroes and the War" (1943) found that pamphlets which aimed to shore up support among African Americans worked especially well in cases where government propaganda resonated with actual experience.[34] Emphasizing the importance of social psychological contexts, *Mass Persuasion* and other mass communication research at the BASR suggested that propaganda was effective primarily if thought of as a two-way communication and as a form of persuasion that acted as a catalyst for already existing predispositions.

Social groups and opinion leaders came increasingly into the purview of communication scholars during the 1940s. Alongside many commercial studies, wartime research helped Lazarsfeld and his collaborators develop what came to be called the "limited effects paradigm," which was most fully expressed in Lazarsfeld and Jehuda Katz's 1955 publication *Personal Influence*.[35] The power of mass media messages was inherently limited, they argued, and audiences were very much active in interpreting information. To be effective, advertising and propaganda had to resonate with audiences and to be interpreted in the context of local groups. Already in 1944, Lazarsfeld and his colleagues had published on the concept of a "two-step flow" of communication that channeled mass media information through local "opinion leaders" who were key to the success or failure of a given message. Such community leaders exercised tremendous influence in determining, for example, the food purchases, movie attendance, or apparel choices of their peers.[36]

Personal Influence, in stark contrast to earlier mass psychology presentations of powerful and manipulative propaganda developed in the wake of World War I, presented a more nuanced but also a more optimistic view of mass media persuasion. Lazarsfeld and Katz rejected the image of "an atomistic mass of millions of readers, listeners and moviegoers prepared to receive the Message," and they were critical of the assumption that media messages were a "direct and powerful stimulus to action which would elicit immediate response."[37] Since the 1930s, as George Gallup observed in the introduction to the volume, research by Lazarsfeld and other scholars in the communication field had helped to elucidate the preferences and behavior patterns of Americans, bringing to light the dynamics through which they change.[38]

Lazarsfeld's more nuanced understanding of media effects was widely accepted in the postwar decades. It served the ideological purpose of making mass media—and by extension advertising—more palatable to a modern and pluralist democratic society. At the same time, it offered to media and marketing specialists a more concrete understanding of how they could effectively reach their audiences. Drawing on panel studies and extensive surveys within one community in Decatur, Illinois, *Personal Influence* analyzed the role of opinion leaders in disseminating and reinforcing information. In the marketing field, for example, "large family wives" with many social contacts acted as central "marketing leaders" within the community, who related their experiences with products to others and thereby influenced the effect of advertising messages. The intervening role of personal relations in communication processes was a recurring theme of the book, which emphasized the complex and overlapping networks of different types of opinion leaders. These leaders did not always conform to traditional, vertical social hierarchies, but within each social group the study found "horizontal opinion leaders" who helped to pluralize communication patterns. Public officials and corporations alike had to account for the role of such opinion leaders in a fundamentally fragmented democratic public.

While the notion of diverse audiences having an active role in communication processes resonates with more recent cultural studies perspectives on mass consumption, the "limited effects paradigm" has also been criticized for its ideological compatibility with Cold War liberalism and American consumerism. Lazarsfeld has been accused of having brought a "marketing orientation" to American social research.[39] Indeed, the Decatur study clearly grew out of wartime propaganda research (which would remain important for Lazarsfeld and the BASR in the context of the Cold War), and the study was financed in part by McFadden Publications, publisher of popular women's magazines. More recently, historians of mass communication have also challenged the simplistic depiction of an interwar media profession naively believing in all-powerful mass media, which could inject the public with their messages like a hypodermic needle.[40] Lazarsfeld, they argue, invented this "straw man" to underscore his own innovations. His overarching interest, however, was the search for more effective means of communicating with citizens and consumers. Analyzing the social context of audience communication was a significant achievement that went beyond seeing consumers as a "black box" as in the behaviorist S-R models of earlier communication studies.[41] Prominent public relations practitioners such as Bernays and Ivy Lee discussed mass communications with reference to S-R behaviorism, as did Harold Lasswell, arguably the most influential early communications scholar.[42] The increased attention paid not only to the psychological makeup of the recipient of communication, but also to the social dynamics influencing this reception, substantially transformed 1930s and 1940s mass communications research.

Postwar audience studies consequently emphasized the importance of small groups in understanding larger social dynamics. Communication between members of the group and between members and specialized opinion leaders within these groups (from church crowds to beauty parlors) was key. The Decatur study could draw on what Lazarsfeld and

Katz—following sociologist Edward Shils—called the "rediscovery of the primary group" in American social studies. Beginning with Elton Mayo's Hawthorne study in industrial relations research and the wartime "American Soldier" project, psychologists increasingly highlighted primary group attachment as a major explanatory variable for social behavior.[43] Here, too, we can make out an important transatlantic strand in the emergence of small group research with respect to marketing and communication. At various points of his research during the 1930s and 1940s, Paul Lazarsfeld drew on the work of social psychologist Kurt Lewin. Much like himself, this German emigré was interested in the role of individuals as "gatekeepers" in communication processes and in-group dynamics as a central element of successful persuasion in the context of wartime research.

Group Dynamics in Consumer Behavior: Kurt Lewin and the Impact of Experimental Psychology

Kurt Lewin's work exemplifies the increasing importance of social and experimental psychology for American marketing thought. Network effects frequently compounded the influence of European emigrés on the social sciences in the United States, as these emigrés helped to familiarize their American colleagues with each other's work. Paul Lazarsfeld, for example, very consciously promoted the social psychological work of his Vienna teachers Charlotte and Karl Bühler.[44] Harvard psychologist George Miller lauded Lazarsfeld for acquainting him and a wider American audience with the ideas of German-speaking experimental psychologists such as the Bühlers and Kurt Lewin, rendering their often inaccessible language and terminology comprehensible.[45] The Vienna group had discussed Lewin's conceptions of decision processes already at the Wirtschaftspsychologische Forschungsstelle during the early 1930s.[46] Ten years later, in discussing wartime survey research in the United States, Lazarsfeld continued to reference insights from Lewin's experimental psychological work, which he now conducted in American exile.

Lewin's research in social psychology appealed to wartime consumer experts. With the looming specter of rationing and wartime shortages, understanding and influencing consumer behavior became of increasing importance. As early as 1940, the National Research Council was charged with the task of setting up a "committee on food habits" to explore the psychological and cultural patterns of nutrition with an eye to wartime needs.[47] Under the acting chairmanship of psychologist Rensis Likert of the United States Department of Agriculture (USDA), the committee initially sought to bundle existing knowledge about food consumption and marketing. They drew on the expertise of commercial market research and invited survey specialist Elmo Roper, advertising experts from McCann-Erickson, and representatives from industry organizations such as the American Meat Institute to share their insights into consumer likes and dislikes and the possibility of changing consumption patterns.[48]

Going beyond a mere survey of consumption practices, the committee's ultimate goal was not simply to understand but to change and engineer American food habits. Under the leadership of anthropologist Margaret Mead, they promoted and coordinated new research at various institutions across the country. Scholars at the University of Chicago studied how children learned to accept socially prescribed foods and how food consumption was interrelated with childrearing, while others looked at possible substitute products in the context of rationing. Paying attention to the cultural needs of consumers, researchers tested emergency rations among groups of nationals from various European countries to whom the United States was expecting to send wartime relief. One of the most prominent research efforts commissioned by the committee in this context was a series of studies by Kurt Lewin at the State University of Iowa. Lewin's group of social psychologists, which included among others a young Leon Festinger, set out to study food habits as they related to rationing: "why people eat what they eat" and the "methods of changing these food habits." They set up experimental designs to explore these questions.[49]

Kurt Lewin's experimental approach to consumer decisions and methods of persuasion was reflective of broader transatlantic transfers in the field of experimental psychology tied to the so-called Gestalt school. In 1933, Lewin had emigrated to the United States from Berlin where he had been associated with the Institute for Experimental Psychology and its influential group of Gestalt psychologists. Gestalt psychology was first developed as a branch of experimental psychology at the beginning of the twentieth century. Its leading figures included Max Wertheimer, Kurt Koffka, and Wolfgang Köhler, with important centers of research at the universities in Frankfurt and Berlin.[50] Gestalt theorists were interested in understanding processes of perception, and they studied the ways in which the human mind compiled different pieces of information into a larger, seemingly coherent "whole" (its shape or *Gestalt*). This "whole," Kurt Koffka famously noted, was different from the sum of its parts. Thus, the brain can perceive a fluid process of dynamic movement, for example, instead of just a series of discrete, individual images with objects in slightly different positions. The brain filled in "the gaps," as it were, and early Gestalt psychologists investigated especially apparent visual illusions. In experiments with geometric shapes, parallel lines, and so on, they found instances in which the mind constructs and perceives a shape (*Gestalt*) that deviates from measurable fact. Perception, from the perspective of Gestalt theory, resulted from complex interactions of various stimuli. In contrast to behaviorist psychologists, who focused narrowly on stimulus and response, the Gestaltists were interested in the broader organization of cognitive processes.

The emphasis on the cognitive processes of perception made Gestalt psychology interesting to the marketing profession, as it could inform, for example, the design of logos or advertising layouts or slogans perceived as especially appealing or memorable to consumers.[51] Leading scholars of the school found their way across the Atlantic. Max Wertheimer, the early innovator of Gestalt theory and the editor of the journal *Psychologische Forschungen*, had worked in Berlin for much of the 1920s before accepting a position in Frankfurt in 1929. As the Nazi party came to power in the winter of 1933, Wertheimer left Germany to join the faculty of the New School for Social Research in New York. Wolfgang Köhler remained in Berlin until 1935 as chair for experimental psychology, when political meddling in his institute led him to leave for the United States. Köhler initially lectured at the University of Chicago before taking a position at Swarthmore College. Kurt Koffka, finally, had left Germany already in 1924 for teaching engagements at Cornell, the University of Wisconsin, and ultimately at Smith College. The emigré Gestaltists never quite became part of the "establishment" of American psychology during their lifetime, and both Wertheimer and Koffka died at a relatively young age during the early 1940s. Still, they left a mark on the discipline, and by 1950 Gestalt psychology had become "an American movement."[52]

Kurt Lewin shared the Gestalt school's experimental approach to psychology, even though he was more of a social psychologist than a pure Gestaltist.[53] Lewin had always been most interested in the social dynamics that influenced perception processes. He completed his doctorate in 1915 in Berlin, where he later taught between 1922 and 1933. Prior to his emigration, he had attended a 1929 conference at Yale and held a visiting position in Stanford in 1932. Still, as an emigré Lewin had to enter American psychology "through the backdoor" with a two-year appointment at the School of Home Economics at Cornell sponsored by the Emergency Committee for Displaced Scholars before moving to the Iowa Child Welfare Research Center at the University of Iowa in 1935.[54] During his early years in the United States, Lewin worked primarily on the psychology of childhood development, gathering a team of young students while at the same time making his prior work in social psychology available to an English-speaking audience.

By the second half of the 1930s, Kurt Lewin had become a prominent voice for experimental and social psychology within the American profession.[55] He advocated for "scientific" psychology, which borrowed from the natural sciences. Even problems of the psychology of "will and needs," he insisted, could be probed in experimental settings and expressed in systematic terms.[56] Promising a look into the "black box" of motivations and decision making, Lewin criticized stimulus-response psychology because it discounted the importance of the "social fields" in which human behavior took place. His field-theoretical approach, by contrast, viewed behavior as a function of person and environment, i.e., of "life-space," which could be expressed in systematic terms. Drawing on Gestalt theory, Lewin suggested that social psychologists needed to look at the interdependencies between individuals and their environment and analyze the dynamic properties of this relationship. Social groups—the unit of analysis at the core of Lewin's experiments—were different from the sum of their parts. How group members perceived the world depended on what he called their social "frame of reference."[57]

Lewin's wartime experiments for the Committee on Food Habits took place against this theoretical backdrop.[58] Inquiring about consumers decisions on food choices, Lewin followed an approach he labeled "channel theory," which traced the channels through which food came to the table and revealed the factors influencing consumer decisions at various points along these channels. The central actor in Lewin's study was the housewife as the family's "gatekeeper," who decided what to buy and what to serve. Housewives' motivations varied by social group, with individual women valuing issues such as "price," "health," "taste," or "status" differently than others: they had different "frames of reference" for making their consumption choices. Consumers, in Lewin's experiments, were not simply a homogenous group but segmented by class, gender, and ethnicity. Any effort to change food habits with regard to wartime conditions, his work suggested, had to focus on the housewife as gatekeeper and on her specific motivations within the social field she inhabited. Persuasion, in other words, required attempts to tweak socially constructed frames of reference.

FROM MASS PERSUASION TO ENGINEERED CONSENT

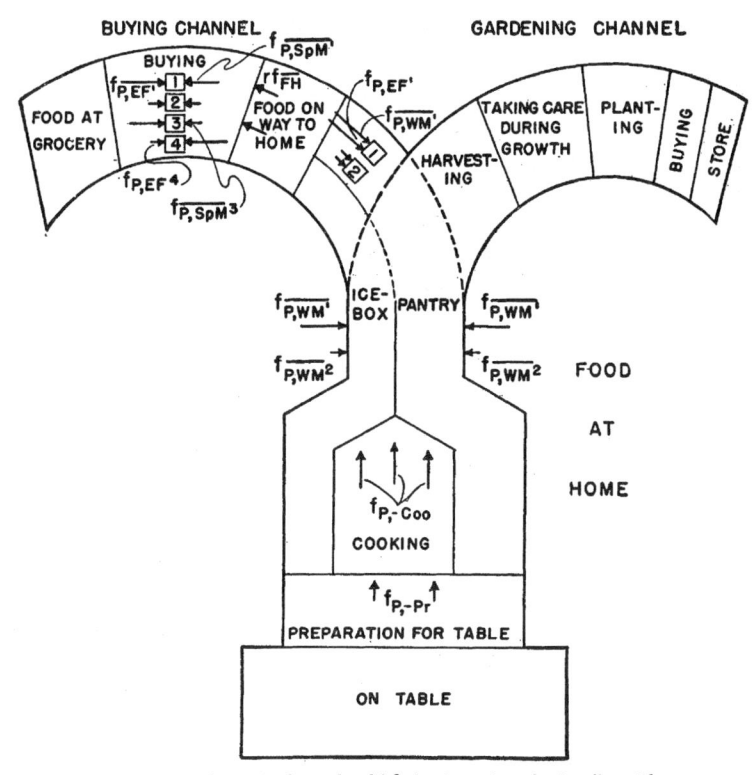

FIG. 1. Channels through which food reaches the family table.

3.1 Schematic depiction of Kurt Lewin's "channel theory" of consumption. Source: Kurt Lewin, "Group Decision and Social Change," *Readings in Social Psychology* (1952), 460.

In response to shortages of specific goods, Lewin's experiments inquired how American housewives could be persuaded to change their food habits and to switch, for example, from high-quality cuts of meat in short supply to less savory items such as chopped liver as a source of protein. Lewin's experiments focused on face-to-face communication and compared the use of lecture formats with a method called group persuasion as the most promising way to induce behavioral change. The researchers provided information about the merits of preparing and consuming less desirable meats such as kidneys to two groups. One group was presented with a traditional lecture by experts including nutritionists and home economics teachers, while the other group engaged in group discussions.

The participatory format offered great promise for mass persuasion efforts. Discussion leaders merely stated some basic information at the outset and then asked the women to discuss their opinions without resorting to "high pressure" to manipulate the outcome. Follow-up interviews revealed that 10 percent of the participants in the lecture group (4 out of 41) had served heart, kidney, or brains following the lecture. In the discussion group, by contrast, 23 out of 44 women (52 percent) served one of these glandular meats—23 percent even served a meat they had never prepared before (as opposed to 3 percent in the lecture group). Lewin's research group conducted similar experiments with different varieties of bread, also finding that student subjects were more likely to choose whole wheat over white bread if they arrived at this decision through group deliberation.[59] The more "democratic" group discussion format, the studies suggested, was better suited to effecting lasting change in behavior among consumers by means of a form of "self-persuasion."

To Lewin, the group decision method was part of "social management." Social "change experiments," which studied the influence of small primary groups on individuals, became a unifying theme of Lewin's research, as sociologist Gordon Allport observed: "the group to which an individual belongs is the ground for his perceptions, his feelings and his actions."[60] Beyond food consumption, Lewin saw social groups as a crucial element of democratic publics, and throughout the 1930s and 1940s he wrote about the influence they exerted on individual behavior in authoritarian societies such as Germany as compared with liberal democracies such as the United States. Democratic behavior, he believed, could be learned. His publications on the social experience of minority groups as well as his concern for effecting behavioral change in postwar Germany recalled his own emigration experience. In 1945, Lewin left Iowa to establish the Center for Group Dynamics at the Massachusetts Institute for Technology (MIT). Engaging questions of communication in democratic societies, the center offered applied research for "social planning" in contexts ranging from community and factory relations to the relationship between "producer and consumer."[61]

While Lewin never directly contributed to postwar discussions about marketing theory (he died in 1947), his work informed a more sophisticated social psychology of sales and communication. "The buying situation," Lewin argued, "can be characterized as a conflict situation," and communication experts such as marketers had to understand what competing social and cultural forces affected consumer decisions.[62] Only then could they effectively employ communication strategies—from radio ads to group discussions—that addressed individuals within their social field and their cultural frame of reference. Getting consumers involved in marketing communication offered greater prospects for effecting lasting change in consumer behavior than media messages delivered to passive audiences. The more tightly knit a group, for example, through joint membership in a social club, the more likely that its behavior could be changed through group decision methods.[63] Defying the notion of a passive mass of consumers, effective consumer persuasion had to pay attention to social dynamics.

These psychological insights from the food habit experiments resonated among consumer researchers. One practical application of group persuasion as explored by Lewin can be found in direct marketing approaches, which gained in popularity in postwar American suburbia. Tupperware parties, for example, relied on housewives as "gatekeepers" who were invited to actively participate in discussions of new products.[64] Here, too, group persuasion and the influence of a group of peers replaced high-pressure sales talk and mass media advertising. More important, however, Lewin's wartime research contributed to the reception of experimental psychology in postwar American marketing thought. Many of the concepts he employed, such as his channel theory or the notion of gatekeepers, informed subsequent consumer psychology. Lewin, finally, was among the first to discuss "levels of aspiration" among consumers. Individuals, he found, often developed socially framed aspirations for goods just beyond their immediate grasp, and they tended to shift these aspirations once specific desires and goals had been gratified. This concept subsequently attained great importance with behavioral economists including George Katona, who recognized this form of psychological framing as a crucial dynamic in avoiding market saturation connected to the phenomenon of "psychological obsolescence."[65] Overall, Lewin's work helped to bring the consumer's frame of reference into sharper focus and revealed the underlying social dynamics that have an impact on consumer motivations and attitudes.

Consumer Attitude Research: George Katona and the Advent of Behavioral Economics

Consumer attitudes present yet another central object of wartime research.[66] The most influential American consumer researcher to come out of interwar Germany's school of experimental psychology was George Katona, who studied consumer attitudes and pioneered the use of consumer confidence measurements. Katona, too, became involved with problems of the psychology of mass consumption during the war as part of the U.S. government's fight against inflation. Katona, furthermore, brought Gestalt psychology into the field of economics. An early exponent of behavioral economics, he challenged the prevailing assumption of consumers as "rational actors" by highlighting the role of psychological attitudes in economic decision making.

George Katona's career was both transatlantic and genuinely interdisciplinary, moving between the fields of psychology and economics. He was born in 1901 in Budapest, where he enrolled in the university in 1918, but he moved to Germany not even a year later after the revolutionary government of Béla Kun had come to power. Katona earned his doctorate in psychology at Göttingen University, developing an interest in sensory perception and the work of the Gestalt school.[67] He subsequently moved to Frankfurt where he continued his research in experimental psychology but also worked for a commercial bank at a time of severe economic strain in Germany. The experience of the 1923 hyperinflation was formative for Katona's pursuit of behavioral economics, because it led him to explore what he saw as an intimate connection between economic developments and the psychology of economic actors. In Berlin, where he had moved in 1926, Katona continued to pursue his dual-track career in psychology and economics: he remained an experimental psychologist studying and working with Max Wertheimer and Kurt Lewin, but he also developed a second career as a financial journalist for Gustav Stolper's *Der Deutsche Volkswirt*. Long before his emigration, Katona had begun to think about the relationship between economic behavior and social psychology.[68]

Following the Nazi seizure of power, Katona immigrated to the United States in 1933 but did not initially achieve a smooth transition into American academia. For a few years, he and Stolper were part of a New York investment office that advised European investors on the U.S. market. With additional income from fellowships and later as a lecturer at the New School's "University in Exile," he also stayed in close contact with Max Wertheimer while working on the psychology of learning from a Gestalt perspective.[69] With the advent of the war, Katona returned to studying the interplay of psychology and economics, publishing *War Without Inflation* in 1942. This book-length essay argued for the importance of utilizing psychological insights to address problems of the war economy, drawing on his prior economic work in Germany as well as on his more recent work in the psychology of learning. The study claimed that it was possible to avoid inflation if the necessity of economic measures was properly conveyed to the public. Favorably received by American economists and marketing experts for its methodological innovations, the book spoke to their overarching interest in shaping consumer behavior to meet wartime needs.[70]

Katona's research dovetailed with the work of Paul Lazarsfeld, Kurt Lewin, and other emigré psychologists studying the U.S. war economy. With its emphasis on the importance of public opinion research, *War Without Inflation* immediately caught the attention of Lazarsfeld. Katona's suggestions about survey methodology with regard to attitudes and expectations, Lazarsfeld observed in a 1942 memo to OFF staff, "go beyond the things we thought of ourselves."[71] During the early 1940s, Katona had been conducting surveys on business and retailer reactions to price controls using detailed interviews with open-ended questions. The inflation study was conducted for the Committee on Price Control and Rationing at the University of Chicago as part of the so-called Cowles Commission under the leadership of emigré economist Jacob Marschak, who had invited Katona to Chicago.[72] Building on these interviews, Katona also published more theoretical interventions on the importance of studying attitudes and expectations for economic policy formation.

As a contrast to stimulus-response models of mass communication, Katona proposed to involve audiences and consumers in an active learning process. Citing Wertheimer and Koffka, Katona advocated the use of Gestalt psychology along with the concept of social frames of reference as developed by Lewin: "All experience is organized within a framework. A stimulus does not give rise to an isolated experience; the meaning of the stimulus changes according to the greater whole of which it is part."[73] The task at hand was to make people think in "appropriate frames" and to bring about a genuine understanding of "changed field conditions." Such a frame could be the conditions of a war economy with shortages and the need for consumers to save. It could also be that of a postwar economy in which, Katona believed, consumers needed to learn spending to ensure continuous growth: "The task of the teacher and the molder of public opinion is, then, to help the public to gain a general orientation for war and for post-war conditions."[74] Katona proposed to use the insights of modern psychology to socially engineer consumer behavior on the macroeconomic level.

As shown above, consumer attitudes toward household spending and saving and toward food consumption were core concerns of the government's home front propaganda efforts. Studies on food and spending formed the focus of the Bureau of Agricultural Economics (BAE) Program Surveys Division as well, directed by the social psychologist Rensis

CHAPTER THREE

Likert. Likert, who had received his Ph.D. in psychology from Columbia University in 1932 for a study on attitude measurements, came to the BAE with experience in commercial market research.[75] During the 1930s, like Lazarsfeld, he had been affiliated with the Psychological Corporation.[76] Indeed, Likert was well acquainted with Lazarsfeld and his survey work, and he even translated the old Viennese research on tea consumption for classroom use in New York. Much of Likert's approach to detailed interviews and open-ended survey questions was developed in exchange with the Vienna School.[77] Likert moved to the BAE in 1939 and pushed for an expansive survey program that soon went beyond strictly agricultural questions. He increasingly employed open-question surveys to study consumer attitudes and motivations more broadly. In 1942, his agency conducted thirty-seven "special" studies on a variety of other wartime issues for agencies ranging from the OWI and the Treasury Department to the War Productions Board and the Office of Price Administration. In that year alone, 80 percent of the office's personnel was utilized for OWI and other "special" reports.[78]

During the war, the BAE thus emerged as a pioneering center for qualitative, psychologically informed consumer research. Survey topics ranged from "Attitudes toward Buying and Shortages of Consumer Goods" (1943) to "What Housewives Eat for Breakfast" (1944).[79] Their studies sought to understand the American consumer's views on rationing and shortages. They asked about future buying plans and about assessments of the current situation. Across the board, the research emphasized the importance of attitudes and expectations and the necessity of "understanding" consumer perspectives toward "sacrifice" and other relevant wartime notions. Predicting consumer behavior increasingly became an aim of government survey work, and the research on the Treasury Department's bond sales conducted by the BASR and by Likert's BAE became a model for the forecasting of buying behavior more generally.[80]

George Katona fit the Bureau's research agenda perfectly when he joined in 1944. He initially worked on studies on the use of wartime incomes, which were conducted at the request of the Board of Governors of the Federal Reserve.[81] These surveys of households sought to capture both economic and psychological data through a mix of statistical area sampling and open interviewing. Katona's surveys asked how savings habits were acquired and how future spending behavior could be predicted. In an effort to break down the consuming mass, he emphasized the importance of segmenting consumers. This meant, for example, paying special attention to affluent households because their attitudes

3.2   George Katona instructing interviewers for the Survey of Consumers, circa 1947. Courtesy of the Institute for Social Research, University of Michigan.

to saving and spending not only differed from those of the rest of the population but had the most significant impact on the war economy as a whole.[82]

With the end of the war, Katona and many of Likert's former staff moved to the University of Michigan's newly founded Survey Research Center (SRC). The group brought with them several contracts from Washington, including the Federal Reserve survey on economic behavior and motives. The survey studies at the BAE thus provided the organizational and methodological foundation for one of postwar America's most influential centers for empirical social research.[83] In 1949, the SRC was joined with the Research Center for Group Dynamics to form the Institute for Social Research. Already during the war, Kurt Lewin had discussed plans for a center for experimental group research with Likert, who after Lewin's death brought his center from MIT to Michigan.[84] George Katona held a dual appointment as professor in economics and psychology at Michigan, and he initiated the "Economic Behavior Program" to pursue a comprehensive program of consumer attitude studies.

Katona became the driving force behind the development of consumer confidence measurements. Building on the "Survey of Consumer Finances" (conducted for the Federal Reserve between 1946 and 1971), the "Survey of Consumer Attitudes" for the Department of Commerce asked representative samples of households about their perceptions and expectations regarding their own finances and the general state of business and of the economy as a whole. Utilizing this data, Katona's team began to calculate an index to predict consumer behavior in the near future. First published as the "Index of Consumer Sentiment" in 1952, it was later included in the Commerce Department's Leading Indicator composite index. Consumer confidence, for Katona, was "a key to the economy," and the empirical data he had been gathering since the war suggested to him that consumer attitudes were rooted in longstanding social and cultural contexts.[85] Consumer sentiments, however, were also highly situational and therefore needed constant monitoring through surveys. Whether in the context of wartime shortages or of postwar affluence, the consumer's "frames of reference" needed to be assessed and adapted whenever necessary.

George Katona's work provided an important link between the theoretical frameworks of interwar German psychology and the study of consumers in the postwar United States. Employing Lewin's concept of "life-space," Katona sought to reconstruct consumer perceptions of the present and the past in order to predict future behavior: "The immediate purpose of psychological studies and of economic surveys as well is diagnostic: we want to obtain as complete an account as possible of the current situation, and this account must include people's expectations, aspirations, plans, fears and many other forward-looking (*ex ante*) variables."[86] On the macro level, his interest in forecasting consumer behavior based on psychological data played into the hands of Keynesian economists concerned with steering levels of inflation or aggregate demand. On the micro level, marketing experts similarly paid attention to his efforts to analyze the psychological dispositions of American consumers. Their attitudes, expectations, and experiences were important variables in understanding purchasing decisions.[87]

Today, Katona counts among the central figures of an early generation of behavioral economists. Not to be confused with the interwar behaviorist psychologists, these scholars sought to bring insights from cognitive psychology to problems of economic decision-making.[88] The conventional "rational actor" models of neoclassical economists, they believed, needed to be augmented with sociological and psychological assumptions about group norms and attitudes and individual motivations.[89] They disagreed with the notion of consumers as cost-value optimizers whose behavior could be easily delineated from a simple set of variables such as income and price incentives. Katona's 1951 *Psychological Analysis of Economic Behavior* argued that attitudes, expectations and aspirations rather than mere physical needs or available income became increasingly important in determining household spending patterns as consumers in affluent societies enjoyed more disposable income. Rejecting stimulus-response models in favor of Lewin's social psychology (and what he called his stimulus—organism—response model), Katona, too, analyzed decision-making processes that had long been a "black box" to economists.[90] Consumers, he emphasized, were imperfectly informed and partially impulse-driven; they were, however, able to learn and to change their behavior. This behavioral perspective increasingly informed consumer activists and government regulators as well as marketing departments at midcentury.

Toward a Cognitive Turn in Marketing: Consumer Psychology and Social Engineering in Wartime and Cold War

Transatlantic knowledge transfers had such a significant impact on postwar American marketing and communication thought that marketing scholar Harold Kassarjian has spoken of the "European roots of American consumer research."[91] Kassarjian identifies several strands of influence. First, much like Roy Sheldon and Egmont Arens in their 1932 *Consumer Engineering*, he refers to the influence of Russian psychologist Ivan Pavlov on the early, behaviorist school in interwar American advertising. More important, however, he points to two further developments in the emergence of academic consumer psychology between the 1930s and the 1950s. On the one hand, Kassarjian notes the influence of individual psychology concepts, particularly from Freud and others from Vienna, on American research on consumer motivations. We have already touched on this strand with regard to the history of the "Vienna School" and will return to it, exploring the rise of motivation research in the following chapter.

Beyond this, European scholarly traditions also contributed to what Kassarjian calls a "cognitive revolution" in American marketing during the postwar decades. The work of the Gestalt theorists—George Katona's economic psychology and especially Kurt Lewin's social psychology—informed those marketing scholars in the United States interested in sensory perception as well as in the social fields and cognitive frames of reference within which such perception took place. They brought experimental research designs to the marketing field and later combined this approach with mathematical models and statistics.[92] Prominent American theorists in the social sciences, furthermore, built their theories in part on an intimate exchange with the European emigrés and their research. Leon Festinger's work on cognitive dissonance, for example, drew explicitly on Lewin's food experiments, which he himself had taken part in.[93] Abraham Maslow similarly developed his concepts of human needs and motivations in part based on his exchange with emigré scholars. Maslow had learned the tenets of Gestalt psychology from Wertheimer and Koffka at the New School and came into contact with European psychoanalysis through emigrés such as Erich Fromm and Karen Horney. In conceptualizing his "pyramid of human needs," Maslow drew prominently on the work of Katona and Lewin among other scholars.[94]

During the 1950s and 1960s, a wealth of publications attested to a broad-based social scientific interest in consumer behavior, increasingly intertwining social science and marketing research. Scholars, especially from the BASR and the SRC, helped to make consumer research a prominent aspect of broader debates on behavioral theory among social scientists.[95] Marketing science opened itself up to new behavioral research first in specialized journal articles and, by the 1960s, increasingly in encompassing surveys and edited volumes intended for classroom use.[96] A massive 1960s anthology, *Consumer Behavior and the Behavioral Sciences*, for example, prominently featured the research of emigré scholars, with contributions from Ernest Dichter, Herta Herzog, George Katona, Paul Lazarsfeld, Kurt Lewin, and Alfred Politz, among others.[97] Noting the still limited role of behavioral science in business school training, Edgar Crane opened his 1965 textbook *Marketing Communications* with an excerpt from Lewin's food habit experiments as a concrete example of the importance of a communications mix in marketing.[98] Advised by Lazarsfeld, the Ford Foundation sponsored a "buyer behavior program" that attempted to synthesize a "theory of buyer behavior" for practical use in marketing management.[99] In part owing to the influence of the emigrés, then, behaviorist S-R models in consumer psychology and mass communication increasingly gave way to a wider range of psychological approaches and to a more complex understanding of consumer behavior in the decades after the war. The "consuming mass" had gained a greater degree of agency in American marketing thought in general; it was increasingly thought of as segmented and broken down into smaller groups with internal dynamics and opinion leaders.

Government-sponsored wartime research acted as an important catalyst for innovations in consumer psychology, but the interrelationship between state policies and social science research did not end with the war. Social scientists were heavily involved in the reconstruction of postwar Europe and especially in the restructuring and "reeducation" of German society.[100] Emigrés like Lazarsfeld, Lewin, and Katona were part of these efforts, and the research institutes they founded became enmeshed with what has been termed the "Cold War social sciences." Both the BASR and the SRC received state contracts and contributed with their research to American foreign and domestic policy making during the Cold War.[101]

Communication research raised the thorny issue of mass persuasion in democratic societies. Lazarsfeld, Lewin, and Katona all reflected on the comparison between U.S. and German society during and after the Nazi dictatorship and on the tools their profession might be able to supply in democratization efforts. Their experience as emigrés sensitized them to the particular challenges of wartime "social engineering" on the part of a democratic government. The German example had delegitimized the use of outright "propaganda," yet the emigrés did not shy away from public debates about the role of political persuasion, arguing for the need to sustain wartime morale.[102] Lewin's theories of suggestibility and Lazarsfeld's work on persuasion by opinion leaders contributed to American controversies about public opinion's role in democracy that had initially been sparked by the use of government propaganda in World War I and later evolved into a debate about differences between authoritarian and democratic societies (and "personalities") during the Cold War years.[103]

Research on mass consumption was part of the Cold War social sciences insofar as mass consumption was presented as a crucial element of the "American way of life." Few scholars were more vocal in their enthusiasm for the American consumer citizen than George Katona, and his studies on consumer attitudes helped sustain the link between state policy and the mass consumer economy.[104] Consumer confidence and the underlying attitudes toward acquiring new goods were vital to American consumer capitalism, as Katona observed in 1960: "Lasting prosperity calls for sustained high demand [and for] general striving for higher standards of living ... Prosperity requires self-reinforcing optimistic attitudes based on sound reasons."[105] His attitude research and the work of fellow behavioral economists advanced this agenda: "Consumer psychology, by providing a better understanding of the factors on which consumer demand depends, will contribute to ironing out excessive economic fluctuations and to assuring a greater rate of growth in our economy."[106]

As it had during the war, postwar consumer research would serve a larger public purpose in the eyes of Katona and of his colleagues.

Many midcentury market researchers and social scientists believed that their profession should not just analyze and understand but promote mass consumption. Ernest Dichter, for example, explained that it was "a task for social engineers" to satisfy consumers and to make American democracy successful.[107] They spent less time, to be sure, worrying about the social costs of expanded material consumption or about the continued inequalities in race, class, and gender. By the end of the war, Edward Bernays wrote confidently about modern capabilities to "engineer consent" and to "sell the American way of life to the American people."[108] A new and improved set of tools including surveys, panels, focus groups, and experimental designs enhanced efforts in "social management" on the part of corporations and government agencies; writing about his research on group dynamics in 1945, Lewin cited President Franklin D. Roosevelt: "The period of social pioneering is only at its beginning. We [need to] bring under proper control the forces of modern society."[109]

As consumers became an increasingly important social force in modern society by the postwar decades, the prevailing image of "the consumer" had changed. Consumers were now seen as embedded in social fields and guided by cultural frames of references and opinion leaders. Despite these intervening forces, consumers still retained some agency that marketing professionals needed to engage: "The American consumer is not a marionette who can be influenced easily," George Katona wrote in 1963; "sometimes he or she acts on impulse; but when it matters he ponders, weighs alternatives, and tries to make an intelligent choice." Influencing people, Katona noted, was a difficult task—it could be done, but only if "the influence is in line with the consumer's wishes and desires" and only, one might add, with the help of competent experts.[110] As much as consumers were able to make informed decisions, the consumer engineers believed, they still needed the help of experts in advertising and marketing. Consumers not only needed information about new products and their purposes but also needed to be given new dreams and ideas. "The consumer," Katona wrote, "is not well endowed with fantasy or imagination; he does not dream up new wants and desires, and therefore new possibilities must be pointed out to him."[111] Postwar consumer capitalism's need to continuously stimulate demand for novel forms of consumption opened up career possibilities for a group of creative knowledge entrepreneurs who, as market research consultants and specialists in consumer motivations, capitalized on the new insights of transatlantic consumer psychology.

FOUR

Hidden Persuaders? Market Researchers as "Knowledge Entrepreneurs" between Business and the Social Sciences

Vance Packard's widely read *Hidden Persuaders* gave a tell-all account of marketing and advertising researchers who employed insights from psychology and the social sciences to analyze and influence consumer behavior. Packard called attention to psychological "probers" at work in American corporations and advertising agencies. They used insights from cutting-edge social science research to unearth the hidden reasons for consumer action, Packard explained somewhat simplistically, because they no longer saw consumers as rational actors in the marketplace. Consumers rarely knew themselves or what they actually wanted, it was now assumed, and their answers to conventional surveys could not be taken at face value. To succeed in selling, advertisers had to appeal directly to the subconscious desires of Americans. "The result is," Packard warned, "that many of us are being influenced and manipulated far more than we realize in the patterns of our daily lives."[1] Much as today's headlines warn about the mining of "Big Data" and the use of psychographic online advertising, Packard and his contemporaries worried about "engineered consent" and about an unholy alliance of marketing and the social sciences that deprived consumers of their free will and agency, undercutting the promises of postwar democracy.

Packard's account can also be read as the fulfillment of Sheldon and Arens's vision of "humaneering." Much as the authors of *Consumer Engineering* had hoped, psychological insights were now used to stimulate consumer demand and sustain consumption-driven growth. A new generation of scientifically trained marketing experts, frequently influenced by Freudian depth psychology, had emerged: "The symbol manipulators and their research advisers have developed their depth views of us by sitting at the feet of psychiatrists and social scientists (particularly psychologists and sociologists) who have been hiring themselves out as 'practical' consultants or setting up their own research firms."[2] Many of those practical consultants that Packard identified personally were actually European immigrants and emigrés or came from institutions heavily influenced by the Viennese group around Lazarsfeld and Herzog. Two of the "hidden persuaders" featured most prominently in the book were Ukrainian-born Louis Cheskin, who directed the Chicago-based Color Research Institute of America, and the Viennese immigrant Ernest Dichter, who had built up the Institute for Motivation Research in New York.[3] Packard, however, paid less attention to the transatlantic background of his protagonists than to the marketing changes their work brought about.

The emphasis on psychological persuasion that concerned Packard and many of his contemporaries was in part born out of increasingly fierce competition in a marketing field where many competitors traded claims of new, scientific capabilities to attract corporate clients. The organizations headed by Cheskin and Dichter were merely two examples of a broader phenomenon that Packard's survey of the postwar marketing industry reveals: the rise of independent consultancies that specialized in various forms of marketing research. Other prominent firms discussed in the book included Burleigh Gardner's Social Research Inc. of Chicago and Alfred Politz Research Inc. of New York. Alongside corporate market research departments and the research divisions of advertising agencies, these specialized institutes were at the forefront of making consumer research a hallmark of America's postwar consumer capitalism.

The "hidden persuaders" Packard described can also be understood as a new type of "knowledge entrepreneur." In the context of an economy that increasingly relied on knowledge and information, the entrepreneurial task of these consultants was to gather, bundle, and analyze market information, utilizing the most up-to-date research methodologies, and then to feed their analyses into the managerial decision processes at their client companies.[4] To succeed, these knowledge entrepreneurs had to bolster the scientific foundations of their research by making expansive claims about the power afforded by their information. They promised to make markets predictable and manageable, providing a solid foundation on which to grow business and launch new products. At the same time, they portrayed themselves as a creative force, coming to the corporate process as imaginative outsiders who could provide both consumers and producers with the knowledge—and, by extension, with the goods and experiences—that they needed. In contrast to Packard's story of manipulation, these experts stressed their role in "market making," i.e., in enabling innovation by overcoming the inherent reservations of consumers toward new products.[5]

Despite Packard's exposé, we know surprisingly little about postwar market research as an emerging industry. This chapter will survey the practice of American market research from the 1940s to the 1960s, focusing especially on the postwar ascendancy of market researchers as independent "knowledge entrepreneurs" outside of corporate hierarchies and advertising agency structures. In part, these new experts were a result of the interplay of academic and commercial research covered in previous chapters. Being "scientific" outsiders to the corporate process provided both advantages and disadvantages within the business world, and research consultants at times struggled to find their proper role. Here, I will highlight two case studies of European emigrés, Ernest Dichter and Alfred Politz, who not only were two of the leading marketing research experts during the 1950s and but also counted among the most influential voices in the field. They stood at the forefront of a growing emphasis on scientific method and on creative innovation in midcentury consumer capitalism.

In their methodology, Dichter and Politz stood at opposite ends of the American market research profession. They were rivals and even open antagonists, and like few others they represented the division between qualitative and quantitative approaches within the postwar discipline. In focusing primarily on the qualitative and psychologically informed depth research represented by Dichter, Packard overlooked the fact that champions of quantitative research modeled on new empirical social surveys also established themselves as marketing advisors with similarly far-reaching claims regarding their ability to analyze and predict consumer markets. Much like the qualitative "Tiefenboys" (referring to the German term *Tiefen-* or depth psychology) the number-crunching "nose counters" equally presented themselves as the future of research-driven consumer engineering.

CHAPTER FOUR

The Glamour of Science? The Expansion of Market Research in American Industry, 1930s–1950s

In 1971, market researcher Alfred Politz drafted a memoir looking back on his experiences in the field of marketing. He entitled it "How to Produce Consumers—Methods and Illusions."[6] Born and trained in Berlin, Politz was a physicist turned marketing consultant who counted among the early champions of random sampling in consumer surveys during the 1940s. Over the course of his career, spanning the 1930s to the 1970s, Politz had seen the marketing profession in the United States become a great deal more systematic in its approach. As the title of his memoir suggests, marketing professionals had become very confident in their ability to understand, influence, and essentially "produce" or "engineer" consumers. This drive toward scientific practice within the profession became especially pronounced during and after World War II. Soon, no marketing executive would approach management and no marketing consultant could hope to win a client without being able to argue that their recommendations were firmly based in (some form of) "scientific" research. "The word 'research,'" Politz observed, "implies a sort of glamorous intellectual sophistication, and marketing research is a symbol of the modernity of the marketer. Marketing research has become a status symbol, and as such it need not perform; it need only exist."[7] As a trained scientist, Politz scoffed at what he regarded as the pseudoscientific veneer that many marketing experts now applied to their work. As a marketing man himself, however, and as the head of Alfred Politz Research Inc., he shrewdly understood that the claim to scientifically established knowledge was at the heart of the marketing profession's rise and of the postwar ascendancy of independent research firms.

Politz also firmly believed in the possibilities of "scientific marketing." He believed in the ability of research to increase "advertising efficiency," to identify the "most important product properties for consumer appeal," and to discover the "most efficient product design." Most important, perhaps, he believed that market researchers had become instrumental in crafting an "image" for products that would make loyalty to brands a part of "daily consumption behavior."[8] Politz and his career, then, are reflective of the rise of a new generation of marketing experts in American industry who influenced the way goods and their meaning in society were communicated and conceived of. They ushered in notions of targeted advertising, market segmentation, and brand identities.

Since the interwar years, marketers had become significantly more systematic in studying markets and consumers. One classic example of the early use of customer research in consumer engineering efforts is the automobile industry.[9] General Motors, already a pioneer of a segmented marketing approach with numerous brands in the 1920s, reacted to the Great Depression by institutionalizing its customer research efforts over the following decade. A new department led by Henry Weaver recorded customer feedback initially through rudimentary questionnaires of consumer preferences and buying habits. Not only were the results utilized in the sales department and for direct-mail advertising campaigns, but Weaver even proposed to use his findings for product planning and body styling.[10] Another industry that pioneered systematic marketing research was the film business. Motion picture studios also increasingly relied on more "scientific" surveys by the 1930s and 1940s, and audience research helped tailor their products more closely to changing consumer demand.[11]

The auto and film industries were early movers in a broader shift away from more informal forms of market observation by "fashion intermediaries" such as retail buyers or art directors.[12] A few firms such as the meat-packing firm Swift & Co. had established specialized market research departments as early as World War I and during the 1920s at a time when advertising agencies, too, began to invest in research capabilities. Following the Great Depression many companies closed their departments again to save on overhead cost or they decided to contract out their research needs to ad agencies and a growing number of independent research firms such as Gallup or the Psychological Corporation.[13] Yet market analyses were increasingly portrayed as indispensable by experts in the field. Industrial consultant Willard Freeland, for example, emphasized in 1940 the centrality of statistics for proper organization and forecasting in all corporations, regardless of size and industry: "Do I hear you say 'That [market research] is all right for the big company'? Bosh! The little company can have it if it has the desire and will."[14] Despite Freeland's emphatic pronouncement, however, market research initially remained an instrument primarily of larger firms particularly in the consumer goods sector.

Over the course of the 1930s and 1940s, systematic research slowly became a more widespread practice. A 1937 survey by the U.S. Department of Commerce, conducted in cooperation with the American Marketing Association (AMA), asked 550 companies about their market research activities. Slightly over a third of the firms surveyed (188 companies) stated that they engaged in market and consumer surveys, but the form

CHAPTER FOUR

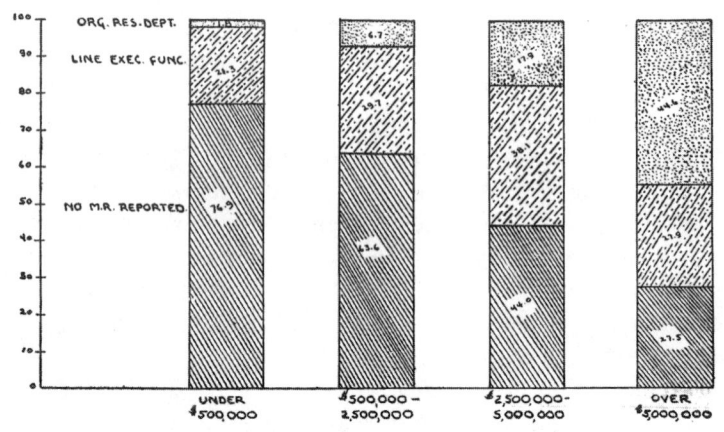

CHART 2. Companies Doing Marketing Research by Size of Company (Net Sales).

4.1 Share of U.S. Companies Doing Market Research by Size, c. 1942/43. Source: Heusner et al., "Marketing Research in American Industry I" (1947).

and intensity of such research use varied greatly. Only 73 companies maintained their own formalized market research departments, most of which had been established only very recently during the 1930s (with the oldest dating back to 1917), and, similarly, only 15 percent of firms surveyed made use of external market research providers. Differences by industry were significant, however. As could be expected, consumer goods firms in sectors such as foodstuffs, personal hygiene, cosmetics, or automobiles were much more likely to employ external market analysts, and market research in general played a much greater role with over 40 percent of consumer goods companies engaged in some sort of research. Their market research departments averaged about five employees, and their average expenditures for research were also above those in non-consumer goods industries with about $20,000 annually for in-house studies and another $12,500 for external research services.[15]

The growing prevalence of market research was confirmed by a second, larger study of nearly five thousand manufacturers conducted by the AMA's Committee on Marketing Research Activities during the mid-1940s. This time, about 38 percent of surveyed firms stated that they were engaged in market research (46% among consumer goods producers). Size still mattered, as not even a quarter of small companies with annual sales under $500,000 conducted research, while that share reached 72.5 percent among the largest firms in the sample with annual sales over $5 million. Large consumer goods firms appeared to heed the call for more systematic marketing research. Nearly half of the largest companies now had market research departments that produced data to predict sales figures and market potential, but also to measure advertising effectiveness or to develop products and merchandising strategies.[16]

Marketing research became a "management tool" with influence on company decision-making processes. In cases where new research departments were installed, the majority of their department heads now reported directly to upper management, while sales departments saw their overall influence shrink. Still other companies had the research unit integrated into larger marketing departments as those became more common by the early 1950s.[17] Regardless of exact organization, a majority of firms with specialized research departments noted in the 1947 AMA study that their "head of marketing research" was part of regular executive meetings or otherwise actively involved in the strategic decision making of the company. Marketing research thus was about more than simply gathering facts and figures but created new forms of strategic knowledge. A majority of surveyed firms consequently indicated that they expected to grow their own research capabilities or to employ more outside services in the future.[18]

How can we account for the growing role of external research consultants and what criteria determined the "make-or-buy" decisions that many firms now faced with regard to market research? Companies took different approaches, with some opting for "lean" research departments with a small, but experienced staff that hired outside help whenever needed. Some large firms, by contrast, handled all sorts of consumer research projects and even the interviewing and tabulation for large-scale surveys in-house, like for example Proctor & Gamble, which had a department of more than 125 people in 1952.[19] In other cases, the choice between in-house or external market research expertise depended on the type of research question at hand. At General Mills an outside advertising agency was primarily responsible for advertising-related research including copy testing and media analyses, as the firm's research director Gordon Hughes explained in 1950. General Mills' own sizeable market analysis department with roughly twenty-four full-time employees, by contrast, was responsible for evaluating the success of advertising work. Most important, their in-house department was also in charge of routine research projects that dealt directly with product development and sales. To Hughes, this was the core business of corporate market research, and in such cases external agencies played only an advisory role.[20] Internal research departments, furthermore, had the advantage of being intimately familiar with the problems of specific industries or products.[21] Still, when General Mills wanted to know in more detail how consumer attitudes toward their Betty Crocker brand compared with attitudes toward competitors, they hired a specialized research firm, Burke Marketing Research Inc., to design a study and conduct the interviews.[22]

Besides advertising agencies, nonmanufacturer market research could come from a variety of sources, including retailers, trade associations, and media organizations.[23] During the postwar decades, upscale retailers such as Bergdorf Goodman, for example, were still involved in surveying consumer taste for the fashion industry, while in industries with many smaller firms, trade organizations often gathered and disseminated pertinent marketing data. Media companies such as the *Chicago Tribune* with its consumer panels and Pierre Martineau's motivation research also played a prominent role. Most important, however, the large advertising firms had expanded their research capabilities since the 1930s to offer a full range of research services. Industry leader J. Walter Thompson began a broad-based consumer panel in 1939 that by 1947 had grown to a national cross-section of 5,800 families keeping diaries of their daily purchases.[24] Its Madison Avenue rival McCann-Erickson, renowned for its research, created a "central research and merchandising department" at its New York headquarters in 1946 after it had hired several academic marketing experts including Herta Herzog and Hans Zeisel. The new department maintained a large library and a research division (for copy tests and media studies) as well as a merchandising division for the implementation of marketing strategies, offering advice to clients on matters from sales to pricing.[25] In the suspicious eyes of many clients, however, agency research appeared biased in favor of agency interests when it came to the implications for marketing strategy.[26]

It was here that independent organizations found a lucrative market niche. Even agencies as large as McCann still "farmed out" specific tasks to specialized market research companies that could, for example, conduct large sophisticated surveys with the necessary staff of trained interviewers and the computing capability of expensive IBM tabulation machines.[27] The expanding postwar demand for market research was thus reflected not only by hires of newly trained researchers coming out of a growing number of college marketing programs but also by an increase in the number of consulting agencies. The largest "complete service organizations" such as A. Politz Research in New York or Alderson & Sessions in Philadelphia had a hundred or more employees and could assemble individual teams for each client project. Numerous smaller, often regional research consulting firms competed for clients by special-

Internal MR within Companies
(e.g., General Mills)
- Product development
- Sales analyses

Advertising Agency MR
(e.g., McCann-Erickson)
- Copy-testing
- Media studies
- "Merchandising": package design, distribution

Specialized MR Firms
(e.g., Alfred Politz, Inc.)
- Large surveys (trained interviewers)
- Tabulation of results
- Specialized studies

4.2 Marketing Research in U.S. Industry, c. 1950.

izing in certain industries (such as research for media organizations). Some focused on certain research methodologies (such as motivation research), or they concentrated on particular marketing problems (such as the design and assessment of packages). Even narrowly specialized consulting firms, finally, could still be quite large, such as the opinion research institutes conducting large, nationwide consumer surveys or companies such as A. C. Nielsen, which offered very specific indices and panels drawing on thousands of households across the United States.[28]

By the mid-1950s, market research had turned into a complex field and a substantial business for some of the more prominent "knowledge entrepreneurs."[29] Yet elaborate consumer research was still not as widespread a phenomenon as its champions hoped and its critics feared. According to one (albeit conservative) 1951 estimate, there were barely eight thousand full-time market research professionals employed in all of U.S. industry. Even some of the largest industrial corporations still lacked full-time research personnel, while service sector companies were hardly engaged in research at all. Nationwide, there were only about five hundred advertising agencies and independent consultancies conducting market studies. The role of market research in product planning, a hallmark of later marketing management, was not yet universally accepted.[30] Nevertheless, consumer researchers had securely ensconced themselves within the world of American business as employees of marketing departments and ad agencies and in independent firms. Especially external research consultants benefited from the increasingly "scientific" nature of the research, as the following case studies will illustrate, because refined survey methodology demanded large samples, and new qualitative approaches required highly trained personnel. A sizeable portion of American corporations thus split their budget for marketing research more or less evenly between their own growing research staff and the services offered by external "knowledge entrepreneurs."[31]

The Drive for "Scientific" Marketing Research: Alfred Politz Research Inc.

Pleading for an integrated approach to professional education in business and the social sciences, Paul Lazarsfeld envisioned a new generation of social engineers as versatile experts with qualitative and quantitative skills: "The government . . . , the sales manager, the leftist organizer—they all wait for a new profession: the research expert on human affairs who combines a variety of skills, just as the medical doctor combines training in basic sciences with clinical imagination."[32] Few marketing consultants during the 1940s and 1950s fit this description of the skilled and imaginative "research expert" as neatly as Alfred Politz, head of Alfred Politz Research Inc. (APR), one of the largest independent market research firms in the United States.[33]

Politz was the ideal type of the market researcher as empirical scientist. In retelling his biography on various occasions during the postwar years, Politz reliably emphasized his professional start with a doctoral degree in physics from the University of Berlin. Here, he later claimed, he had studied with Max Planck and Albert Einstein. Politz had also minored in psychology and in speeches referred to contacts with the experimental Gestalt psychologist Max Wertheimer. Much like Lazarsfeld, Politz cultivated the persona of the European academic, but, to fit his particular purposes, he was careful to associate himself with the natural sciences. In 1937 he left Nazi Germany even though he was neither Jewish nor politically active, but Politz's emigration experience was still typical in so far as it required him to reorient his career and to find a new niche in which he could advance professionally.

When he came to New York, Politz would later joke, he had looked for the profession that "makes the most money with the least intelligence. The answer was advertising." And since his language skills at the time were not good enough to write ad copy, he explained, he went into advertising research, which suited his scientific background.[34] In reality, Politz had been in the advertising business in Germany since the late 1920s. In 1927 he helped create a new advertising agency for the well-known Rudolf Mosse Annoncen-Expedition (part of the German Jewish Mosse publishing house, publisher of the influential liberal daily *Berliner Tageblatt*) to acquire commercial ads for print publications. He had published articles on "advertising psychology and technique" in German trade journals and in 1933 established his own firm "A. Politz."[35] His resumé lists accounts with several large German companies, which he consulted on sales and advertising in Germany and Europe during the mid-1930s, including Kienzle clocks, Emil Busch optical instruments, Vorwerk carpets, and Zündapp motorcycles. He even had some international clients such as the London-based Abdullah Cigarettes and the American Parker Pen Company.[36]

Thus, Politz's career in the United States was hardly a sharp departure from his prior work in 1930s Germany, and economic opportunity was a strong motivation for his move across the Atlantic. His initial plan was to sell Spalt headache tablets in the United States and the United Kingdom. He had worked on advertising for the Much Aktiengesellschaft, which succeeded with the brand in Germany, and Politz had acquired the production titles for these countries. He had also established a marketing test laboratory for "Spalt-Tabletter" in Sweden that provided him with some funds during his initial stay in the United States. As the American Spalt business failed, however, Politz soon found himself in New York with little money and few social contacts outside of the emigré community.[37]

At this moment, Politz reinvented himself as the European expert scientist with a knack for marketing. He initially got a job as research director with the polling firm Elmo Roper through his ties to the Parker Pen company and subsequently worked for the advertising firm Compton

4.3 Market researcher Alfred Politz (1902–82), c. 1952. Source: Harris & Ewing Studio, Washington, D.C.

during the early 1940s. This relatively brief engagement, however, ended in his dismissal at the request of a client, Proctor and Gamble, whose in-house representatives apparently objected to Politz's new, unorthodox research methods.[38] Remaining in Compton's office space and drawing on the help of his former co-workers as well as his wife Martha, he started his own advertising research business in 1943, which he incorporated as Alfred Politz Research Inc. (ARP) in 1947. Politz quickly gained a reputation for his novel sampling methodology, and his firm attracted numerous large clients, including Mobil (Socony Vacuum) Oil, U.S. Steel, Bristol Myers, and DuPont.[39] ARP eventually included a prominent media division that studied advertising performance, as well as a products and advertising division that advised on product development and marketing based on large surveys. By the early 1960s, his Madison Avenue firm had grown to 220 employees in the central office (85 of whom were highly trained professionals) and an additional 600 full-time field researchers.[40]

Politz was well regarded within the profession for the "scientific" rigor of his statistical methodology. The American Marketing Association presented him with several leadership awards, and in 1953 he was inducted into the Marketing Hall of Fame.[41] Listing him as one of the eight "Founding Fathers" of advertising research in the United States, the *Journal of Advertising Research* singled him out as the "sampling innovator." Politz recalled that he found approaches to sample selection to be "rather naïve" when he first entered the research field in the early 1940s.[42] With his mathematical training he developed alternatives to the widely used "quota sampling," which aimed to approximate the overall population in the relative composition of the sample. Quota sampling was statistically flawed, Politz argued, preferring instead an approach that utilized a large, randomized sample of interviewees within carefully selected geographic areas. First developed within the context of the U.S. Census Bureau, such area sampling minimized traditional sampling biases. Politz was among the first to apply this technique to commercial research, and in 1945 the advertising trade publication *Tide* ran an in-depth article on the "revolutionary" yet expensive "Politz plan."[43]

Alfred Politz Research Inc. catered to a small set of very large clients that the firm advised in product development and marketing. To Politz, the success of market research depended on its ability to isolate and analyze the specific marketing problems of a particular company and to employ statistically sound data gathering to furnish solutions.[44] He credited his survey research on drinking habits and package perception, for example, for Coca-Cola's decision to introduce larger bottle sizes in 1952 and new cans in 1955, for which he had predicted sales increases to the tune of $40 million.[45] Besides packaging design, research also influenced product development. When Bristol-Meyers introduced stick deodorants, Politz's surveys suggested that consumers disliked the sensation of friction during the application process. In collaboration with the engineering department, a roll-on container was developed based on ballpoint technology, which, Politz boasted, led to the introduction of

Ban as the first brand of roll-on deodorant.[46] Detesting hunches or speculation, Politz demanded rigorous research that quantified and weighed various variables in explaining consumer behavior.

The product Politz sold to his clients was knowledge, or better, the illusion of reliable "scientific" knowledge. Companies needed to know their product and their consumers, he believed.[47] Professional market research, he wrote, provided "relevant knowledge" obtained by standard scientific methods of "questioning" and "observation."[48] Only scientifically trained experts such as himself were able to produce useful, predictive knowledge based on empirical survey measurements.[49] Market research, Politz consequently believed, should be provided by specialized external firms rather than internally generated, and he argued for the benefits of long-lasting relationships between clients and research firms.[50] Like many advertising agencies, Politz took on only one client from each industry to avoid conflicts of interest. Acknowledging the complexities of each individual field and product, he believed that contracts needed to establish a one-year minimum to allow for thorough investigation. This, he hoped, would become part of a code of professional ethics for research consultants.[51]

Between the 1930s and 1950s, the market researchers increasingly viewed themselves as experts on a par with scientists, doctors, and engineers. While early marketing men had learned their trade in the business world, in sales or retailing or advertising, postwar marketing experts came to the field with formal academic training. Frank Coutant, president of the American Marketing Association, offered cautious optimism in 1936: "On bringing science into marketing, only a start has been made. On marketing research, about two or three millions of dollars a year are being spent . . . . It is only a drop in the bucket, of course . . . , but it shows that we are on our way to greater efficiency and better profits in marketing."[52] To contemporaries, the scientific character of research legitimized marketing's growing role in American business.

By the early 1950s, marketers proudly surveyed the growing body of knowledge in professional journals, and many quite confidently spoke of "marketing science" as a new and independent area of scientific inquiry.[53] Consumer psychology, as already discussed, had benefited greatly from an exchange with academic psychology.[54] At the same time, quantitative survey research and statistical sampling techniques underwent a similar revolution. The large survey institutes like Gallup and Roper vastly improved the traditional "quota sampling" approach while others such as Politz championed new randomized "area sampling" methods.[55] Consequently, survey institutes became more specialized and diverse.

Advocates of "scientific marketing" enthusiastically anticipated a new and unified "theory of marketing" that would make it a science in its own right.[56]

Few marketers saw themselves as pure scientists, however; especially those with practical experience were inclined to see marketing as a practical science that, like engineering, heavily drew on scientific methodology. Marketing scholar Kenneth Hutchinson, for example, saw marketing as "an art or a practice, and as such much more closely resembl[ing] engineering, medicine, and architecture." According to Hutchinson, "Engineers and physicians are trained to approach their problems in this spirit of scientific inquiry; marketing men are learning rapidly to follow their examples."[57] Such analogies to doctors and especially to engineers were a recurring theme in midcentury debates about the marketing profession: "Most of us are so close to the area of engineering, that there is surely little need to labor the point," another observer noted as he discussed the engineering profession as a model for marketing.[58] Herbert Hess even proposed: "Marketing engineering which is in reality human engineering has the task of getting the right article into the possession of the right individual at the right time and at the right price."[59] Such an understanding of marketing as a technical science easily meshed with notions of consumer engineering.

Claims to scientific expertise opened doors in the corporate world. As marketing professor Edmund McGarry observed, scientific credentials created high expectations among businessmen: "[businessmen] are prone to look upon a scientific expert as one who has remarkable and mysterious powers of foresight. . . . He must be a prophet who can foretell where profits are to come from. He is expected to know the unknown, to foresee the unforeseeable."[60] Such attitudes promoted the close proximity between commercial marketing and academic research observed already with regard to the Vienna School, with careers that frequently crossed the permeable lines between business and academia.[61] Companies increasingly paid attention to the academic qualifications of their research personnel, and two forms of degree qualifications were especially sought after during the early 1950s: extensive statistical knowledge and a background in psychology.[62]

New experts were not always welcomed with open arms, however, especially when academic culture collided with business culture. Innovations in scientific marketing suggested by research consultants, for example, sometimes faced opposition from middle management who regarded such changes as a challenge to the "tacit knowledge" of sales department personnel. Other times market researchers simply came across as overly technical. According to one contemporary commentator, experts who needlessly "belabored the techniques" or presented "riddles instead of results" had a difficult time getting heard by management.[63] Language and the use of jargon was a frequent point of contention between academics and businessmen, as Henry Bund warned: "scientific and pseudo-scientific mumbo-jumbo" was likely to scare away rather than attract new clients and corporate executives.[64] Scientific facts and the most elaborate methodology, finally, were not always appreciated by commercial customers or accepted within corporate organizations, especially when they became too costly and stood in the way of financial interests.

To be successful as knowledge entrepreneurs, Politz and other consultants thus made it clear that they were businessmen first and academics second. Politz was adamant that marketing consultants needed not only to thoroughly research the economic challenges faced by each individual company, but also to provide clear and concrete solutions to relevant marketing questions in a language that management could understand: "The production of figures is really very simple. Anybody can produce figures. To produce figures of known accuracy is another story. . . . Research is finished when the interpretative predictions are finished, ready for the use of management. . . . Research must show what will happen if a decision goes this way, and what will happen if it goes the other way."[65] Science and its ability to provide "accurate" figures provided Politz with a crucial competitive edge, and his firm proudly advertised the fact that they had established what Politz called an "experimental marketing laboratory" that employed and trained "people from the exact sciences."[66] To succeed as "knowledge entrepreneurs," however, they needed more than the data alone; they further needed to provide concrete and innovative marketing solutions.

Desire as Marketing Strategy: Ernest Dichter's Institute for "Motivation Research"

According to sheer numbers, Politz may well have been the most successful of the 1950s research consultants, but by the second half of the decade his statistical approach was increasingly overshadowed by the type of psychological research that Packard's exposé focused on. A 1956 *Fortune* article on "motivational research" highlighted the fierce and public competition between two emigré experts, the "ex-German" Alfred Politz and his antagonist Ernest Dichter, the "ex-Austrian" moti-

Table 4.1

	A. Politz Research, Inc. "nose counters"	**Inst. for Motivation Research (Dichter)** "tiefenboys"
Size:	220 employees (+600 field interviewers)	50 employees (c. 200–300 field)
Sales:	$2,500,000 (1955)	$750,000 (1955)
Clients:	11 (1955)	c. 30 (1955)
Method:	statistical, large-n surveys	qualitative, small-n panels
Cost:	$100,000+	$6,500–$20,000

Based on: Stryker, "Motivation Research" (1956).

vation researcher. While Politz's firm had grossed $2,500,000 from a group of eleven major clients in 1955, Dichter's institute had made a still quite respectable $750,000 providing research to over thirty firms. The two rivals publicly traded barbs, accusing each other of "pseudoscientific" research and flaws in methodology, but both consultants drew their authority from claims to scientific respectability. In the end, they were quite similar in the way they sold knowledge about markets and consumers and what they called "creative research" to American companies.[67]

Today, Ernest Dichter is certainly the most well known of the postwar market researchers, and his penchant for publicity helped him to establish his own legend. Historical accounts have frequently overstated his role as an "innovator" in qualitative research techniques, celebrating him as a "pioneering change agent" without acknowledging the tremendous intellectual debt that he owed among others to the Vienna Forschungsstelle and the group around Lazarsfeld, which helped establish his career.[68] Still, Dichter's career not only is exemplary for the impact of European emigrés on American marketing, but also deserves special attention for the way in which Dichter conceptualized the role of the market researcher as a knowledge entrepreneur in consumer capitalism.

Dichter's transatlantic career and his role as a pioneer in "motivation research" are too well known to be discussed here in detail.[69] The son of a Jewish immigrant family from Galicia, Dichter grew up in Vienna, and during the late 1920s and early 1930s, he studied psychology in Paris and Vienna, where he earned his doctorate in 1934. He subsequently set up a practice as a psychoanalyst but became interested in market research, to which he was introduced through work for Lazarsfeld's Wirtschaftspsychologische Forschungsstelle. Dichter claimed that

4.4 Ernest Dichter (1907–91). Courtesy of Hagley Museum and Library.

in 1938, even prior to his emigration, he told a U.S. consular officer in Paris that he intended to "revolutionize" consumer research in the United States. Indeed, he quickly became part of Lazarsfeld's New York network of Austrian emigrés engaged in various forms of consumer research, starting on Lazarsfeld's recommendation at a small research firm called Market Analysts Inc.

Dichter, who changed his first name from Ernst to Ernest and was eager to "Americanize" and blend in with the Madison Avenue professionals, soon developed a reputation as an expert on psychological research. He did a much-noted study on bathing habits for Ivory Soap at Compton Advertising, which during the early 1940s had also employed Politz. His first professional break came with market study on the psychology of car buyers in 1939 for the Chrysler Corporation and its advertising agency Stirling Getchell. Dichter's psychological technique of detailed interviews, which he likened to the "diagnoses of physicians or psychoanalysts," was met with initial skepticism at the agency. The study, however, which famously likened convertibles to mistresses in the phantasy of male automobile buyers and suggested ways to exploit this mental connection in advertising, was well received by both agency and client. Dichter joined Stirling Getchell, one of the largest advertising firms during the early 1940s, where he became director of psychological research and went on to conduct over thirty psychological studies on products ranging from breakfast cereal to gas stations.[70]

While his commercial consulting career was taking off, Dichter continued to strive for academic recognition of his psychological methods. Throughout the early 1940s, he remained in close contact with Lazarsfeld, whose approval he still sought, even as he was beginning to develop his own, more psychoanalytical methodology. While the relationship between the two emigrés became increasingly tense, Lazarsfeld's networks proved again valuable for Dichter as the Stirling Getchell agency faltered and he found himself searching for employment in late 1942.[71] He was quickly hired as a research psychologist and a special consultant on daytime programming by Frank Stanton at CBS, where he was to devise audience research in collaboration with Lazarsfeld's Bureau and Elmo Roper's polling firm.

Dichter's employment at CBS proved to be a formative time as he refined his own approach to consumer studies and learned to defend it against objections from more quantitatively minded colleagues. In meetings and memos, Dichter argued for the increased use of in-depth qualitative interviews with radio listeners to reveal underlying motivations. Effective programming could address the everyday frustrations of listeners and appeal to their inner desires. With input from Lazarsfeld, and cognizant of the work Herta Herzog had done on the subject of daytime listening, Dichter developed a complementary questionnaire to analyze and categorize listeners by fundamental attitudes and psychological dispositions. Questions probed into political attitudes, asked opinions about religion, surveyed aesthetic preferences, and inquired about "life style" (self-categorization as optimists or pessimists).[72] Dichter sought to develop a psychological profile of listeners to establish motivations but repeatedly clashed with more statistically minded researchers in-house and especially at Roper, which sent Alfred Politz as a representative to several of the meetings. In the end, Dichter's psychological case studies were supposed to form "pretests" in the research process at CBS to build hypotheses for a subsequent nationwide mass survey of about six thousand interviews conducted by Roper. He helped legitimize small-n qualitative studies as part of a mix of qualitative and quantitative approaches.[73]

In 1946, Dichter set up his own market research firm in upstate New York, the Institute for Motivational Research (originally Institute for Research in Mass Motivations). With his reputation on the rise, his relationship with CBS had devolved into a more external consulting role, while he produced reports for other media outlets and worked as an advisor to the Klein Institute for Aptitude Studies. His own institute began with a few mass media clients but quickly grew to serve numerous large companies across the consumer goods sector. By the mid-1950s, Dichter's consultancy had a staff of fifty employees at its New York headquarters, including over twenty specialists in fields ranging from psychology and anthropology to statistics and merchandising. In addition, his organization maintained several smaller offices across the country and abroad as well as a field staff of several hundred trained interviewers, which it could draw upon for its studies. In 1954, the institute listed about five hundred studies for various clients from consumer goods producers to public organizations. About fifteen advertising agencies had regular consulting arrangements on a retainer basis with Dichter.[74]

Research consultants such as Dichter offered a way to reduce the uncertainties of a dynamic marketplace. Understanding the hidden motivations and desires of consumers, he asserted, required in-depth, psychologically infused analysis of what goods meant and of how they were perceived and used. Such knowledge did not come cheap. Politz charged hefty prices for his elaborate statistical analyses according to a 1954 *Wall Street Journal* report. A full nationwide survey of up to twenty thousand consumers that gave a detailed picture of the composition of consumer markets according to household size, income level, sex, social status, among other things, could cost the client over $100,000. By comparison, motivation research was more affordable with an average price tag of $15,000 for a full study. Costs for a Dichter study ranged from $750 to $4,500 for simple advertising or packaging tests. A complete study of the advertising and sales problems of a given product could run anywhere from $6,500 to $19,500, depending on the number of respondents and questions involved.[75] Dichter also offered half- or full-day consultations for $250 and $500 respectively as well as a test study at just a few thousand dollars.[76]

Dichter's appeal lay not just in his comparative affordability, however, but also in his willingness to offer unorthodox interpretations. A typical Dichter investigation began by interviewing a small group of consumers in-depth to reveal the perceived "personality" of a given product and at the same time allowed for the psychological categorization of different types of consumers. A 1950s "Creative Research Memorandum on the Psychology of Hot Dogs," for example, was based on focus group sessions with 27 people of various backgrounds. The hot dog's appeal, the report explained, was based on the fact that it was an "adult food" that was still easy to eat and could be "easily conquered" even by children. On a "deeper psychological level, [hot dogs] provide an oral satisfaction comparable to cigarettes." Respondents, the client was told, "came up quite readily with the comparison to a phallic symbol. There is no doubt that some sort of oral sex is connected with hot dogs." Such sexual allusions were par for the course with the Freudian-inspired Dichter, but it did not simply boil down to selling sex. The hot dog's "democratic nature" and its evocation of "homemade" memories were similarly part of the product's complex personality, according to the memorandum. To connect to the "hot dog lover," Dichter advised his client, Swift & Co., to personalize their advertising for different types of hot dog consumers: "the nibbler," "the gulper," the "sensuous hot dog eater," and, finally, the "delicate hot dog eater," "possibly a woman," holding the hot dog between two fingers like "a jewel." In this last case, Dichter told the advertisers, erotic parallels were "unavoidable and probably desirable as long as they are done in good taste."[77]

To underscore his credentials as a psychological expert, Dichter sought academic and professional audiences as well. Before the Market Research Council, for example, he elaborated on his methodology in "depth interviewing" in 1943, arguing for qualitative approaches in early stages of consumer research and for the use of detailed, probing interviews.[78] In the trade magazine *AdQuiz*, he explained his use of so-called "psycho-panels," permanent panels of about 250 households, in which participants were classified not only by demographic and social factors, but also by psychological dispositions. Was a family governed by the authority of the father, were they secure or insecure, resigned or ambitious, overspending or miserly? Knowing these traits, Dichter suggested, allowed marketing to target specific segments of the population.[79] For the journal *Sociometry*, Dichter expanded on his ideas regarding the personality of products and the role of "commodities as intersocial media." He understood every purchasing act as a "highly dramatic event, full of spontaneity and emotion." In the "psychodramic performances" that unfolded on the stage of the store or the advertising tableau, Dichter proposed, products and consumers appeared as "co-actors" whose interrelationship the marketing psychologist had to reveal.[80]

Dichter's primary audiences were marketers and corporate executives, however. His question to them was: "Do you really know your customers?" Traditional, quantitative market research, he claimed, could say very little about the "true" reasons why people bought a certain product or brand or about the personality of customers.[81] Companies, Dichter suggested, often actually knew next to nothing about how their products were perceived by consumers, and only through collaboration with psychologists could they learn more about their consumers and their "innermost wishes." Motivation research offered a plethora of innovative approaches from the depth interviews and psycho-panels mentioned above to content analysis and projective techniques such as inkblot tests and word association methods. "Like a medical practitioner systematically examining several hypotheses," Dichter noted, describing his consultant work, "the researcher translates the clinical picture into . . . a questionnaire and makes a quantitative check-up of the various buying reading and listening mechanisms . . . involved."[82]

Dichter promised to elucidate for his clients "the real reasons why people buy goods." Using the example of an ice cream company, he illustrated how conventional surveys helped consumers rationalize their decision for a specific brand with references to rational factors such as preferred taste or price, indicating they bought an ice cream because they liked the flavor or because it was affordable. Buying motivations, however, were mostly more complex and could not be teased out with a "ready-made formula." The "true reason" for brand preference, one of his case studies suggested, could be the imagery of a bowl overflowing with ice cream, used in a specific brand's advertising, which evoked memories of childhood connected with emotions of joy and abundance in individual interviews.[83] Only a specialized psychologist could untangle this complexity of needs and motivations, Dichter suggested to companies and advertising agencies.[84] The trade press helped his business with flattering portraits in journals like *Business Week* and numerous articles on motivation research as the "new fad" in marketing; *Printer's Ink* even celebrated Dichter as a "Doctor for Ailing Products."[85] Dichter's institute began to publish its own monthly newsletter, *Motivations*, which counseled companies on changing consumer trends, new marketing challenges, and the advantages of employing cutting-edge research.

Knowledge of tastes, knowledge of "untapped demands," knowledge of sociological trends at home and abroad: *Motivations* advertised what Dichter and motivational researchers could provide. Using generalized insights from institute studies, the newsletter suggested to companies how to create a new line or product. A March 1956 report, "The Changing American Taste," advised producers that Americans were becoming more individualistic and cosmopolitan in their outlook and companies had to adapt.[86] The newsletter warned large corporations of being complacent and of losing their dynamism if they took consumers for granted. Consumers were constantly changing, and especially women, *Motivations* noted repeatedly, had become more "mature" and "self-assured" as consumers, balancing household work and a career at the same time. "Know your customers!" was the recurring mantra of *Motivations* articles.

The newsletter also helped clients to become familiar with the language and concepts of psychology. Articles taught readers what they "should know about perception" as understood by contemporary science and what Gestalt psychology could do to give a "new look" to an old product.[87] Business, Dichter urged, had to follow academia in abandoning the simplistic notions of behaviorist conditioning in favor of new understandings of human learning based on social psychology, which "leave room for human intuition and creativity in a wider and more meaningful context."[88] Dichter and Politz saw consumer capitalism as increasingly dependent on the expertise of independent knowledge entrepreneurs. In a culture where needs and desires were seemingly in constant flux, they sought to forge brands and new consumer loyalties. They saw themselves as the rising creative force in American marketing.

Image and Brand: Market Research as Creative Consumer Engineering

Creativity was as important as scientific exactitude when it came to promoting products and brand images. Dichter represented but the tip of the iceberg of a much larger consulting industry specialized in motivation research that emerged after World War II. In 1954, according to the *Wall Street Journal*, about eighty firms claimed they had the qualified expertise to provide motivation research. Companies such as Lever Bros. even opened their own psychological research departments, which were "loaded with Ph.D.'s."[89] While some scoffed at the new trend as "hocus pocus," others saw psychological studies and profiling of consumers as the ideal way to make their products and advertising unique and distinctive. Affording a "personality" to products became a common quest as brands, their "image," and their symbolic meaning received increasing attention.

Early motivation research took on a variety of different forms. In his 1957 handbook, Richard Crisp identified several methods in common use. Besides depth interviews, he discussed various forms of projective techniques that aimed to reveal "repressed" attitudes and feelings about a product by projecting them onto other people, such as neighbors. Images of particular brands were tested with sentence completion and word association tests. Thematic apperception tests, by contrast, asked consumers to "tell a story" about shopping experiences or about decisions a hypothetical person had to make when weighing, for example, the purchase of ready-made cake mixes against that of more basic ingredients such as flour.[90] To help laymen, i.e., the businessmen and marketers not properly trained in psychology, to make sense of the "turbulent" new field of motivation research with its "esoteric devices," the Advertising Research Foundation published an extensive bibliography of texts that discussed basic concepts and specific research techniques in an accessible, "non-technical" language.[91]

A 1956 *Fortune* article distinguished "three main schools" of motivation research in the United States.[92] The first school, of which Dichter was the most prominent representative, utilized Freudian concepts and individualized interviews and relied heavily on psychoanalytical methods to get at consumer motivations and product perceptions. A second school, which the article identified with Paul Lazarsfeld of Columbia University and Herta Herzog of McCann-Erickson, was deemed more open to quantitative methods. It combined several approaches: conventional surveys to identify different consumer segments, qualitative interviews and projective tests with a small subset of those groups, a larger "structured questionnaire" survey of 1,200 to 3,000 consumers to test the conclusions of the qualitative phase, and finally the qualitative pretesting of psychological advertising appeal based on the result. A third school, finally, emphasized the role of group behavior as well as the impact of environmental and social factors on consumer motivations. This "psycho-sociological" approach was most prominent with Chicago-based researchers such as the sociologist Lloyd Warner, Pierre Martineau of the *Chicago Tribune*, and the commercial researchers associated with Social Research Inc. (SRI), Burleigh Gardner and Sidney Levy.

The Chicago group was particularly influential in popularizing the connection between psychological research and brand images. Pierre Martineau's 1957 *Motivation in Advertising* became a standard book on the subject, bringing notions of "product personality" and "product image," lifestyle and the symbolic quality of goods to prominence among contemporary marketing specialists.[93] Already in 1955, Gardner and Levy had argued for the importance of qualitative motivation research in uncovering the symbolic dimensions of "the product and the brand." Brand names, their sound, and the psychological associations they conveyed were deemed crucial for the "public image" of a given product. Often regardless of actual properties, the authors argued, the perceived image of a product had to match the personality of the buyer—a car or cigarette had to be "right for me" from the consumer's perspective. Understanding consumer attitudes and motives thus helped manufacturers and advertisers to develop different images for specific market segments and durable, long-lasting brands.[94]

There was a keen sense of rivalry among the various motivation research camps. To Chicago knowledge entrepreneurs such as Levy at SRI, the roots of motivation research lay within the Chicago intellectual milieu and a group of social scientists including Reinhard Bendix, Ernest Burgess, Erving Goffman, David Riesman, and Edward Shils. Levy even claimed to have coined the term "motivation research," suggesting it to Martineau for use in his monograph title.[95] Such claims, however, not only willfully ignored the work of competitors such as Dichter and Herzog but also obscured their common interest in overcoming purely quantitative approaches to marketing research. As marketing historian Ronald Fullerton has observed, motivation research was part of the broader movement toward consumer behavior research among marketing scholars during the 1960s, as discussed in chapter 3.[96]

However, while qualitative methods were definitely on the upswing during the 1950s, they never fully replaced quantitative research, as the example of Politz's success has shown.[97] Politz attacked the psychological approaches as a flimsy fad. Results based on a small number of interviews, he argued, lacked empirical validity. The main fallacy of psychological motivation research, to Politz, lay in the assumption that there even was such a thing as the one, true motivation, which could somehow be "uncovered." Instead, he asserted, many factors influenced and determined buyer behavior, and such multicausality demanded careful statistical measurement to determine the relative importance of each individual factor.[98] Dichter, by contrast, called the emphasis on statistics an "outdated" and "unpragmatic" "belief in Aristotelian empiricism."[99] Responding to Politz's insistence on the importance of large-scale, projectable samples at a marketing conference, Dichter allegedly quipped: "But, Alfred, ten thousand times nothing is still nothing!"[100]

Despite such public differences over methodology, the market research consultants of the 1950s still agreed on the general importance of sophisticated research for modern mass marketing: being "scientific" had become central to their own marketing efforts. The leading consultancies all offered the full complement of research designs. In all cases, too, they now went well beyond more traditional interwar attempts at market research, which had simply "counted noses" by collecting demographic information about customers. Their research—whether through depth interviews or complex statistical cross-tabulation—went beyond observable acts of stimulus and response to "find out what consumers' lives are like" and to craft brand images and product personalities to match these lives.[101]

Dichter called his product "creative research," which combined systematic scientific analysis with his own "creativeness and intuitiveness": "You can't research without knowing what you are searching for. Thus you need hypotheses, and *that* means creativity."[102] Creative research, to Dichter, drew on a complex understanding of consumer motivations derived from Freudian and Gestalt psychology (which left room for human intuition) and on the realization that postwar marketing was "facing an entirely new type of consumer." The new consumers, Dichter thought, were more individualistic than ever before, more "mature" in their understanding of marketing practices, had a stronger desire to participate in marketplace decisions, and—most important—they had the financial means to make demands and to choose between options.[103] With the deprivations of the Depression and World War II safely left behind, consumers now sought goods with personalities that they could relate to, that were symbols for their desires, and whose qualities expressed individuality through "newness" or sensuality.[104] In a buyer's market, these new consumers had a much greater degree of power, and only creative research and marketing could assure a continuous satisfaction of consumer demand.

Politz, too, made creative research a recurring theme in his writings on statistics and science as tools in the creative process. "Creative research," he argued, separated the important from the unimportant and enabled businesses and advertisers to anticipate and think ahead.[105] Politz, to be sure, was more of a traditionalist when it came to his understanding of effective advertising, and he was dismissive of early advances of what has been described as the "creative revolution" in advertising by the 1960s. He disliked the rise of humorous, unconventional, or in other ways attention-getting ads that brought carnivalesque elements back into advertising after decades of efforts by professional admen to distance themselves from the boisterous and sometimes shady beginnings of their profession.[106] This kind of "creative" ads, Politz felt, served only to highlight the ad executive or agency but distracted from the product; entertainment and wittiness were counterproductive in the long run.[107] Evoking Max Wertheimer and Gestalt psychology to different ends than most of his colleagues, he argued that physical properties of a product should retain precedence over its psychological "image."[108] But while Politz might have missed a major shift in the advertising business in this particular respect, he still believed in the innovative force of scientific marketing.

Creative experts, Politz believed, were crucial to further product innovation and ultimately social improvement.[109] He attacked what he saw as the misguided "belief that marketing research exists to find out what the consumer wants."[110] While knowledge of explicit consumer wishes was of course useful, these wishes constituted only a small part of what marketers needed to know in order to create and successfully distribute a new product. "Consumers," Politz asserted, "do not know what they want and why they act. If the uninfluenced opinions of consumers in the year 1800 had determined the development of lighting systems, we would have today merely vastly improved kerosene lamps. The electric light bulb . . . was not part of consumer thinking in 1800. Consumers do not solve problems by answering questions. Research has to solve these problems by intelligent experimental design, and by drawing intelligent conclusions from the raw material of consumer opinions and behavior."[111]

In such sentiments, Politz channeled the self-perception of postwar consumer engineers: They understood markets and consumers better than anyone, because they had the scientific tools and creative ingenuity to reveal problems and preferences, needs and desires. If given the proper authority within corporations or advertising agencies, they could be a creative force aiding companies to succeed in a competitive marketplace and aiding consumers by providing them with the goods they needed. Since more and more needs in America's affluent "consumer's republic" were thought to be of a psychological nature, even new product personalities and brand images served a real function in the eyes of most consumer engineers.[112] Knowledge entrepreneurs such as Politz and Dichter, finally, saw themselves as furthering the public good by helping consumers and companies realize the potential of postwar mass consumption and by providing the necessary creative input to assure that American consumer capitalism, instead of falling into affluent complacency, did not lose its innovative dynamism.

Consumer Engineering and the Limits of Hidden Persuasion

Vance Packard and other contemporary critics correctly observed that marketing efforts in postwar consumer capitalism had become more systematic and more subtle than in previous decades. Market research stood at the center of this transformation with new psychological and statistical methodologies informed by transatlantic exchanges. Corporations got to know their customers in more detail, and they used this knowledge to their advantage. They designed products that appealed to hidden consumer desires, and they created product personalities and brand images through their advertising, which resonated with specific, carefully targeted segments of the buying public. The advent of computers swung the pendulum back toward quantitative, data-driven approaches by the 1960s, but qualitative, psychological studies with focus groups and other devices were here to stay for decades to come. People, consumer engineers believed, could be swayed by repressed memories, sexual allusions, prejudices, and emotional associations.

Even more fundamental, historian Stefan Schwarzkopf has recently discussed Dichter as an agent of cultural "disembedding." Psychological research helped companies to target previously inhibited impulses and further removed social norms from market relations.[113] The therapeutic ethos of consumer satisfaction and psychological needs fulfillment found especially in motivation research, it could be argued, advanced a more individualized, hedonistic, and ultimately unsustainable mass consumer culture during the final decades of the twentieth century. The combination of "European" psychological traditions and "American" professionalization in marketing would cast a long shadow.

With "creative research" it was possible to manipulate consumers, Dichter was the first to admit. He would argue, however, that since the experts often knew consumers' true needs better than they themselves did, well-meaning nudging through marketing measures could not be condemned. It would indeed be shortsighted to reduce consumer engineers such as Politz or Dichter to agents of corporate profit maximization. Their efforts stemmed in part from a genuine belief in the potential of scientific marketing to optimize economic processes and to produce "a new age of prosperity" with continuous growth through mass distribution. They were agents of a technocratic "growth liberalism" at mid-century that equated ever-growing consumption with the promise of a peaceful and democratic society.[114] Dichter and others were part of a Cold War culture that advertised not just goods but an entire political and economic system whose appeal was intricately tied up with its capability to satisfy the inner needs of its citizens.[115]

Like many of his colleagues, Dichter regarded marketing as a "task for social engineers" who employed psychology to help American democracy and capitalism succeed.[116] While he took Packard's accusations of manipulation as a testament to his skills as a research consultant, Dichter also believed that his critics overestimated his power over consumers. Much like Katona and other consumer researchers, he believed that consumers—especially in the longer term—could be swayed to go down a certain path or to make certain decisions only if there was some perceived benefit. The hypnotic force of a sales pitch could go only so far, Dichter explained, but consumers liked to blame the advertising when they found themselves with buyer's remorse.[117] It would be more honest, he felt, to admit that people are persuaded at all times and for various reasons, that their motivations are often contradictory, and that emotions govern our behavior as much as reason. Marketing experts like him, Dichter argued, simply wanted American consumers to embrace a modern world of goods to the benefit of all.[118]

Another, more mundane, response to Packard's critique of mass manipulation would be to stress the limits of consumer engineering even more strongly than Dichter would have been willing to do. For all their "scientific" sophistication, postwar consumer engineers produced flops as grandiose as the Ford Edsel. They frequently misread the market, misunderstood consumers, or pushed for misguided innovation. Unsurprisingly, the stories of their failures have survived the times to a much lesser degree than tales of success. Competing for customers as independent consultants and knowledge entrepreneurs, market researchers like Politz and Dichter talked up their scientific prowess, their creative genius, and their abilities to sway consumers. Yet one would be mistaken to take their claims at face value. The discrepancy between expert claims and reality became most obvious with marketing psychologist James Vicary, who claimed to be able to psychologically manipulate consumers into buying certain goods by splicing hidden, symbolic images into advertising films. Such hypnotizing, "subliminal" advertising resonated with 1950s fears of propaganda and manipulative mass persuasion, but the experiments turned out to be a hoax.[119] Vicary did not have that kind of power over consumers, and neither did his more legitimate colleagues. As marketers had become more sophisticated, behavioral economist Raymond Bauer argued in 1958, so had consumers, whose power to resist psychological appeals created by new techniques was aided by the fact that every consumer engineer faced stiff competition from other brands and products similarly vying for the consumer's attention.[120]

Still, by 1960 we can observe a marketing profession in which psychologically infused research played a vastly greater role than three decades before, when Sheldon and Arens called for a program of "consumer engineering." European emigrés and the methodological toolkits they developed within transatlantic contexts played an important role in this transformation, as the previous three chapters have demonstrated. As academics within the American marketing profession such as Lazarsfeld and Katona or as "knowledge entrepreneurs" with independent consulting firms such as Dichter and Politz, they not only reflected the growing "scientization" of American marketing but actively promoted that change as well.[121] In the process, they helped bring about the postwar world of goods with "product personalities" and "brand images." Market researchers, however, were not the only self-proclaimed agents of creativity within midcentury American consumer capitalism. Increasingly, they cooperated and competed with a new generation of commercial designers who had similarly evolved into a profession with claims to innovative potential and unique consumer knowledge over the course of the 1930s to 1950s. Transatlantic exchanges facilitated by European immigrants and emigrés had an outsized impact on the field of design as well.

SECTION TWO

Designing for Sustained Demand

"Tastemakers" or "Wastemakers"? Commercial Design at Midcentury (1930–1960)

Just a few years after his attack on the "hidden persuaders," Vance Packard took aim at another aspect of consumer capitalism: "planned obsolescence." Consumer products, Packard claimed, were built to quickly go out of style and to create continuous replacement demand.[1] With this, commercial and industrial designers, another relatively new profession in American industry, now came into the crosshairs of his critique. The practice and profession of industrial design had "exploded" in the United States since the 1930s, with the promise of raising aesthetic standards and bringing "good design" to mass markets. Nationally known designers fashioned themselves as "tastemakers" for the American way of life. Yet, as historian Lizabeth Cohen has noted, industrial design frequently "fed into the strategy of planned obsolescence, with designers regularly restyling mass consumer goods."[2]

Like market research, design became an increasingly crucial element of midcentury marketing with regard to product, promotion, price, and place. Department store displays and later the new consumer environments of supermarkets and shopping centers required professional design expertise. Product styling became more systematically integrated into the production process. Modern art forms found their way into graphic design for advertising, packaging, corporate public relations, and other forms of promotion. By allowing for both subtle distinctions and wide gradations in price and appearance, design marketing furthermore advanced practices of market segmentation. Designers helped to develop brand images and corporate identities, and they promised a vision of consumer modernity that would fulfill physiological and psychological needs.

Several parallels between consumer research and commercial design are instructive for understanding the rise of midcentury consumer engineering. In both cases, a new profession emerged with new organizations and educational programs. Both design and consumer research worked through advertising agencies, specialized corporate departments, and independent consultancies. Many industrial and graphic designers, too, retained an outsider status as artists within the corporate world, and both fields saw lively debates over questions of academic and artistic integrity, professional standards, and the demands of commercial marketing. Several designers also came from interwar reform movements and social planning efforts into the world of corporate capitalism. They shared a genuine interest in understanding the psychology of consumer markets and saw themselves as creative entrepreneurs within the structures of large corporations.

Midcentury design and consumer research, finally, shared the fact that European immigrants and emigrés had an outsized impact on both fields. The chapters in this section trace both the expansion of professional design as an integral part of consumer marketing and the transatlantic transfers that shaped that process. Chapter 5 discusses the rise of design professionals as marketing consultants beginning in the 1930s and shows the influence of European modernist designers on American notions of "good design" in numerous areas ranging from industrial to graphic design and from textiles to housewares. The influence of German interwar design on American design education stands at the center of chapter 6, which also explores the tensions inherent in adapting the social reform ideas of the Werkbund and Bauhaus movements to the demands of American industry. Postwar design consultants increasingly relied on practices of market and consumer research, as chapter 7 demonstrates using the case studies of Raymond Loewy and Walter Landor. Designers increasingly looked beyond individual products as they created brand systems and entire ensembles of goods. While market researchers touted their scientific and creative research, designers boasted of engineered creativity tailored to consumer needs and desires.

FIVE

The Designer as Marketing Expert: European Immigrants and the Professionalization of Industrial and Graphic Design in the United States

In 1955, designer Raymond Loewy addressed an audience of professional industrial designers in Essen, Germany. An American citizen born and educated in France, Loewy was likely the most famous member of his profession on either side of the Atlantic. In his speech, he reflected upon the development of industrial design since the 1930s. Working alongside engineers and sales and advertising executives, he noted, product and graphic designers had become an integral part of industrial development processes. Increasingly, they were employed as marketing experts in support of America's largest manufacturers and corporations.[1] Industrial design had moved well beyond its origins as a form of applied arts in the early decades of the twentieth century. "As you see, gentlemen," Loewy continued, "our highly complex profession is now a far cry from the early days of the justly famous BAUHAUS. The happy days when an artistic gentleman could give his undivided attention to the fine handicraft of a lamp shade or the precious twisting of aesthetically bent wires are over. Nevertheless, such

133

early activities . . . as those of the BAUHAUS as well as those of the DEUTSCHER WERKBUND were the sparks that started the present chain reaction. To these men, we like to express our appreciation for what they did so early to assist industry."[2] Loewy was well aware that many in his audience of German designers were reluctant to embrace the role of the designer as a commercial marketing consultant. He therefore carefully connected corporate design in the United States to aesthetic traditions such as the Bauhaus that were deeply rooted in interwar European artistic reform movements.

Indeed, this connection between interwar European avant-garde modernism and the development of the design profession in the United States was hardly far-fetched. While the rise of "styling," "streamlining," and specialized industrial design offices has been typically described as a specifically American response to the expansion of consumer markets and the challenges of the Great Depression, artists born and educated in Europe played a significant role in this process.[3] They occupied leading positions in industrial design offices and advertising agencies, they helped institute new programs and professional institutions, and they transferred modernist design approaches to the United States. This chapter sets out to trace the rise of commercial design while systematically highlighting the contributions of European emigrés. Aside from work on individual immigrant artists, there is no comprehensive account of the impact of European emigrés on commercial design in the United States. Yet transatlantic exchanges significantly shaped the appearance of American consumer goods and commercial aesthetics by midcentury. Much like Loewy, moreover, some of the emigrés became prominent champions of a "new type of artist" with little inhibition in engaging in commercial art. The resulting "romance of commerce and culture" became a crucial aspect of modern consumer capitalism.[4]

A New Profession: Industrial Designers as Consumer Engineers

Several new handbooks surveyed industrial design as a young profession during the 1930s and 1940s. Design departments emerged in corporations, independent design studios became successful business ventures, design professionals published professional guides and journals, and they established courses of study. Whether as company stylists, art directors at magazines and advertising agencies, independent consultants, or as academics freelancing in the private sector, designers became a significant force in the business world. Their professional identity oscillated between art and commerce. Industrial designer Harold Van Doren had published what he called the first "practical guidebook" for the field in 1940. Van Doren believed that the designer should be "primarily an artist" but also a marketing specialist: "Stripped of hocus-pocus, the goal of design is sales—at a profit. For ten years, the author has practiced this new profession with no thought but to create merchandise that the public would buy."[5] Industrial design, to Van Doren and his colleagues, concerned the creation and styling of mass manufactured goods for broad-based consumption. To him, machine-made metal and plastic goods for everyday use (the form of which once used to be determined primarily by utility and price) were the true domain of the industrial designer. He observed that "planned obsolescence has become almost a routine in stoves, radios, refrigerators, out-door furniture, and other lines." The purpose of design, he wrote, "is to enhance . . . desirability in the eyes of the purchaser through increased convenience and better adaptability of form to function; through knowledge of consumer psychology; and through the aesthetic appeal of form, color, and texture," producing a seemingly endless stream of novel shapes and aesthetically appealing consumer goods.[6] This, of course, directly echoed the sentiments of consumer engineering.

In industrial design as much as in graphic design for packaging and advertising and in the more traditional fashion goods industries, American business turned its attention to matters of aesthetics and artistic design. Business did not merely coopt artists by shaping them into marketing men and blunting their agenda of design modernism, however.[7] Sociologist Andreas Reckwitz has argued that advertising, fashion, and design served as pioneering sectors of a new creative economy. In Reckwitz's reading, an ideal of artistic creativity won out over hierarchical rationalization in the self-understanding of many business actors.[8] A closer look at the growing importance of the design profession in the United States suggests that aesthetic considerations indeed became an integral part of the American consumer goods sector several decades before the much-discussed "creative revolutions" of the 1960s.[9] Already the post-Depression era witnessed a large transformation with regard to the role of aesthetics in American capitalism.[10] Michael Augspurger has traced the aesthetic sensibilities of a new generation of "corporate liberal" business executives, including *Fortune* founder Henry Luce, who demanded not only an economy of abundance but an "economy of abundant beauty." The magazine signified the businessman's embrace of high culture: "Business and art, *Fortune* argued, were not only compatible but were also symbiotic."[11] Not only had artists made their way into the major advertising agencies of Madison Avenue, but college educated managers increasingly populated the boardrooms of American corporations, bringing with them a new openness toward modern forms of aesthetic representation.

American manufacturers were "intensely alive to industrial design values," as a 1935 survey of the design field observed.[12] In industries from textiles and furniture to upscale luxury goods like pianos, small batches production had long entailed frequent model and design changes attuned to shifting consumer demand.[13] As consumer markets became increasingly saturated by the late 1920s, marketers became more attuned to design changes in the mass production sectors of American industry as well. Product differentiation by style began to provide a contrast to "Fordist" uniformity, and automobile companies, for example, expanded their styling departments.[14] One of the stars of the new profession was industrial designer Norman Bel Geddes, who was among the first to open an independent design studio in 1927. Inspired and mentored by stage designer Joseph Urban, an Austrian immigrant who created the scenes for the Metropolitan Opera and the Ziegfeld Follies in 1920s New York City, Bel Geddes had a background in theater design. Credited as the "father of streamlining" by *Fortune Magazine* in 1934 for his book *Horizons*, Bel Geddes and his design studio soon designed nearly everything from radios and refrigerators to automobiles and locomotives. His most acclaimed achievement was perhaps General Motors' famous "Futurama" display for the 1939 New York World's Fair, which placed GM automobiles in a futuristic vision of urban mobility. The exhibition connected his work as an industrial designer directly back to his roots in stage and theater settings. Bel Geddes's work as a designer, historian Christopher Innes suggests, shaped the look of American consumer modernity and helped spark the imagination of future material abundance.[15]

While Bel Geddes was a trailblazer for streamlined design, he was part of a larger group of design professionals whose early careers were closely connected to large advertising agencies. Among the most well known were Raymond Loewy, Walter Teague, and Henry Dreyfuss. Teague, too, had started an independent design consultancy in the late 1920s, and while Bel Geddes initially worked closely with the J. Walter Thompson advertising agency, Teague collaborated with Calkins & Holden. His early designs included work for Eastman Kodak as well as for Steinway pianos and Corning Glass. Henry Dreyfuss was a stage designer as well with connections to J. W. Thompson before opening his own design studio. Some of Dreyfuss's most influential designs include the Western Electric desk telephone for Bell Laboratories and the 1936 Hoover vacuum cleaner.[16] Others, including Harold Van Doren and Joseph Sinel, round out a group of pioneering design all-rounders, which Sheldon Cheney and Martha Candler Cheney in 1936 referred to as the "renaissance-men" of early U.S. design.[17] To them, the decade of the Great Depression produced a sea change in American industry as functional design was introduced to help expand sales.[18]

Alongside "consumer engineer" Egmont Arens, these designers counted among the fourteen founding members of the Industrial Designers Society of America, which was created in 1944. They represented a larger group of prominent designers. In New York City George Sakier and Donald Dohner created influential design work in affiliation with the Brooklyn-based Pratt Institute. Architects such as William Lescaze and Richard Neutra were part of the emerging design field, as were interior designers such as Donald Deskey, Kem Weber, Gilbert Rohde, and Robert Heller.[19] Few of these designers were affiliated with any one company, but instead they acted as independent consultants for design jobs of various kinds, from fairs and exhibitions to corporate public relations.[20]

A slew of new design education programs helped to foster a sense of professionalism. As late as 1935, New York architect Ely Jacques Kahn had bemoaned the problems of American design schooling while lauding the "superior schooling" in Europe, from the Kunstgewerbeschule in Vienna to the Bauhaus in Dessau: "Much has been done in Europe in recent years," he concluded, "that represents distinct progress."[21] A few years later, by contrast, Harold Van Doren could point to several centers of modern design education in the United States. In New York, the Pratt Institute offered education to aspiring industrial designers, as did the Laboratory School of Industrial Design directed by Gilbert Rohde. Elsewhere in the country, too, design programs were established. In Pittsburgh, Peter Muller-Munk headed the industrial design department at the Carnegie Institute of Technology, and in Los Angeles, Walter Baermann had helped build up the California Institute of Technology's graduate department of industrial design engineering.[22] In addition to the Rhode Island School of Design, directed by Alexander Dorner between 1937 and 1941, Van Doren also pointed to the existence of several new Bauhaus offshoots in the United States. We will return to American design education and the influence of European-born designers on it in chapter 6. For now, it suffices to note that design education in the United States had expanded significantly over the two decades following the Great Depression. A 1946 Society of Industrial Designers bulletin detailed twenty programs for professional education in industrial design, preparing new generations of commercial designers in aspects ranging from artistic modernism to business psychology.[23]

During the 1940s designers passionately debated their professional identity between artistic freedom and commercial constraints. A November 1946 conference, "Industrial Design: A New Profession," at the New York Museum of Modern Art provides a unique glimpse at these debates.[24] Organized by the Society of Industrial Designers (SID) and chaired by Joseph Hudnut, dean of Harvard's School of Design, the four-day conference assembled many of the leading designers in postwar America. Egmont Arens, Walter Baermann, Raymond Loewy, Gordon Lippincott, George Nelson, George Sakier, and Walter Teague all participated, as did corporate designers and representatives of the American Designers Institute, the Pratt Institute, and the Institute of Design in Chicago.

Professional standards were the focus of the conference's first session.[25] Participants agreed on the need for shared educational standards and a common code of ethics. Joseph Hudnut called upon the profession to define a special set of skills that would set the profession apart, a "special competence" in a "well-defined field." Walter Baermann of Caltech's Graduate School of Design summarized the "basic training" of industrial designers very much in terms of marketing and consumption:

> An industrial designer coordinates consumer demands, merchandising demands and the demands of advertising. With these, the industrial designer also has to coordinate production methods, available and appropriate materials—and, last but not least, he must consider the ultimate function and behavior of the product in our daily life. He may go beyond that—I wish he would—and also consider the product as a social element.[26]

Critics, however, questioned if commercial design work could truly be reconciled with serving a greater social good as Baermann hoped. What were the aesthetic lines that artists should not cross? And, Hudnut wondered, should "a man's conscience" be troubled a little "when he puts out a line of hideous lamps or chairs"?

The search for a professional code of ethics focused on the relationship between designers and their industrial clients. Raymond Loewy and Walter Teague introduced the discussion with a role-play sketch between Loewy as the design consultant and Teague as his fictional client, America's largest "widget manufacturer."[27] Loewy wins the skeptical manufacturer as a client by emphasizing his attention to the company's needs. He addresses the concerns of the internal engineering and design departments, he proposes a detailed study of the "widget" market, and he asks to consult with sales managers and retail outlets. No leading professional designer, Loewy asserted, would offer sketches or design ideas without a clear grasp of the market conditions first. Like Politz and other market research consultants, Loewy and Teague modeled their view of ethical standards for the new profession on the advertising industry. For any corporate account, all information was to be treated with complete confidentiality so as to establish the necessary trust on the part of companies when it came to their business conditions and market strategies. Designers would have only one client per industry and an exclusive relationship, which would help avoid any conflict of interest. Design consultants, Loewy explained, strove to establish long-term relationships with manufacturers. Depending on the nature of the project, clients could choose between a nominal retainer fee and compensation based on royalties or a fixed-fee arrangement. This level of professionalism in business conduct, the short skit suggested, was key to making the design artist a reliable and successful partner to business.

Others, however, worried about the constraints that arose from the designer's relationship to industry. Constraints of cost, engineering requirements, and especially of consumer demand could produce aesthetically inferior products. Particularly with regard to fashion goods, "lack of taste" on the part of the consumers and—as Egmont Arens and Walter Baermann pointed out—the "taste of buyers at the large department stores" or the power of advertising often held more sway over the eventual shape of goods than designers themselves.[28] As much as they were willing to pay attention to market demand, design professionals still regarded themselves as the ultimate arbiters and experts on matters of taste and style. Utilizing consumer surveys and market research, they listened to consumers, but they were reluctant to abdicate their aesthetic authority to them or to their corporate clients. Many designers bemoaned the restrictions put on their work by retailers with conservative tastes or by the cost cutting in materials demanded by producers. But others stressed these limitations as a wellspring of opportunity. "All great art springs from limitations," cheered Teague, and Loewy added that cost-cutting for a still reasonable design "allows a manufacturer to sell his product at a much lower price, [reaching] a mass market of poor people who could never afford that sort of unit if it were just a few dollars higher."[29]

The division within the profession between the business-oriented consumer engineers and those who prioritized artistic authority was especially pronounced with regard to notions of styling and obsolescence.

"Why must we change styles each year?" the Chicago Institute of Design's James Prestini asked, and Walter Baermann complained about the "false sentimentality" in advertising that used images of beautiful women rather than sober forms to sell cars by indirect association. Advocates of outright consumer engineering, by contrast, defended the practice. In industries such as automobiles, Walter Teague argued, change was an economic necessity as production tools lasted for only about two years, allowing companies to incorporate improvements and changes into new model years. Manufacturers, Egmont Arens added even more bluntly, had a concrete reason for changing designs, as this stimulated demand and created new interest with consumers: "Newness and difference happen to be the things that people respond to: newness, difference, interest."[30]

However, was it legitimate for the professional designer to follow consumer desires in this way? The designer as expert, many conference attendees believed, had a social responsibility to educate consumers and elevate public taste. Some suggested this could be done by simply making "better design" (usually defined as simple designs of high-quality goods appropriate to intended use) widely available to consumers. Others saw the need to involve schools and other educational institutions. The importance of training retail buyers and sales staff was raised by several attendees—George Sakier and Walter Baermann pointed to the example of Swedish cooperative stores where salespeople (and consumers) were taught in special forums to appreciate the merits and beauty of a commodity. By eliminating waste and directing sales, designers would be acting as true professionals, "because they will then be serving society, rather than just serving business."[31]

The notion that the design profession ultimately served the "greater good" of society by creating good design was at once a convenient self-legitimization and effective PR. Several designers at the conference—and prominently those with a European background—described a peculiar American design culture that arose from this larger context. Bernard Rudofsky, for example, warned that U.S. mass production systems had lost sight of genuine human needs—creating standards for kitchens and bathrooms, for example, that, he felt, were not adapted to the complexity of human physiology and cultures of use. Hungarian-born Bauhaus artist László Moholy-Nagy argued that American abundance created a propensity for throwaway designs and obsolescence. To Moholy, the concept of "novelty" was merely a simulation of a desire rather than "true need," and he scoffed at American advertisements driven by status anxiety or "the fear of body odor or halitosis." He contrasted an

American economy built on a cycle of expanding domestic consumption and full employment with postwar European economies, which commanded few resources and were dependent on trade and exports. Thus, Moholy-Nagy surmised, Europeans preferred durable goods that saved resources by lasting a long time, and they produced quality goods that proved internationally competitive for export.[32]

Drawing on this European perspective, Moholy-Nagy and others like him argued for lasting standards rather than "style-cycles" dictated by advertising as the true ideal of a design profession. Pratt's Dean, James Boudreau, by contrast, spoke for those consumer engineers who did not share this idealistic vision. "Accelerated obsolescence," he shot back, formed the cornerstone of the American economy: "You cannot impose upon America the European philosophy of goods, any more than we can impose our philosophy upon Europe. Our entire economy is based upon accelerated obsolescence." Long-term sustainability of economic development was not a primary concern for the adherents of the gospel of growth. However, "good design," they felt, would still prevail in the long run, aided by the power of the market: "Design just can't be obsolete, if it is good."[33] Thus, while skeptical voices persisted, the new profession as a whole looked confident going into the postwar years. Backed by new education programs and new professional standards, designers positioned themselves as key players in American industry and consumer capitalism.

Transatlantic Modernism: European Immigrants and American Commercial Design

The MoMA conference also documents the influential role of European immigrants and emigrés in establishing the American design profession. Among the select group of designers present, Walter Baermann, Raymond Loewy, László Moholy-Nagy, and Bernard Rudofsky all had been born and educated in Europe, and many others had studied there. In part, this reflected a longstanding deference to European art and design that architect Ely Kahn observed: "an object to be interesting [had to] be made in Paris, Berlin, or some exotic corner in Europe." While American design had made great strides especially with consumer durables, he cautioned: "I venture the . . . premise that many of the designs made in America for goods also made here were executed by Europeans, or Americans trained in Europe. . . . We have profited of course by the assistance of our guest conductors, but a moment arrives when self-reliance might have some virtue."[34] By the 1930s, Americans increasingly began to challenge notions of European cultural superiority.[35] Still, even after World War II, historian Jeffrey Meikle notes, many American designers felt like "second-class citizens" in their own country, because the postwar American debate about "good design" was superficially cast in terms of European "quality" versus American "chrome" and because prominent emigrés such as Walter Gropius or Ludwig Mies van der Rohe seemingly dominated public design in the United States.[36]

Such a dichotomy between Europeans and Americans or between "European" and "American" design is somewhat misleading, however, because of the numerous transnational transfers in the design field. At times the emigrés brought an outsider's perspective to bear on the American economy, but generally migrant designers were actively engaged in developing the design profession in the United States from within. One central contribution of the European immigrants and emigrés was the promotion of various strands of modernism on both sides of the Atlantic. Design historians argue about what constituted "true" modernism in design, with some favoring the strict functionalism associated with the German Bauhaus over more organic forms and the streamlining trend. While such distinctions informed contemporary debate, it appears difficult from a historical perspective to neatly separate an "authentically" modern form from mere commercial "styling." The popular organic streamlining of the 1930s was very much rooted in the ideas of technocratic "high modernism," as historian Christina Cogdell has argued.[37] From the angularity of French-inspired Art Deco to the more severe forms of functionalism, a variety of design traditions all aimed to create simple and clean forms to match a new "machine age."[38]

Many shades of "European" modernism crossed the Atlantic and influenced American design education during the interwar years.[39] While they have—with some justification—garnered much attention among contemporaries and historians alike, we need to look beyond the Bauhaus emigrés for a more complete understanding of transatlantic traditions in design education. Alexander Dorner, for example, studied in Berlin and was director of the State Museum in Hanover before he headed the Rhode Island School of Design from 1937 to 1941. Dorner believed in the "spirit of functional design" for all areas from furniture to typography.[40] Peter Muller-Munk took over the design program at Pittsburgh's Carnegie Institute. He had studied as a silversmith and designer with Bruno Paul in Berlin before moving in 1926 to New York, where he first worked at Tiffany's and exhibited his work at a Macy's exhibition in 1928 as well as at several Metropolitan Museum of Art shows. His industrial design work focused on an upscale Art Deco designs including the 1937 chrome Waring Blender and his Normandie Pitcher for Revere Copper and Brass. Muller-Munk established his own design firm in Pittsburgh, but his biggest contribution to American design came through his academic work and through his leadership in professional organizations.[41] He was part of a broader modernist design reform movement that swept from interwar Europe to the United States. Bruno Paul's state school for applied arts in Berlin was connected to the tradition of the Munich and Berlin Secession as well as to the Werkbund and Muller-Munk was one of several of the school's students who came to America with strong notions of modern design without embracing the angular radicalism of the Bauhaus.

The immigrant designers thus hailed from various European design traditions. Walter Baermann, another important immigrant educator at the California Institute of Technology, had studied architecture and engineering in Munich. Baermann came to the United States in 1929 and pursued first and foremost an academic career. Before doing a wartime stint as chief of graphics for the Office of Civilian Defense, he moved from California to Bloomfield, Michigan, to head the design department of Cranbrook Academy.[42] Since the 1920s, Cranbrook's design program had been developed by another European immigrant, Finnish-born architect Eliel Saarinen. Saarinen, whose work drew initially on the English Arts and Crafts movement as well as on Jugendstil designs and the Vienna Secession, was president of the Cranbrook Art Academy from 1932 to 1946. Under his direction, the school became an important training ground for designers with a modernist outlook akin to the interwar European reform traditions.[43]

Other immigrants came out of engineering and technical drafting. Railroad designer Otto Kuhler, born in Remscheid, Germany, in 1894, had studied electrical engineering and technical drawing in Belgium and worked as a designer for a Berlin car manufacturer. After immigrating to the United States in 1923, Kuhler worked as a freelance industrial artist in Pittsburgh before opening his own New York design studio in 1928 with railroads as his major client.[44] Peter Schlumbohm, born in Kiel in 1896, the son of a chemical manufacturer, was also particularly interested in the technical aspects of design. After studying chemistry as well as Gestalt psychology with Wolfgang Köhler at Berlin University, Schlumbohm moved to New York in 1936 where his most acclaimed works included the utilitarian Chemex coffee maker.[45]

Several other immigrant designers, finally, came from the field of modern architecture, including Swiss-born William Lescaze, who had trained in Zurich and France, and Vienna-born architect Frederick

Kiesler.[46] The lines between architecture and design were blurred, and architects often designed furniture and other items while architectural firms also employed designers. California architect Richard Joseph Neutra was one of the early advocates of applying industrial finishes and materials to residential housing. Born in Vienna in 1892, Neutra worked in Switzerland and Berlin before immigrating to the United States in 1923. Neutra became a pioneer of "international style" residential architecture and interior design on the West Coast.[47]

Interior design was one of the few fields in which female designers were able to carve out a significant professional space for themselves.[48] Their work frequently drew on that of other women designers working in textiles and in household wares from ceramics to glass. Here, too, several prominent immigrant artists helped to bring modernist design impulses to the United States. Bauhaus designer Anni Albers's work in textiles influenced a new generation of American artists. Her weaving workshop, along with the workshops taught by Finnish immigrants Loja Saarinen and Marianne Strengell at Cranbrook and by German emigré Marli Ehrmann at the Chicago Institute of Design, helped bring constructivist ideas in textile design across the Atlantic.[49] Noted particularly for her work in ceramics, industrial designer Eva Zeisel (née Striker) had been educated in Hungary and came to Vienna by the mid-1930s, where she connected with her future husband, market researcher Hans Zeisel (of Lazarsfeld's "Vienna School"). The couple immigrated to the United States in 1938 where the designer and the market researcher would pursue careers in the two fields that most defined American consumer engineering at midcentury.[50]

To fully appreciate the transnational dimension of midcentury American design, we also need to consider the manifold ways in which American-born designers exchanged ideas with European colleagues as students, mentors, or collaborators. As we have seen, interwar Americans generally held European design aesthetics and design training in high regard. Walter Teague, for example, was strongly influenced in his work and in his ideas about design for the "machine age" by the writings of modernist architect Le Corbusier.[51] Designer George Nelson counts among those Americans most influenced by the architects of European avant-garde modernism. He had studied in Rome, and by the 1930s he was writing for *Pencil Points* and the *Architectural Forum*, introducing the design concepts of Walter Gropius, Mies van der Rohe, and Le Corbusier to a broader professional audience in the United States.[52] Whereas many American designers such as Nelson had traveled, studied, or worked in Europe, others learned from immigrants in the United States. Charles

Eames, for example, studied at Cranbrook Academy, while Eliot Noyes, who would head the industrial design department at the Museum of Modern Art in New York between 1939 and 1946, studied with the emigrés Gropius and Marcel Breuer at Harvard's School of Design.[53]

By midcentury, the emerging American design profession was thus profiting from a wide cast of European immigrant designers. The Bauhaus emigrés and other representatives of radical modernism allied with the Congrès Internationaux d'Architecture Moderne (CIAM) left a significant mark on transatlantic design debates dating back to the 1920s.[54] To reduce European influences on the American design profession to the Bauhaus, however, would be misleading, as the above examples have demonstrated. French and Italian designers informed interwar American design in areas from glasswork to metals, as did the Central European craft movements, most notably immigrants from cities such as Vienna, Munich, and Berlin.[55] Some had come already during the 1920s, including Kem Weber—born Karl Emanuel Martin Weber in Berlin—who had been singled out by Arens and Sheldon as "a new type artist" in *Consumer Engineering*.[56] Weber was another student of Bruno Paul's in Berlin who first came to the States in 1915 and who later was among the first designers to bring streamlining to California during the 1930s.[57] Others, such as Munich-born and London-educated packaging designer Walter Landor, arrived as late as 1939 just before the outbreak of the war. No single style or design ideal unified this diverse group, but the vast majority of emigrés subscribed to modernist traditions broadly defined, which proved suitable to mass production and the aesthetic demands of consumer capitalism.

The Designer as Expert Consultant: Raymond Loewy, French-Born Star of "American" Industrial Design

Raymond Loewy was the paradigmatic example of the successful immigrant designer. Although he was no radical functionalist, Loewy's work adhered to basic ideals of modernist design. While often out of reach of average consumers, his designs still helped to set the standards for mass market goods in the United States between the 1930s and the 1960s. His career not only shows the developing function of the designer as consumer engineer but also underscores the comparative advantages that European immigrants had over their American-born colleagues.

Throughout his career, Raymond Loewy crafted a public image of himself as a European immigrant entrepreneur. Speaking before design students at Harvard in 1950, for example, he treated his audience to an immigrant story of success by his own bootstraps: "The only thing I had when I landed in America thirty years ago, broke, without friends, unable to speak English, weakened by combat wounds, was enthusiasm."[58] Save for the enthusiasm, however, none of this was exactly true. Loewy was born in 1893 into a wealthy Parisian family with Viennese roots. In Paris, he had studied electrical engineering and technical drafting before serving in the French Army as a liaison officer to the American Expeditionary Forces during World War I. When Loewy came to the United States in 1919, planning to work as an engineer at the General Electric Corporation, he not only spoke English but also had family connections and a few professional prospects.[59]

Instead of becoming an engineer, Loewy established himself within the commercial design world of 1920s New York City. As with many in the field, one of his early jobs was employment as a window dresser for stores like Saks Fifth Avenue with preference for Art Deco designs. He had also brought with him a recommendation letter for the magazine publisher Condé Nast, and for much of the decade Loewy produced fashion designs and other illustrations for *Vogue* as well as for other magazines such as *Harper's Bazaar*.[60] His design background was thus in the kind of sales aesthetics (including retail decoration and fashion drawing) that set the visual standards for interwar American consumer culture.[61] In 1929, Loewy moved into the industrial design field with a remodeling job for a Gestetner duplication machine. Through the advertising agency Lennen & Mitchell he acquired the car manufacturer Hupp as a second prominent account, winning prizes for "elegance" at automobile shows across Europe during the early 1930s.[62] His real breakthrough came with the 1934 redesign of the Coldspot refrigerator sold by Sears & Roebuck. Loewy created a streamlined design for the appliance that helped boost sales in subsequent years. His independent design studio expanded tremendously by the late 1930s with offices in New York and London, creating streamlined locomotives for the Pennsylvania Railroad, outfitting "stratoliners" for TWA airlines, and styling a host of other goods. In 1937, he claimed that a total of $400 million in Loewy design goods had been sold.[63] Reflecting on the quick success of the industrial design movement, Loewy concluded as early as 1938: "the theory of well-styled and good-looking merchandise is now an accepted tenet of industrial life."[64]

How did Loewy define well-styled design and what role did he envision design consultants would play in the American economy? Despite his overtly commercial understanding of design, Loewy fell squarely within the modernist spectrum by consistently advocating "simplicity" and "functionality" in designing mass manufactured goods while avoiding any sort of unnecessary ornamentation. He lauded the interwar "Arts in Industry" movement for fostering a growing appreciation for functionalism among American consumers. While designers improved profits and the appearance of products, Loewy felt, "the competent industrial designer [also] exerts a great influence on improving the standards of public taste."[65] Design consultants had a social role to play in the minds of Loewy and his colleagues. On the one hand, they provided "a sort of public service," he claimed, "by supplying the American public with goods which are simple, that are reduced to their simplest expression without adding unnecessary decoration or what is called 'Gingerbread.'"[66] On the other hand, the "social responsibility" of designers included a commitment to lowering the costs of manufactured goods: bringing goods "within reach of the underprivileged class—this is democracy in action."[67] Modern design for mass production, Loewy felt, furthered the promise of democratic material abundance and still increased the cultural acceptance of what he saw as "improved" aesthetic standards.

Most of all, however, modern design for consumer goods promised expanded sales for businesses. Initially, modernist designs were marketed primarily to high-end consumer segments with goods from furniture to silverware that were meant to signal social status and ("European") sophistication. Over the course of the 1930s, the angular forms and dynamic streamlined shapes came to signify an orientation toward the future and a sense of "newness" that appealed to an increasingly broader market. As versatile marketing experts, design consultants needed to be attuned to these trends, Loewy insisted. Addressing Harvard design students as a guest speaker, he encouraged his audience to push the limits with regard to style but to remain keenly aware of what was aesthetically acceptable to producers and to consumers.[68] The "true function of the industrial designer," he explained to his colleagues at the Society of Industrial Designers in 1953, "far transcend[s] the mere solution of a given design problem, clearly stated and well defined by the management. [It reaches] into the field of product determination and market evaluation. The designer's duties become, in many cases, executive duties, and the client expects such attitudes from the designer."[69] Anticipating what the market would bear, successful design consultants needed to assume an executive mindset to understand merchandising conditions and economic limitations.

At the same time, Loewy wanted designers to be a creative force in marketing that could transcend these same market constraints. His involvement in engineering and marketing innovations across several industries allowed him to present himself as an arbiter of new trends. In speeches before industry associations Loewy speculated about future possibilities. In predicting market developments, he drew on the insights of academic sociology and other fields, but especially on his experiences with manufacturers. Loewy knew what was going on in the research laboratories at General Electric or at Westinghouse, he knew what improved types of Pyrex glass Corning had just brought to the market, and he knew what type of calculation machines IBM had created that could potentially revolutionize railroad ticket sales and future station design. "Speaking as a design consultant to 52 large producers in all fields of manufacturing—from lipstick to locomotives," Loewy assured the attendees of a 1944 marketing conference that "extraordinary things" were about to happen in the postwar economy.[70] Yet Loewy scoffed at exuberant predictions such as $300 automobiles or flying cars. Such unrealistic ideas, he felt, mainly came "out of the brains of copy writers who know little or nothing about automotive design."[71] Compared with advertising agencies, design studios cast themselves as the superior consultants for creative marketing with a clear grasp of engineering possibilities.

To fulfill their creative role, designers needed some degree of independence. "The trouble starts when the client tries to influence us in one way or another," he told his team. Whether it was against a conservative engineering department or a sales department guided by "false" notions about what the public demanded, designers had to defend their artistic vision.[72] Despite wartime achievements in engineering, he argued, for example, automobiles were still too big, too bulky, or too gaudy because of sales executives with their misguided ideas and because of advertising executives who manipulated status anxieties among consumers.[73] Speaking before members of the Society of Automotive Engineers in Detroit about his experiences working for Studebaker, Loewy defended the creative freedoms of the "stylist" who could not abide by the 9-to-5 workday of the corporate bureaucracy. The role of the styling office was, on the one hand, to develop designs meeting economic limitations in close collaboration with the engineering department. On the other hand, however, it needed to "constantly originate other designs of a more advanced nature." This included so-called "blue-sky design" projects that were purely experimental and unhampered by economic and engineering constraints. It was the responsibility of the design consultant, Loewy believed, to "maintain a state of alertness and creativeness" within the department that would benefit the corporation as a whole.[74]

In their demand for constant innovation, the design consultants

THE DESIGNER AS MARKETING EXPERT

5.1 Raymond Loewy designing (with Robert Bourke) for Studebaker in South Bend, Indiana, 1953. Source: Library of Congress, Prints & Photographs Division. Visual Materials from the Raymond Loewy Papers.

resembled the "knowledge entrepreneurs" in market research. They saw their task as introducing a constant creative urge to innovate and update into the hierarchies of American consumer goods corporations. This engineered creativity and perpetual marketing innovation became a crucial characteristic of midcentury consumer capitalism and a core element of what Reckwitz identifies as the inner dynamic of creative economies, the drive for institutionalized creativity.

Loewy had no patience with critics who felt that obsolescence demonstrated the "artificiality of our American way of life."[75] He acknowledged that market saturation had contributed to a culture of obsolescence among Americans since the late 1920s. Pointing to the automobile industries of Nazi Germany and Stalinist Russia, however, he regarded any form of standardized government design purporting to know "what's best for the people" as doomed to failure.[76] Loewy was confident that American consumer goods were superior to those of all other countries. Yet he was also convinced that Americans could still learn from Europe with regard to style and design. As an immigrant from Europe, he claimed a special awareness of the design shortcomings of American industry, and he shared his first impression of "shock" on numerous occasions: "I remember that, arriving in the United States about which I had heard so much . . . I was stimulated by the movement, the energy, the enthusiasm, the force of spirit there. But, possibly because I was French (and came from a country where a graceful archway [or a] subtle water color were more highly prized than the mass produced sink) I could not understand why so many marvellous machines and city streets were so ugly."[77] Loewy, the French immigrant, styled himself as a European on a "crusade against ugliness" to improve the aesthetics of the American machine age.[78]

Presenting oneself as a designer from Europe brought a certain cachet in the American market that Loewy systematically used to his advantage. He knew American business, but he had retained his European sensibilities—he spoke "French with a Chicago accent" as he liked to observe.[79] He had a villa in Saint-Tropez, a French subsidiary of his business (the Compagnie de l'Esthétique Industrielle founded in 1953), and he divided his time between households in Paris and New York throughout the postwar decades. The beginning of nearly every public speech he gave to American audiences had him "just" returning from Europe with fresh observations on trends on the continent.[80] Every time he returned to America, Loewy told his design staff, he felt the need to talk about design philosophy, about the "finesse" that he had grown up with in France, and about his overarching goal to fight "schmaltz and overabundance," the gaudy design trends he observed in the United States. In countries like Italy, by contrast, he found simplicity and restraint: "There again it was a case of finesse revisited. Again, the underlying reason was economy of means and restraint. Designs weren't so much functional for reasons of intellect or logic."[81] In contrast to the cerebral central European design reformers, Loewy invoked the cliché of a kind of natural modernism in Southern Europe: men with taste but with limited means creating well-designed goods under conditions of postwar scarcity.

To the design consultant Loewy, the very consumerist abundance he helped to create in America carried a dual threat, both aesthetically in exuberant overabundance (as in the notorious tailfins) and economically in a complacency that no longer pushed for visionary design innovations beyond current sales. The growing success of European cars by the mid-1950s underscored Loewy's contention that "Europe has much to offer America, design-wise."[82] As professional designers emerged as a new kind of creative "knowledge entrepreneur" within American industry at midcentury, expertise in the artistic traditions of interwar Europe or simply appearing "European," Loewy's case suggests, could provide a distinct competitive edge.

Between Art and Commerce: A "New Type of Artist" in Graphic and Advertising Arts

Corporate employers, from consumer goods manufacturers to advertising agencies and magazines, increasingly looked for immigrant artists with a modernist bent. Graphic design work for advertisements or corporate identity programs composed another expanding sector of design after the 1930s. Here, too, European emigrés were able to build transatlantic careers. Dutch-born graphic artist Leo Lionni, for example, succeeded as an art director at the advertising agency N. W. Ayers and later at *Fortune* magazine. Like many of his colleagues, Lionni had to overcome initial skepticism vis-à-vis his avant-garde designs but eventually became part of a group of advertising artists who understood themselves to be "artists of a new type," simultaneously attuned to the creative world of modern art and to the commercial demands of the business world.

Advertising design as a field had straddled the line between art and commerce since the early twentieth century.[83] Illustrators had become a growing profession with the rise of mass market magazines such as *McClure's* and *Collier's*, and since the 1920s ad agency art directors who saw art predominantly as a means to generate desire had gained in prominence. Some art directors, such as Ayer's Charles Coiner, however, pushed to include modern art in their commercial campaigns, and by the 1940s the line between "art-art" and "commercial art" had become blurred.[84]

Lionni's career provides a useful lens into American graphic design as it increasingly embraced modernist elements.[85] He had been born in Amsterdam in 1910 into family of Jewish diamond traders, but as a teenager he moved to Genoa, Italy, where he studied economics and during the early 1930s began to paint and draw, heavily influenced by the style of Italian futurism. When Lionni came to the United States in 1939 he initially had trouble finding employment in American advertising, which prized polished, professional designs and technical perfection. Whereas some European businesses were willing to take a chance on experimental poster designs by artists such as A. M. Cassandre or Henri

Henrion, American agencies typically copy-tested every slogan and produced more conservative graphic design.[86] Yet the advertising business, too, was interested in creative challenges, and Philadelphia-based N. W. Ayer, the third-largest agency at the time, took him on as an art director working under Charles Coiner and Leon Carp. By the early 1940s, Lionni was directing the artwork for several major accounts, including *Ladies Home Journal,* General Electric, the Container Corporation of America, and the Ford Motor Company. His work proved successful enough that in 1946 the J. Walter Thompson agency asked Lionni to take over a prestigious automobile account and to provide a critical review of the direction the agency's entire art department was taking.[87]

Large advertising firms increasingly sought the kind of modernist creative input that Lionni and other Europeans offered. He had developed strong ties to the modern art community on the East Coast and was able to win "real artists," as he noted, including immigrants and emigrés such as Jean Hélion, Willem de Kooning, and Fernand Léger. Lionni, too, at times struggled with his dual identity as an artist and a commercial art director. After a postwar return to Europe, he recounts in his memoirs, he felt "ashamed" of his commercial success in advertising and of the artistic restrictions this career path brought with it. He left Ayer to pursue freelance consulting work at the Museum of Modern Art, for consumer goods firms like Proctor & Gamble, and media corporations including the broadcasting network CBS, and he taught classes at the Parsons School of Design. In 1952, Lionni designed a MoMA exhibition on the design of the Italian Olivetti company, which he had been consulting for on product design and advertising art.[88] Between 1948 and 1960, his primary source of income came from *Fortune* magazine, which had played such an influential role in pushing for the aestheticization of American industry.

The world of U.S. magazine publishing and advertising was heavily populated with European-born art directors at the time.[89] At *Fortune*, Lionni had been preceded by Will Burtin, an emigré designer who had studied and worked in Cologne before fleeing to the United States in 1938.[90] Lionni hired former Bauhaus student Walter Allner as a cover designer who would eventually become art director at *Fortune* himself.[91] Peter Piening was yet another Bauhaus student turned art director within the Luce publishing empire. Upon coming to the United States during the mid-1930s, Piening first worked for Condé Nast's *Vogue* and advertising agencies including Ayer before becoming art director first at *Life* and later at *Fortune* magazine during the later 1930s and 1940s.[92] These immigrant artists styled magazines and advertisements that helped shape the American consumer imagination. From French poster artists such as Cassandre and Léger to Bauhaus designers such as Piening and Allner, immigrants and emigrés had a tremendous influence on commercial graphic art and on the broader field of visual arts in general.[93] "All that talk about advertising design being an 'American form' or an 'American art' is cheap superstition," Art Institute of Chicago director Daniel Catton Rich observed in 1946 while pointing out that many of the more notable contemporary advertisements were created by European-born artists.[94]

Few emigrés illustrate the midcentury ties between art and commerce better than graphic designer Herbert Bayer, who had headed the Bauhaus's workshop for advertising art from 1925 to 1928.[95] Bayer's lasting achievements include the so-called "universal typeface," which he developed in 1925 and which later became commercially adapted as the ITC **Bauhaus** font in use to this day.[96] Bayer's work and teaching at the Bauhaus was geared toward the psychological effects and commercial potential of graphic design, emphasizing the importance of new technologies. He especially valued the photomontage for its psychological and suggestive properties, and his collage work won acclaim at the first exhibition of European advertising art in New York in 1927.[97] Bayer extolled the virtues of standardized mass production, arguing that achieving "good form" was an aesthetic as much as an economic balancing act. The aesthetics of function and of standardized production, he believed, were applicable to graphic design and typography in advertising as well.[98] After collaborating with the German Association of Advertising Professionals, he left the Bauhaus in 1928 to work as an art director for Condé Nast's *Vogue* and the American-owned advertising agency Dorland in Berlin. In his advertising work for Dorland, Bayer applied his interest in modern art to campaigns for new consumer products such as Telefunken records, Boehringer cold medication, or Venus underwear.[99]

Building on his personal ties to former Bauhaus colleagues and on his professional connections through Dorland and *Vogue*, Bayer left Germany for the United States in 1938, even though he had experienced few personal problems under the Nazi regime and even engaged in design work for state propaganda exhibitions.[100] Upon coming to America, he was involved with a Bauhaus exhibition at the Museum of Modern Art, and he soon made a good living as an independent design consultant in New York City. One of his first professional contacts was Robert Lincoln Leslie, a Brooklyn-based designer who helped several emigrés get their start in commercial New York.[101] Bayer created posters for the New York Subways Advertising Inc., worked at the Pennsylvania display

CHAPTER FIVE

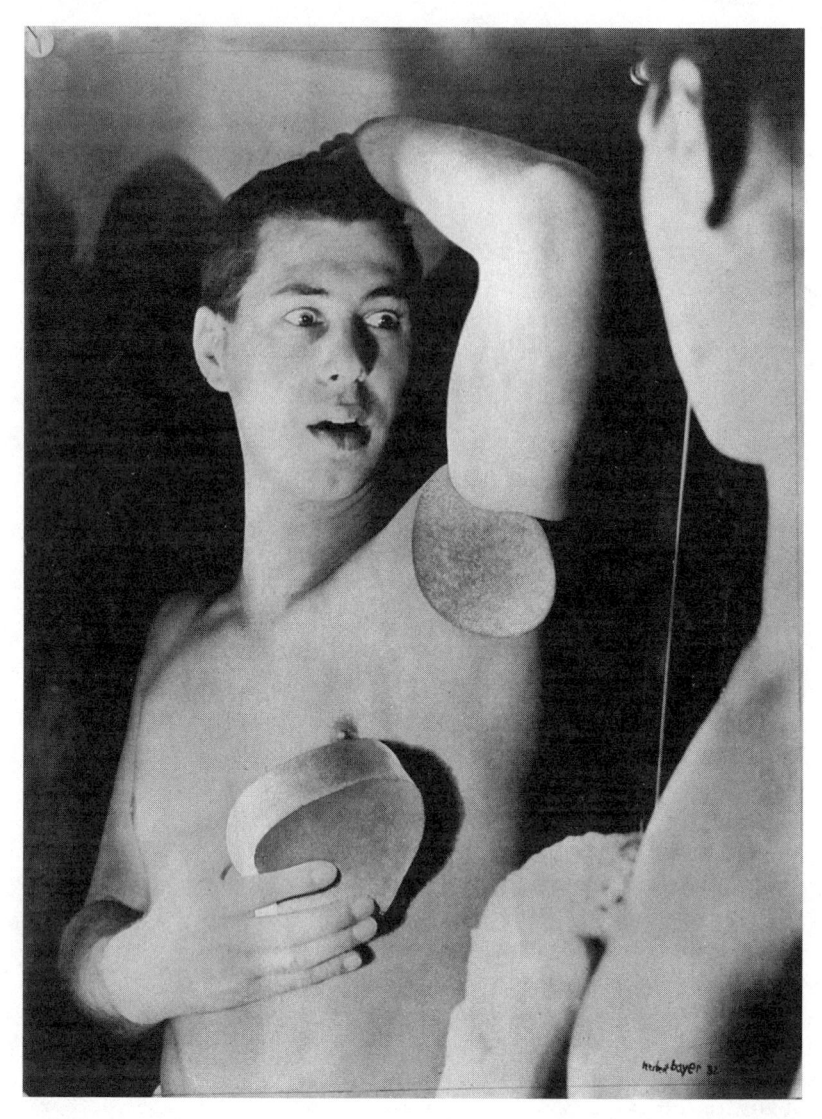

5.2 Herbert Bayer (1900–1985), self-portrait, 1932. Courtesy of Bauhaus-Archiv, Berlin.

at the 1939 New York World's Fair, designed displays for the Museum of Modern Art, and did freelance work for Wanamaker's department stores as well as for magazines including *Life* and *Fortune*.[102] When Peter Piening left *Life* to become art director at *Fortune* in 1941, Bayer got his friend Walter Gropius to write Henry Luce on his behalf so that he could secure a more permanent position with *Life*. Bayer, Gropius wrote, would be a "most ingenious and versatile art director," full of innovative ideas for the magazine.[103]

Bayer brought to American graphic art the same attention to consumer psychology and its commercial ramifications that had infused his work in Germany since the 1920s. During the early 1940s, he conducted workshops sponsored by the American Advertising Guild in which he taught the use of montages and other techniques such as jarring pictorial contrasts in magazine ads. Bayer tried to "influence his students to aim for the emotions of people," Percy Seitlin wrote in the trade journal *A-D*, insisting that they must understand the "nature of these emotions" and the psychological interplay between stimulus and response in order to create attention and to make a subtle appeal to consumers' "anxiety." Bayer's techniques, Seitlin concluded, "were progressive and modern because they try to utilize a body of knowledge about psychology as it relates to design. They are not art for art's sake."[104] By the end of the war, during which he designed several home-front posters for the U.S. Department of Agriculture, Bayer had attained national renown as a modernist advertising artist and art director, and he returned to the Dorland advertising agency as a vice president at their New York office.

Still, Herbert Bayer's modernism pushed the boundaries of accepted commercial design, as he found out during his short employment with J. Walter Thompson in 1944. He experienced the leading Madison Avenue ad agency as rather conservative. Bayer had initially been brought into the company on a freelance basis. When he became a regular employee, however, working on an account for Douglas Aircraft among others, Bayer quickly came into conflict with his co-workers in the creative department. "Bayer's type of art was new to this country and was entirely different to anything existing here," one of his JWT colleagues recalled in a 1950s interview, and "his form of art was not accepted" within the creative department. "I will say that Bayer's form of art is generally accepted now," he was quick to add; "[Bayer] was years in advance of our thinking at that period of time."[105] Several other former JWT co-workers mirrored this sentiment that during the mid-1940s Bayer had been "too advanced" for the agency and that the quality and commercial potential of his artistic style had not been adequately recognized.

Like many of his fellow emigrés, Bayer straddled the line between modern art and commerce, two worlds that became increasingly enmeshed through new institutions and companies such as the Container Corporation of America (CCA). Bayer's most recognized contributions to American advertising included widely acclaimed ads for CCA's "Great

CHAPTER FIVE

5.3 "Paperboard Goes to War," promotional booklet for the Container Corporation of America designed by György Kepes, c. 1944. Courtesy of University Archives and Special Collections, Paul V. Galvin Library, Illinois Institute of Technology, and Ms. Juliet Kepes Stone.

Ideas of Western Man" campaign that Lionni had initiated at Ayer. The campaign, which eventually ran for several decades, helped to solidify that firm's public image as a pioneer in modern industrial and graphic design.[106] The company specialized in shipping boxes and crates and supplied the packaging for many of America's leading consumer goods manufacturers. During the 1930s, CCA had built up a strong design department, run by designer Egbert Jacobson, who traveled to Europe to recruit modern artists such as A. M. Cassandre from France and Toni Zepf from Germany for their advertisements. Under the leadership of longtime chairman Walter Paepcke, a German immigrant with an affinity for modern art, the company supported the emigré art community in Chicago, including György Kepes and others from Moholy-Nagy's "American Bauhaus."[107] Paepcke was instrumental in organizing a 1945 Art Institute of Chicago exhibition, "Modern Art in Advertising," which featured Bayer prominently alongside Jean Carlu, Cassandre, and other luminaries of interwar European poster design.[108] Bayer and Paepcke connected over a shared interest in art and design in industry, and in 1946 the designer accepted Paepcke's invitation to come to Aspen, Colorado, to help him develop the elite ski resort into a center for international cultural exchange. In 1950, Bayer became regional director of the International Design Conference in Aspen.[109]

"Design as a function of management," the motto of the 1951 design conference in Aspen, reflected a guiding theme of Bayer's corporate design work at CCA, as he explained in the German trade journal *Gebrauchsgraphik*. To attract customers and personnel, corporations had to create a unifying "personal touch." Uniform typography was essential to such efforts, as were recurring symbolic forms, which tied a company's advertising to its delivery trucks, its stationary, and even its office cafeteria.[110] This modern co-dependence of art and commerce demanded a different, novel type of artist, Bayer believed. This "new" artist would shed all reservations vis-à-vis the world of material production and commercial advertising, and he now saw artists of this mentality taking hold in the design departments of American corporations and ad agencies. Artists should be regarded as an essential component of business no different from the banker or the worker. Because material goods and commercial advertising played an increasing role in modern consumer society, designers played a crucial moral and aesthetic role. The artist in industry, Bayer observed by the early 1960s, now worked alongside professional market researchers and motivation analysts and could not ignore the insights of modern social science and engineering, carefully balancing art and marketing.[111]

Bayer's sentiments were reflective of the wider design profession, as the Aspen conferences demonstrated. Initiated by Paepcke and Jacobson at the CCA as well as by a group of prominent artists like George Nelson and businessmen such as Frank Stanton of CBS, the conferences stressed the importance of the designer in boosting sales and in engineering economic growth as well as a perceived civilizing aesthetic mission of design in modern consumer capitalism.[112] In an echo of the call of Arens and Sheldon's "consumer engineering" from thirty years earlier, artists were called upon to enter a fruitful cooperation with business that would give companies a competitive edge but would ultimately also uplift the moral and aesthetic sensibilities of consumers. Among the attendees at the conference, few argued more vigorously for an aesthetically improved consumer capitalism than Edgar Kaufmann Jr. of the Museum of Modern Art, another key institution in promoting an "American" design profession informed by "European" modernism.

Emigrés on Display: "Good Design" and the Aestheticization of American Consumer Capitalism

Many of America's leading art institutions were actively involved in fostering a synthesis between art and commerce. When the Art Institute of Chicago showcased the advertisements of the Container Corporation in 1946, curator Carl Schniewind praised CCA's efforts as the "most progressive advertising arts program ever to have been instituted by a large industrial enterprise." He optimistically raised the prospect that commissioned artworks for advertising could financially support an entire new generation of artists in their artistic endeavors.[113] In New York, the Museum of Modern Art (MoMA) led this effort of promoting "Good Design" in industry to the benefit of both business and artists. At midcentury, the museum systematically provided a platform for the new design profession and its European emigrés.

MoMA sought to directly influence the development of professional design and the aesthetic standards of American mass consumption. The museum's interest in modern design began in the late 1930s with a series of exhibitions of contemporary "useful" household objects. By 1939, MoMA had established a department of industrial design headed by designer Eliot Noyes, who had previously worked with Bauhaus emigrés Walter Gropius and Marcel Breuer in Boston.[114] Under his direction, the museum initiated industrial design competitions, with exhibits ranging from furniture and lamps to fabrics, as well as putting on specialized exhibitions, for example on the role of modern textiles in interior design.[115] Against the application of superficial "style" or "spurious art" in industrial design, Noyes advocated a strict utilitarian ideal, and his successor in the design department, Edgar Kaufmann Jr. of the Pittsburgh department store family, similarly pushed for a purist understanding of "modern design." Artificial style changes and merely decorative "streamlining" were not "good design" to Kaufmann, and MoMA sought to put a check on any design excesses that had resulted from the new enthusiasm for commercial design among American marketers.[116] This was not a fundamental attack on commercial forms of art, however. To the contrary, from the beginning the museum developed close ties with industry and retailers. The annual "Useful Objects for under Ten Dollars" exhibit, for example, aimed to encourage manufacturers to produce "well-designed useful objects at prices within the reach of the average person." The goods on display were available in large department stores throughout New York.[117] In 1945, the museum even introduced annual design awards for mass-produced goods in everyday use. The expressed intent of the program was to "stimulate consumer demand for really good modern design in manufactured objects, thus assisting manufacturers and retailers to produce better modern design."[118]

European-born artists featured prominently among the commercial designers displayed in U.S. museums. The catalogue for the Chicago "Modern Art in Advertising" exhibit featured a list of 41 artists on display, of which 23 artists had a European background, including Carlu, Cassandre, György Kepes, and Richard Lindner.[119] Since its inception, the Museum of Modern Art had had a close relationship with European modernists and was among the strongest supporters of Bauhaus design in the United States. The 1938 Bauhaus exhibition in cooperation with Herbert Bayer, Ise Gropius, and Walter Gropius was only part of MoMA's wider and sustained interest in the emigré art community that had come to the United States and especially to New York City over the course of the 1930s.[120] Anni Albers and other immigrants featured prominently in the modern textiles exhibition; Eva Zeisel had an exhibition of her modernist ceramic designs for Castleton China.[121] Bernard Rudofsky, Marcel Breuer, and Alvar Aalto were among the large number of immigrant designers whose work was on display as part of larger shows or as individual exhibits at MoMa throughout the 1940s.[122]

The annual "Good Design" show, launched in 1949, presented the culmination of MoMA's efforts to define the aesthetic standards of American mass consumption in terms of "European" modernism. Guiding the style choices of both marketing experts and consumers was the core idea of the exhibitions, which Kaufmann organized in cooperation with the Chicago Merchandise Mart.[123] With 850 market tenants and thousands of employees, the Merchandise Mart was the largest American wholesale operation, putting on semiannual shows of home furnishings and consumer goods for retailers from across the United

CHAPTER FIVE

5.4 MoMA "Good Design" exhibition, photograph 1952. Digital image © The Museum of Modern Art/Scala, Florence.

States and abroad.[124] The two institutions agreed to install a design show on the premises of the market, featuring new outstanding designs every six months. A selection of the best designs were shown at MoMA in New York, chosen by a committee of design experts from among thousands of entries sent in by manufacturers and retailers.[125]

The "Good Design" exhibits brought together the leaders of the American design profession and prominent industrialists. Every year, notable designers such as Charles Eames and Alexander Girard were chosen to create the exhibition space.[126] The selection committees included among others Florence Knoll and Russian emigré Serge Chermayeff, who had succeeded László Moholy-Nagy at the Chicago Institute of Design. Among the designers on display were many of the immigrant modernists discussed in this chapter, along with leaders of commercial design such as automobile stylist Harley Earl, George Nelson, or members of Raymond Loewy Associates.[127] Behind the scenes, MoMA had assembled an impressive cast of sponsors ranging from architect Mies van der Rohe to Paul MacAlister of the Industrial Designers Institute, CCA chairman

Walter Paepcke, and *Chicago Sun-Times* publisher Marshall Field to the National Retail Dry Goods Association and original "consumer engineer" Egmont Arens as president of the Society of Industrial Designers.[128] When the first exhibit opened, the opening luncheon brought together three hundred guests from the worlds of manufacturing, retailing, and design. Immigrant designers such as Jens Risom and William Pahlmann of the American Institute of Decorators mingled with CCA's chief designer Egbert Jacobson and Anne Swainson, director of product design at Montgomery Ward. Design consultant Alfred Auerbach was the principal speaker at the event. He observed that while America had been ashamed to submit its goods to the Paris Exposition of 1925, today it "leads the world in good design."[129]

Addressing consumers, retailers, and producers as well as elite "tastemakers" and "fashion intermediaries," the Museum of Modern Art presented itself by the early 1950s as an intermediary between art and industry. Exhibitions of graphic design and advertising presented the commercial work of designers as art, worthy of display in a museum. These exhibits heavily featured emigrés such as Herbert Matter, Erik Nitsche, Herbert Bayer, and Will Burtin.[130] The "Olivetti: Design in Industry" exhibition organized by Leo Lionni even went so far as to portray the modern corporation as a "total" work of art. Reflecting corporate identity concepts developed by Bayer, Loewy, and others at the time, the museum displayed Olivetti's product designs and its retail and advertising displays, together with company posters and technical pamphlets as objects of an art exhibition.[131] To some critics, to be sure, MoMA was an elite institution with an air of superiority and only limited influence on mass market taste.[132] This verdict, however, ignores the degree to which the museum and the designers it represented strove to combine modern functionalism with an appeal to retailers and manufacturers. It also overlooks the fact that more commercially oriented designers from Arens to Loewy were not only involved in the museum's efforts but also shared broadly reformist notions of "uplifting" the consuming masses.[133] Along with MoMA they aimed to redefine the aesthetics of American consumer capitalism, an effort in which European-born designers played a central part.

New Experts for America's Midcentury World of Goods

Midcentury America witnessed a broad aestheticization of consumer capitalism. For the design profession this process promised both economic opportunity and a sense of a larger cultural mission. Reflecting on the relationship of modern art and contemporary industry, French emigré artist Fernand Léger expressed the hope that industry and commercial work would be able to reconcile modern art and popular taste in a democratic fashion.[134] Many in the design profession understood their work as contributing to and providing aesthetic guidance to a new era of material abundance in the United States.

The industrial or graphic designer had become a new creative force in American industry who could combine marketing and engineering competence with aesthetic expertise. Whether as independent designers, as company stylists, or as art directors in the advertising and publishing business, design professionals gained in prominence in the corporate world. In 1953, Raymond Loewy spoke not only in Essen but also to the Society of Industrial Designers in Bedford Springs, Pennsylvania: "We no longer have to sell the idea of design," he assured his colleagues. "Most of the new people we meet now come to us because they take it for granted that a progressive business needs design; that it is as much part of their operation as a sales or accounting department."[135] The challenge before the profession was now to expand the areas in which design was relevant with an even more systematic integration of design into the marketing process.

A significant number of the midcentury design professionals had a European background. These immigrant artists contributed to debates about the professional role of designers between art and industry, and while many of the immigrants saw themselves as artists of a "new type" and open to work with business, others sought to balance out what they perceived as the stylistic and commercial excesses of American consumer capitalism. They tried to reconcile the social and aesthetic visions of the modernist reform movements in interwar Europe with the commercial demands they experienced in the United States. The tensions and new impulses that arose from the Atlantic crossing particularly of the Bauhaus and Werkbund ideals in particular will be at the center of the following chapter.

SIX

The Commercialization of Social Engineering? Adapting Radical Design Reform to American Mass Marketing

In a short May 1945 article, the *New Yorker* magazine featured emigré designer Ferdinand Kramer as their "architect of the week." Kramer was singled out by the magazine for his innovative designs in what was then called "knocked-down" furniture—mass-produced furniture that could be easily assembled and disassembled. Because such dressers, beds, and cabinets could be efficiently stored and transported in bulk, several European governments had expressed interest in Kramer's designs in anticipation of postwar reconstruction efforts. In the United States, his low-cost furniture designs had already been marketed for several years to meet the housing needs of war workers who moved frequently and often lived in temporary dwellings.[1] While in Britain the *Daily Mirror* celebrated Kramer as a wartime exponent of the "American genius for mass production," in the United States the *Wall Street Journal* recognized him as one of the many Europeans who had fled Hitler to "set up shop" in the United States. In furniture manufacturing and other fields, the paper observed, "European 'know-how'" was helping war production and offering postwar solutions.[2]

CHAPTER SIX

Ferdinand Kramer's career exemplifies the tremendous transatlantic divides that many designers coming from reform movements within interwar Europe had to overcome.[3] Kramer's designs originated in public design efforts in Frankfurt that centered on providing affordable housing to working class families through standardized mass production. Here he had first anticipated the low-cost, demountable furniture that would later become the object of interest for American media. Kramer had been involved in social reform projects that were far removed from the visions of affluent mass consumption developed by "consumer engineers" in the United States. Still, he was able to engage with the world of commercial design after his emigration: he learned to understand the conditions of mass marketing, the psychology of mass retailing, and the appeal of novelty to consumers. His furniture, while low-cost and focused on basic needs, still foreshadowed a consumer interest in do-it-yourself assembly and eventually the success of Ikea-style peg-and-board furniture. For postwar consumer households there was room to grow, as he told the *New Yorker*: "My cabinets come in sections. You find you need more brandy in the house, or more toys, and you add a section."[4] In a sense, Kramer adapted principles of the German Werkbund to the new contexts of American consumer capitalism.

Squaring the reform visions of avant-garde European designers with the demands of American industry was a complicated and conflict-laden process. Kramer was connected to groups such as the International Congresses of Modern Architecture (CIAM) as well as to the Bauhaus.[5] Their strict functionalism and social reform ideals were not easily reconciled with trends such as "streamlining" and frequent style changes aimed at creating continuous demand. Many American observers such as industrial designer Harold Van Doren were therefore skeptical whether Bauhaus emigrés like Walter Gropius and Joseph Albers could really contribute to American consumer capitalism. In 1940, Van Doren wrote: "The Bauhaus approach [lacks] the realistic qualities that we Americans, rightly or wrongly, demand. Much of the writing of the group is vague to the point of complete unintelligibility. . . . It will be difficult, I believe, to acclimatize the esoteric ideas of the Bauhaus in the factual atmosphere of American industry."[6] Indeed, many of these artists kept a critical distance from commercial design.

At the same time, however, there was a great deal of interaction between these emigré artists and American commercial design. This chapter will trace both the inevitable conflicts between radical design visions and corporate America and the surprising degree to which interwar European reform traditions in design informed American "consumer engineering." After a closer look at Ferdinand Kramer's career between Frankfurt and New York, I will inquire more generally into the transatlantic commercial impact of the Bauhaus school and of European design modernists organized by CIAM. Bauhaus emigrés, furthermore, established themselves in prominent positions in American design education, including Walter Gropius at the Harvard School of Design and Josef Albers at Black Mountain College in North Carolina. The most comprehensive of these educational ventures was the so-called "American Bauhaus" established by László Moholy-Nagy in Chicago. Its history most clearly illustrates the conflicts as well as the potential in this meeting of reform visions and commercial demands. Some historians of industrial design have criticized the "elitism" inherent in the reformers' goal of "uplifting" the popular taste of the consuming masses. Others have bemoaned the loss of a more radical social vision as designers cooperated with industry.[7] The reformers' ideal of providing democratic access to well-designed standardized goods, I argue by contrast, held surprising appeal to "consumer engineers." In fact, their careers point to the lively connection between "social engineering" and marketing at midcentury.

Ferdinand Kramer: From Standardizing Working Class Homes to Marketing Novelties

Ferdinand Kramer's work is exemplary for the convergence of marketing and design reform. Born in Frankfurt in 1898, Kramer grew up in a family that owned one of the city's leading fashionable hat stores. He received his political socialization in the turbulent years following World War I, when he became involved with Social Democratic reform circles and with Frankfurt intellectuals organized in a group called the "Society of 1918." In the spring of 1919, Kramer moved to Munich, a city even more heavily caught up in revolutionary fervor, where he began his studies in architecture at the Technical University alongside Sigfried Giedion, intent on finding solutions to Germany's social housing crisis.[8] His interest in the social aspects of architecture and design brought Kramer briefly to Weimar, where Walter Gropius had just opened his progressive Bauhaus School. Despite what he described as a stimulating mix of architecture and politics, however, Kramer quickly moved back to Munich, and after graduating in 1922, he returned to work in Frankfurt. Here, Kramer started his career by designing small pieces of furniture for a Werkbund exhibit. A first larger contract followed in 1924 when he was commissioned to design the travel office of the HAPAG steamship line at the Frankfurt rail station. The modern execution of the sales area designed with unadorned counters and fluorescent lighting received controversial reviews in the local press, but it brought Kramer to the attention of Frankfurt's new city planner Ernst May.

Frankfurt was a center for social housing reform in 1920s Weimar Germany. The city hired Ernst May in 1925 to initiate an ambitious housing and urban redevelopment program entitled the "New Frankfurt." The program entailed the construction of entire neighborhoods with twelve thousand modern housing units for working-class renters. Roughly at the same time that American department store mogul and progressive philanthropist Edward Filene called for the construction of "houses like Fords," Frankfurt officials, too, were engaged in the serial mass production of living space. May offered Kramer a position within the housing administration, supervising the city's standardization efforts. Kramer designed standardized door handles, hinges, windows, and various other elements in a comprehensive effort to set new norms for everyday items. His interest in lowering cost was matched by an emphasis on efficiency and rationalization in interior design that also informed the famous "Frankfurt kitchen." Its emphasis on rationalization, minimum standards, and partially centralized infrastructures (e.g., communal laundries) made the "New Frankfurt" a case study for social democratic planning efforts and interwar Germany's "municipal socialism."[9]

Kramer's work was driven by an overarching interest in improving the standard of living for low-income, working-class consumers. Anticipating his later U.S. efforts in knock-down furniture, Kramer designed flexible shelving and storage units that could be rearranged by consumers in different ways. All this was part of a larger vision of reform architecture that came together in the 1929 CIAM exhibition entitled "The Apartment for Minimum Standards of Existence" and focused on ideal housing units for those consumers who had but the bare minimum. Responding to Walter Gropius's call for "housing rations" for consumers, Kramer had organized the exhibition along with architect Mart Stam, graphic designer Hans Leistikow, and his friend Sigfried Giedion, who served as CIAM's first secretary general.[10] Kramer, too, believed that housing was a good fit for mass production. From kitchens and bathrooms to handles and doors, ovens, furniture, and illumination devices, "the smallest things in apartments were rethought in terms of function and form," he later recalled; "Frankfurt was devising new standards for living."[11]

His reformist vision did not preclude Kramer from pursuing commercial applications for his designs. To the contrary, he was keenly interested in the possibilities of commercial mass production and the machine age.

He expressed his fascination with mass production in a 1929 article on serial production at the Thonet furniture company. Kramer viewed the firm as exemplary for integrating the organizational lessons of the American automobile industry into modern furniture making, and he even designed a chair for Thonet.[12] His greatest commercial success was a simple, affordable oven (later marketed as *Volksofen*) that he created for the Buderus Company during the 1920s.[13] Kramer chided German industry for its conservatism when it came to mass marketing new designs to consumers. As the country descended into the Depression era, he observed a reluctance to innovate among German companies fearful of competition. Both business considerations and the needs of consumers, Kramer believed, required rational design and modern functional forms (*Zweckform*) that saved resources and were simple to produce.[14]

Kramer's interest in creating affordable mass market goods resonated with various efforts to create "people's products" in 1930s Germany. The *Volkswagen* (people's car) and the *Volksempfänger* (people's receiver) became two of the most notorious examples of the Nazi regime's propaganda efforts, which extended the (largely empty) promise of making new consumer goods available to the German population.[15] Kramer himself did not participate in such developments, however. After May left Frankfurt to pursue urban and industrial planning projects in the Soviet Union, Kramer opened a private architectural practice in Frankfurt.[16] He quit the Werkbund in 1933 to protest the exclusion of Jewish members from its ranks, and in 1937 he himself faced expulsion from the architectural profession by state regulators, presumably for his political leanings. Like many of his colleagues involved with CIAM, the Werkbund, and particularly the Bauhaus, Kramer and his first wife, textile designer Beate Kramer, left Germany to pursue careers in the United States.[17]

Arriving in New York in 1938, as he later recalled, Kramer like many of the European emigrés was at once fascinated with what he perceived as the dynamism of American society and weary of its heavily commercialized consumer culture.[18] Yet he quickly succeeded in establishing himself in the areas of furniture and interior design during the late 1930s and early 1940s. Drawing on his social housing background in Frankfurt, he initially found employment with architect Ely Kahn, designing a trailer camp. In 1939, he helped design two FHA housing developments in Westchester. They were owned by the Frankfurt Institute for Social Research, whose members, including Max Horkheimer and Theodor Adorno, were prominent in New York's larger emigré community.

Kramer's design network, however, was not limited to a narrow circle of emigrés but quickly extended to the centers of New York commercial

CHAPTER SIX

design. In 1939, he worked for the design studio of Norman Bel Geddes on the General Motors account and helped prepare the World's Fair's "Futurama" exhibit. In 1943, Kramer and his associate, Czech emigré Fred Gerstel, set up their own design studio, the Products Marketing Corporation on Park Avenue in New York City. They worked primarily as design consultants of the Lehman Brothers investment bank, which at the time was the financial force behind several large national department store chains including Allied Purchasing Corporation and the Federated Department Stores (Bloomingdale's). With these stores as customers, Kramer devised new lines of mass-produced, demountable furniture meant to cater to a highly mobile, flexible lifestyle.[19] Like the knock-down furniture that would win the attention of the *New Yorker*, all Kramer designs emphasized practicality and transportability. Produced by a furniture company in Sumter, South Carolina, the designs were heavily advertised and sold across the country.[20]

Kramer's utilitarian designs fit the demands of wartime consumption. An advertisement for a European-style demountable wardrobe emphasized not only its low price and convenient transportability but also its easy assembly with ten wooden pegs and no tools needed: "Wonderful for apartment dwellers! Service wives! Defense workers!" the ad exclaimed.[21] Not entirely coincidentally, Kramer and Gerstel's Park Avenue office shared an address with the General Panel Corporation of emigré architects Konrad Wachsmann and Walter Gropius, which conceived ready-made houses for postwar production during the 1940s.[22] This suggests not only the close ties within the community of immigrant architects and designers, but also the degree to which former members of reform-minded design schools tried to comprehensively influence American mass consumption. Kramer's "knock-down furniture" was designed exactly for the type of functional environment represented by Wachsmann's "ready-made houses" for mobile wartime workers and low-income postwar consumers.

Kramer's designs did anticipate postwar affluence, however. Among the more successful designs of his office was a line of foldable lawn and garden furniture made of aluminum pipes and synthetic Koroseal strapping that could even be spotted on the White House lawn.[23] His demountable designs and flexible modular storage systems offered possibilities to grow and expand for families and households as they moved into bigger dwellings. By the late 1940s, Kramer's products increasingly catered to new forms of leisured domesticity.

Department store interiors, furthermore, became a growing field of activity for the Products Marketing Corporation. Kramer recognized the need for fast turnover and merchandise-centered forms of display.[24] The

6.1 Advertisement for "knock-down" furniture designed by Ferdinand Kramer, c. 1944. © Kramer Archiv, Frankfurt a.M.

CHAPTER SIX

modern department store had become a "machine for selling," he believed, and designers had to engineer the efficient movement of goods from the loading dock through storage and sales areas to the customer. Based on sales research and analyses of consumer behavior, he advocated new forms of "visual selling" to bring customers into direct contact with the merchandise.

In keeping with general trends in the design profession, Kramer believed that designers had to pay attention to consumer psychology. Factors from lighting and temperature to traffic patterns between floor displays informed shopping behavior, he observed.[25] He created a model store for the Alden mail order business outside of Chicago that emphasized open displays and removed visible barriers to attract interest in merchandise for "better sales" and "more sales per square foot."[26] To achieve these effects, Kramer developed Vizual, a comprehensive display system centered on simple and slanted Plexiglas display stands that could be flexibly assembled and rearranged. He collaborated with Swiss-born graphic artist (and his neighbor in suburban Port Chester, New York) Xanti Schawinsky on the visual representation for Vizual, which was meant to amplify trends toward self-service in retailing.[27] Only a few stores adopted Vizual at the time, but it anticipated subsequent efforts to open up retail sales areas and eliminate traditional counter space.

Despite his preference for functional design, Kramer was not oblivious to marketing considerations, and he was not above catering to American consumer society's penchant for novelty. His most famous design, the "Rainbelle," was an umbrella made almost entirely of folded paper coated with vinyl.[28] Marketed explicitly as a novelty product and designed for mass production in a wide range of color variations, the Rainbelle was a media hit in 1951. The product was featured in *Look* magazine and in the display windows of prominent department stores across the United States. Kramer had designed the umbrella as a disposable product for the "throw-way society" of the 1950s after market research had indicated a need especially among commuters hurrying from their cars to downtown offices and stores. While the original target price of less than one dollar could not be realized by Bristol-Myers, which produced the Rainbelle, the company still sold 58,000 units in 800 stores in the first six weeks. The Rainbelle drew on Kramer's interest in simple, standardized products that could be mass produced at low cost. The design, later on display at the Boston Institute for Contemporary Art, met reform aspirations of "good design for mass production," yet at the same time it was a reflection of American consumer capitalism's fascination with novelty, disposability, and rapid turnover.[29]

THE COMMERCIALIZATION OF SOCIAL ENGINEERING?

FERDINAND KRAMER: *he folded a piece of paper . . .*

6.2   Ferdinand Kramer (1898–1985) presenting his "Rainbelle," 1952. © Kramer Archiv, Frankfurt a.M.

Kramer understood the dynamics of American consumer society, but he was—at best—a reluctant consumer engineer. When he moved back to Frankfurt in 1952, it was with a clear grasp of commercial "industrial design" in the United States. American design dynamics, he told his European colleagues after the war, were dominated by competition between producers and "conspicuous consumption" among consumers.[30] This led to fashion-driven designs and a fast turnover in mass produced goods. The design studios of Loewy, Bel Geddes, and others, he reported, relied on market research and advertising, and, he claimed, they had become the most important instrument of corporate sales departments. Yet Kramer cautioned that the sales manager had become the real "dictator of taste." Like many contemporary design critics, he thought sales numbers were driving the rise of unnecessary "streamlining," and he warned against commercialized aberrations from modern functionalism. Still, his commercial success in the United States drew on his earlier work in Germany in often surprising ways: he was a "social engineer" turned marketing expert, and his expertise in functional design for mass production was in high demand.

Radical Modernism in the United States: Commercial Applications of Social Engineering

Ideological and aesthetic differences over obsolescence notwithstanding, there was a sustained interest in European design reformers among American commercial designers. As early as 1936, Sheldon and Martha Cheney's account of the emerging design profession in the United States noted that the Bauhaus in particular brought together theory and experimental practice "as has never been done in America." American designers from Donald Deskey and Henry Dreyfuss to Norman Bel Geddes, George Sakier, William Lescaze, and Frederick Kiesler, the authors asserted, "all speak the language of the Modern Art studio [and] they acknowledge their indebtedness to Moholy-Nagy and van der Rohe and Lissitzky, though not audibly in the presence of industrialists who are fearful of aesthetic theory."[31] While these European artists were known only to professional circles at the beginning of the 1930s, by the end of the decade even prominent industrialists were interested in what Bauhaus modernists had to offer to American consumer design.

The emigration of the Bauhaus designers was part of a larger transatlantic shift in the world of modern architecture and planning. To interwar avant-garde artists design was but one element of a broader vision to transform the built environment that ranged from the shape of everyday goods to architecture and urban planning. The International Congresses of Modern Architecture (CIAM), which besides Ernst May and Walter Gropius included leading modernists from across Europe such as Josep Lluis Sert and Le Corbusier, had provided a key forum for such ideas since the 1920s. Over the course of the 1930s, the center of CIAM activities shifted gradually from the continent to the United Kingdom and to the United States as many central European protagonists of the CIAM network moved into exile where they engaged with local urban planning traditions such as the garden city movement. In addition to Gropius and Sert, the group's wider American exile network also included Sigfried Giedion and Herbert Bayer, who created the cover art for CIAM publications. Drawing on ideas of "organic neighborhood communities" in Gropius's work or on Luis Sert's notions of a "human scale" in planning, CIAM thinkers devised community spaces and modern civic environments that were hardly as sterile or functionally divided as some critics have alleged.[32] These comprehensive visions to completely recast the material environment provide an important frame of reference for the work of Bauhaus artists and for designers from related reform traditions as they put the design of individual goods within a larger "social engineering" framework.

Such reform visions had informed the Bauhaus school since its inception amidst the social upheaval following World War I. In the recollection of emigré designer Alexander Dorner, who later became head of the Rhode Island School of Design, Gropius's new school had offered a medley of different "revolutionary" art traditions from German Dadaism and Russian Constructivism to French abstractionism and the Dutch "De Stijl" movement. The Bauhaus crossed the boundaries of traditional fine art by offering workshops in painting and photography next to more craft-oriented fields such as printing, weaving, carpentry, metalworking, and architecture. Reconciling art and industrial society had been a central mission for Gropius, who counted among the youngest representatives of the German Werkbund. As early as 1910, Dorner notes, Gropius and German industrial design pioneer Peter Behrens had pushed American-style mass production with a holistic synthesis of art and technology in all areas of life from the individual household to the community level. Against the handicraft romanticism of the earlier Arts and Crafts movement, Gropius and his disciples wanted to provide functional solutions in design and architecture to the benefit of both society and industry.[33]

Especially after the Bauhaus had moved from Weimar to Dessau in 1926, the school developed direct ties to industry. The school, as historian Frederic Schwartz has observed, was by no means "opposed or hostile to the mass market for consumer goods." Herbert Bayer and members of the Bauhaus print workshop cooperated with commercial advertisers, the school marketed its own products, and artists produced commissioned work for companies. In contrast to Kramer's work in Frankfurt, critics have noted, many interwar Bauhaus designs from furniture to lamps found their way only to an upscale, elite market.[34] Still, ultimately the school shared Kramer's aim for a democratization of goods in society through mass production. Especially under Hannes Meyer's leadership after 1928, the Bauhaus pursued a social-democratic agenda of affordable mass production, and the commercial application of design became part of its curriculum. Students received schooling in the psychological ramifications of design. Occasional courses on the basic principles of business and marketing similarly allowed students to gain more than just an artistic perspective on their field.[35]

Several of the protagonists of Bauhaus design were architects, but in keeping with the school's holistic vision, fields were not clearly separated. Marcel Breuer was among the most commercially successful Bauhaus artists, marketing his bent-tube chairs and other furniture designs for serial production already by the late 1920s.[36] Breuer headed the carpentry workshop and, together with Mart Stam, created pioneering designs in tubular steel furniture and modular unit cabinets. Breuer's chairs, emigré architect Peter Blake noted in 1949, became "the model for thousands of copyists the world over."[37] Like Kramer, Breuer was fascinated by American mass production methods and applied them to the serial production of interior furnishings. While he was open to change and to "fashions," Breuer believed that designs had to anticipate a basic set of "true" human needs. This was a vision of mass production that stopped short of the demand-stimulating, accelerated marketing of 1930s "consumer engineers," but quite early on Breuer called for the collaboration of artists and engineers to find economical design solutions.[38]

Applicability to industry and a focus on "rational" design became an important part of the Bauhaus's appeal in the United States once Breuer, Gropius, and other former protagonists of the school made their way into exile. The Dessau Bauhaus had experienced severe political pressure from 1930 onward and was shut down by the local National Socialists in 1932. Many of the artists affiliated with the school eventually went into exile. Marcel Breuer traveled across Europe during the first half of the 1930s, while some left-leaning artists and architects spent time in the Soviet Union.[39] Walter Gropius was one of several Bauhaus members who moved to London, where he set up a private architectural practice with the help of British modernist Maxwell Fry in 1934. The forced emigration of Bauhaus artists and other modernists aided the transcultural dissemination of modernist design concepts across the Atlantic world and elsewhere.[40]

Eventually, however, most of the Bauhaus emigrés settled in the United States. Joseph and Anni Albers and the graphic artist Xanti Schawinsky were among the first former Bauhaus members to cross the

Atlantic. They took up teaching positions at Black Mountain College, an experimental liberal arts college in Asheville, North Carolina, founded in 1933. The Alberses were attracted to Black Mountain College by the school's dedication to reform pedagogy.[41] Commercial application was of little direct interest to Joseph Albers, whose teaching primarily concerned the basic experience of color and form. He did not see himself as the standard bearer of the Bauhaus at the new institution, but still built on his experiences to further develop notions about sensory awareness in artistic production and the pedagogical impact of practical work. His wife Anni, by contrast, headed the school's weaving workshop and had much more direct interest in the design of materials for commercial purposes. At Black Mountain College the Alberses managed to establish a new meeting place for artists interested in modernism for much of the 1930s and 1940s. Especially the summer schools they offered drew on the exile community for instructors as well as on other modern artists from across the United States.[42]

Other former Bauhaus members followed into American exile after 1933. In August of 1937, Breuer and Herbert Bayer crossed the Atlantic on the steamship *Bremen* to meet up with Gropius, who had recently accepted a position at Harvard's School of Design. While Bayer became a freelance design consultant in New York, Breuer followed Gropius to Harvard, and the two launched a private architecture firm together. Gropius's appeal to Harvard stemmed from his international reputation as a member of CIAM and of the European architectural avant-garde. A further appeal, however, was his technocratic vision of social engineering, which emphasized methods of rationalization and standardization and an orientation toward design.[43] Over the course of the 1940s and 1950s, Gropius became an influential educator and facilitator of modernism and managed to build wide-ranging networks in American society and government, ranging from cultural institutions such as the Museum of Modern Art to media tycoons like Henry Luce.

Gropius, too, was active as a commercial artist in the United States. His letters to emigré friends like Bayer and Breuer, to be sure, betray a certain detachment from the New York arts scene with its commercial ties.[44] The stereotype of the elitist European emigré is reflected in a letter in which Bayer apologizes to Gropius who had apparently been "bothered" by inquiries from a mutual acquaintance, an industrial designer employed by Raymond Loewy.[45] Gropius, however, who himself had done design work for railroads, Adler automobiles, and other companies back in Germany, was not generally dismissive of commercial design professionals. He wrote in glowing terms about the graphic artist Paul Rand, for example, who he saw as genuinely interested in improving the profession, appreciating his "very direct approach": "[Rand] has 'zip' and does not sell his soul." In early 1944, Rand even helped to bring Gropius on as a consultant at the Weintraub & Co. agency, a New York advertising firm that counted the later advertising star and "creative revolutionary" William Bernbach among its creative personnel in the early 1940s.[46]

The "General Panel Corporation" was Gropius's most extensive commercial venture. With emigré architect Konrad Wachsmann as partner and Herbert Bayer and others as investors, the firm planned to market Wachsmann's "packaged house concept." The prefabricated housing elements and modules were conceived as easy to ship and to assemble, an ideal solution for wartime and postwar housing shortages.[47] While the "factory-made home" never went into serial production, it provides a telling example of how as emigrés, Gropius and other Bauhaus artists adapted their social engineering interest in functional design for industrial mass production to the needs of American consumer capitalism.

The American Bauhaus: Between Experiment in Totality and Design for Industry

The most comprehensive attempt to bring the reform vision of the Bauhaus to bear on American marketing and design was the "American Bauhaus" in Chicago. On June 6, 1937, Sibyl Moholy-Nagy received this telegram intended for her husband, László, at their London residence:

Plan design school in Bauhaus vision to open for fall. Marshall Field offers family mansion Prairie Avenue. Stables to be converted into workshops. Doctor Gropius suggests your name as director. Are you interested?—Association of Arts and Industries, Chicago

Hungarian-born László Moholy-Nagy, a renowned Bauhaus designer and graphic artist, had been living in exile in the United Kingdom for some years and was at the time traveling in France. Sibyl forwarded the cable to Paris and added a word of caution of her own:

Forwarded Chicago cable today. Urge you to decline. German example shows Fascist results when *field marshals* take over education. Stables and prairie sound just like it. Love.—*Sibyl*[48]

Only a subsequent telegram clarified the reference to Marshall Field III, a wealthy Chicago philanthropist and heir to the department store fortune. The short exchange marks the beginning of László Moholy-Nagy's attempt to bring Bauhaus-style design education to Chicago and its business community.[49] The communication also underscores the challenging transatlantic gap that the couple needed to bridge. Sibyl, daughter of Werkbund architect Martin Pietzsch, student of sociologist Max Horkheimer, and accomplished Berlin actress, was a highly educated and cosmopolitan woman (and her initial telegram may well have been tongue-in-cheek). Still, she and her husband were complete strangers to the world of American business, or for that matter to the famous Chicago department store that would come to play such an influential role in their lives and careers over the coming decade.[50] They set out from Europe for a professionally and culturally foreign world.

Some historians of American design and contemporary design professionals have described the foray of Bauhaus design philosophy into the hard-nosed world of American business as a venture doomed from the start.[51] The careers of individuals such as Gropius and Herbert Bayer, by contrast, suggest that Bauhaus and American industry could very well collaborate fruitfully. The telegram cited above demonstrates that it had been U.S. business interests that initiated the project in Chicago. The Chicago Association of Arts and Industries represented a group of philanthropists and businessmen interested in modern art and its role in commercial culture. The initial contributor list of the association reads like a Who's Who of Chicago area civic and business leaders including about three hundred individual, institutional, and corporate sponsors such as Marshall Field, Harold McCormick, Harold Swift, and the deceased William Wrigley.[52] In the late 1930s, the association, under the leadership of Mrs. Norma K. Stahle, set out specifically to find a Bauhaus emigré to head their envisioned design school. They first approached Walter Gropius, who declined but recommended his friend and former Dessau colleague, Moholy-Nagy.

It was more than a philanthropic interest in modern art and architecture that moved the Chicago Association of Arts and Industries to bring a European "modernist" to their city. Business considerations were in play as companies across the United States increasingly turned to product and graphic design to market their goods and incite consumer interest. With design education programs emerging all over the country, Chicago did not want to lag behind. Home to the corporate offices of major corporations such as United Airlines and Kraft Foods, as well as to advertising agencies and big department stores such as Marshall Field's, the city and its business community hoped for tangible benefits in opening a cutting-edge design school in Chicago. The association saw itself

CHAPTER SIX

6.3 László Moholy-Nagy (1895–1946), photograph, 1925. Source: National Portrait Gallery, Smithsonian Institution / Art Resource, NY.

as a "liaison" between the manufacturer and design education, intent on promoting "good design in industry" across the Midwest.[53]

László Moholy-Nagy was among the most radical design theorists in interwar Europe. After the collapse of Béla Kun's communist regime, he had fled Hungary for Berlin imbued with the ideas of artistic constructivism. In 1920 he met Walter Gropius and soon after settled in Weimar, where his ideological radicalism quickly translated into an experimental approach to photography and painting.[54] Emphasizing a holistic approach to art education, Moholy-Nagy became one of the leading Bauhaus theorists as author of several Bauhaus books to appear during the interwar years. His idea of Bauhaus art and education entailed a comprehensive artistic and social vision (or *Gesamtkunstwerk*), a "totality of vision" that did not mesh easily with business demands for quick sales and profitability.[55] However, Moholy-Nagy had also engaged in design and advertising work during the early 1930s as a consultant to Schott glassworks.[56] Their heat-resistant Jena glass products, many of which were designed by Moholy-Nagy's student Wilhelm Wagenfeld, were considered path-breaking examples of consumer product design at the time. Thus, the hopes placed by the Chicago association in Moholy-Nagy and his Bauhaus concept did not seem entirely out of place.

The school began with high expectations on all sides; the hope was to replicate the success of the original Bauhaus in a new Chicago setting. Once Moholy-Nagy had arrived in America, he quickly set out to draft a comprehensive program for the school and began to recruit faculty both from the United States and from among former Bauhaus affiliates. In letters, Sibyl had encouraged him to target specifically middle-class Americans as students and to focus on design for commercial purposes.[57] As the school opened in the fall of 1937, the faculty included several former Bauhaus colleagues and other emigrés including Hin Bredendieck György (George) Kepes as well as Ukrainian-born Alexander Archipenko. In its first semester, the "New Bauhaus" enrolled thirty-five students.[58] It soon became clear, however, that student enrollment alone could not financially support the school and that further donations would be more difficult to come by than anticipated. By the summer of 1938 the association had decided that the school would not remain open for a second year. Parting ways with the Association of Arts and Industries, Moholy-Nagy began a new institutional venture, the School of Design, which in 1942 became the Institute of Design. While the Bauhaus name was officially dropped to signal a new beginning, the continuities in staff, program, and vision were substantial, and the phrase "American Bauhaus" remained in use for much of the 1940s.

One can easily tell two very different tales of the New Bauhaus and its successor organizations. One conforms to the view of the Bauhaus vision as incompatible with American business culture. The schools remained small, financially struggling, and continuously on the verge of closure. A second story, however, shows an educational enterprise that for all its problems garnered a great deal of attention and support in the art and design world and that had a perhaps surprising degree of successful interaction with corporate America.

Lost in Translation: Moholy-Nagy's Struggles with Corporate America

After a brief, initial honeymoon, relations between Moholy-Nagy and the Association of Arts and Industries quickly soured. In part, this was because the corporate sponsors realized that the much-desired Bauhaus education with its emphasis on basic and abstract training did not immediately yield concrete and practical results. At the same time, the association ran into fund-raising troubles. Many of their initial donors could not be tapped again, and new sponsors were hard to find under the economically depressed conditions of 1937/38. The school soon had problems paying its staff. Moholy-Nagy was frustrated because he was not involved in any fund-raising efforts and generally felt that the board of directors kept him out of the loop.[59] He had managed to persuade three more emigré designers, Herbert Bayer, Jean Helion, and Xanti Schawinsky, to join the staff for the second year of the school, but just as they had made arrangements to come to Chicago, news broke of the school's imminent closure. A lawsuit ultimately awarded some compensation to Bayer and the others, and the settlement kept the association from further using the Bauhaus name as Moholy-Nagy was rallying the staff to reopen the school independently.[60]

Many corporate sponsors were concerned that the theoretical approach of the school did not lend itself easily to commercial application. To compensate for the lack of donations, László Moholy-Nagy took a consulting job with the Chicago mail order firm Spiegel Inc.[61] The job paid about $10,000 a year, which helped support the school's payroll while leaving Moholy enough free time to tend to his teaching and administrative duties. Spiegel hired Moholy to consult on their catalogue design, but the cooperation was not hugely successful. Moholy sketched designs of products for the catalogue and worked on its typography, but the client was ultimately not satisfied.[62] In a 1976 letter, Modie Spiegel recalled his collaboration with "Holy Mahogany," as he called him: "He was supposed to be innovative and creative, which he was, but thoroughly impractical for the times. I tried to get our various executives to take him seriously and did meet with him and them occasionally to see if there was anything usable. . . . After about a year or two, I terminated

Moholy, against which he protested on the theory that in Hungary one couldn't dismiss a person so lightly.... Naturally I told him, we weren't in Hungary and I could do what I wanted, and I thought I had gone pretty far to try to help support the Bauhaus without getting anything usable in return."[63] The episode underscores the dual character of the Chicago community's support for the New Bauhaus enterprise, which was derived from both practical business interests and philanthropic considerations. However, it also reinforces the notion of a cultural and professional transatlantic gap between Bauhaus emigrés and American businessmen.

The Chicago business community remained skeptical of the American Bauhaus and Moholy-Nagy's professionalism. A December 1944 letter from the Chicago Association of Commerce to Walter Paepcke, who had emerged as the institute's principal patron and advocate, listed "typical questions" they received about the school, including "is it ... the personal hobby of Mr. Paepcke?," or does it represent "a particular sect of art?"[64] Sears Roebuck similarly inquired if the Institute of Design was "a school or a one-man show" and requested concrete data about enrollment, retention, and placement of students to decide on future support for the school.[65] Experiences with night classes offered to professionals were mixed. Zay Smith, an art director at United Airlines, suggested that the faculty at the school was "setting up a cult." In a memo to United Airlines' president, Smith related the experiences of five airline employees with their classes at the school. While they generally found the experience "stimulating" and worthwhile in developing "an awareness of a new sense of form and color," they found the courses to be "unorganized" in methods and thinking, a deficit Smith attributed to "the European influence." He concluded: "They have a good product to sell, but, we feel, they do not adequately sell it."[66]

Proponents of the new design profession who saw themselves primarily as marketing specialists frequently did not accept the artistic vision of the school. Tensions with commercial art directors were palpable in a letter from a General Outdoor Advertising Co. executive who related their impressions of their art director. Moholy-Nagy, this critic alleged, had barely survived with his school for five or six years by hustling companies for donations, while his methods were "questionable." None of the students, he claimed, had become "outstanding," and he himself had quit a course ("a complete flop!") in "disgust."[67] Walter Paepcke responded to the letter by pointing to Moholy-Nagy's international renown, to his ties to the B&O railroad and Parker Pen company, to the school's reciprocal credit arrangements with leading institutions

CHAPTER SIX

including Harvard, Yale, and Princeton, and to various prizes and competitions won by institute students.[68] Privately, however, he shared his concerns over Moholy-Nagy's lack of professional leadership with Bauhaus allies such as Gropius and Herbert Bayer.[69] Corporate fund-raising for the school remained a challenge, and after Moholy-Nagy's premature death, a merger with the Illinois Institute of Technology (where Bauhaus emigré Ludwig Mies van der Rohe headed the architecture school) appeared increasingly attractive to ensure the long-term continuity of the program.[70]

The tension between commerce and art was by no means one-sided, however. Moholy-Nagy and his staff retained an idealistic commitment to the Bauhaus vision and at times were highly disdainful of commercial work. In a 1945 letter to Herbert Bayer, Sibyl Moholy-Nagy vented her frustration with well-meaning advice for the school from fellow emigrés: "Not a single friend of the European Bauhaus group . . . has found it worth his while to give up the flesh-pots of commercial New York for the hard work and financial risks of an experimental school."[71] Bayer and Gropius were upset about the charge that they had been anything less than supportive of the Chicago school, but they shared Paepcke's assessment that Moholy-Nagy's personality and management style had contributed to the problems between the Institute of Design and the business community.[72]

In their opposition to streamlining and to the influence of market research on professional design, Moholy-Nagy and many of his faculty were at odds with the commercial mainstream of the profession. Looking back at the program, Sybil wrote to György Kepes in 1948: "The more I analyze the success and failure of the Chicago school, the more I see that the unholy alliance between experimental education and 'big business' blocks the road to all genuine achievement."[73] This attitude remained even after Moholy-Nagy's death and Sibyl's departure from the institute soon thereafter. Finding a successor for Moholy-Nagy proved to be difficult. In light of his deteriorating health, Moholy-Nagy had suggested several candidates to Paepcke, most of them emigrés with ties to the Bauhaus including Marcel Breuer, Josep Lluis Sert, and György Kepes. In case the director "would have to be an American," Moholy-Nagy further suggested Cranbrook-trained designers Ralph Rapson and Charles Eames, who in many ways came closest to the design and educational tradition of the Bauhaus.[74] Ultimately, Serge Chermayeff, a Russian emigré designer who had worked in London, San Francisco, and New York, became the compromise choice.[75] As Chermayeff himself left in 1952, the Illinois Institute of Technology (IIT) had again tremendous problems filling the position. When the IIT administration finally announced the appointment of Jay Doblin in 1955, a full-blown controversy erupted. Doblin had served as an administrator and teacher of industrial design at the Pratt Institute but most recently had worked as an executive designer with the Raymond Loewy Associates design firm. Where the school's administration saw a candidate with practical experience and good connections to the commercial design profession, many of the faculty saw an outsider who was "too commercial" in his designs and lacking the background training in Bauhaus education.[76]

This crisis of succession with its underlying tension between the radical educational ideals of the Bauhaus and American "commercial design" suggests that the Atlantic gap between the more radical Bauhaus emigrés and their students on the one hand and corporate America on the other indeed never completely closed. Longtime faculty member George Keck invoked the school's original "idealism" and cited Moholy-Nagy's decades-old resignation letter from the Dessau Bauhaus where similar tensions had played out: "We are now in danger of becoming what we as revolutionaries opposed: a vocational training school which evaluates only the final achievement and overlooks the development of the whole man."[77]

Fruitful Cooperation: Business Ties of the Institute of Design

Despite such conflicts, the records of the American Bauhaus during the late 1930s and 1940s also support a second, more optimistic story about successful interaction between European emigrés and corporate America. No one promoted fruitful cooperation more vigorously than Walter Paepcke, president of the Container Corporation of America (CCA), who in many ways emerged as the savior of Moholy-Nagy's educational experiment in Chicago. Paepcke, a second-generation immigrant entrepreneur, had a longstanding interest in modern art, and his company had built up a decidedly modernist design department in the 1930s under the leadership of Egbert Jacobson.[78] As Moholy-Nagy's school in Chicago was struggling, Paepcke helped to keep it afloat through several generous donations and active fund-raising efforts among his business connections.

The American Bauhaus built a sizeable network of business support during the 1940s. Chicago socialite Inez Cunningham-Stark had advised Moholy-Nagy and Paepcke to forgo dilettante patrons in favor of corporations like Eastman Kodak or Oscar Mayer who could engage in a mutually beneficial relationship with the school and provide the kind of credibility that would ultimately attract more money from foundations.[79] In May of 1940, Paepcke thus sent out invitations for a luncheon at the Chicago Racquet Club, an event to explore how the Institute of Design could "best meet industrial and business needs of Chicago." The school, Paepcke explained to fellow industrialists, was unique in its methods and offered promising potential at a time when "Chicago is seeking new industries [and] product design is of primary importance."[80] The luncheon laid the groundwork for a wide network of support with representatives of companies such as Kraft Cheese, Rand McNally, Sears & Roebuck, and Walgreens. In its wake, a group of "Friends of the Institute of Design" emerged who regularly contributed small sums but also several hundreds or even thousands of dollars to the school.[81] Beginning in 1942, Paepcke acted as the Institute of Design's chairman of the board of directors, which also included other influential businessmen such as William Patterson of United Airlines, John Kraft of Kraft Foods, Herbert Johnson of S. C. Johnson & Son, Don Mitchell of Sylvania Electric, and Edgar Kaufmann of Kaufmann Department Stores in Pittsburgh.[82] Corporate contributions, to be sure, did not fully end the financial difficulties for the school, which had to rely on further fund-raising and grants from the Carnegie Endowment and the Rockefeller Foundation to meet operating costs.[83] Still, Paepcke's fund-raising efforts attest to the potential that Chicago business elites saw in the school.

The school itself did its part to offer the Chicago community what it was expecting: "good design for mass production." Moholy-Nagy promised a "new education" that aimed to eliminate "the gap between the economic and cultural potentialities of the industrial age."[84] In the school's brochures, Alfred Barr of the Museum of Modern Art touted the Bauhaus's global reputation for tackling the problem of good design in the age of mass production.[85] The curriculum Moholy-Nagy had devised built on the Bauhaus model of basic and specialized workshops in product design, textiles, painting, sculpture, and display, augmented by general courses in the humanities and social and natural sciences. The workshops were headed by specialized artists such as György Kepes, who organized the light workshop. By the late 1940s, the bachelor of arts degree in product design combined Bauhaus methods with a great deal of practical application. Students began with the first-year foundation course and by the third semester tackled problems of applied design. After learning about various materials and methodologies from weaving to pressure molding, third-year students were acquainted with methods of mass production. In their final year, students focused on product design and research and on

6.4 Workshop of the Institute of Design, Chicago, 1947. Courtesy of Bauhaus-Archiv, Berlin.

developing projects in cooperation with industry.[86] The school's promotional literature was imbued with Moholy-Nagy's vision of art education as suited to the demands of the machine age, integrating art, science, and technology into a holistic concept, a "vision in totality."[87]

The school's curriculum reflected wartime concerns with rationalized mass housing. Gregory Keck and Moholy-Nagy conceived a course in "prefabrication" and "factory-built" housing during the 1940s that explicitly drew on Gropius's and Wachsmann's ideas.[88] Especially during the war, the school reached out to government partners to receive funding. Their most prominent such initiative was a new course in camouflage design, taught by György Kepes. The "industrial camouflage course" was offered in 1944 with support from the Office of Civilian Defense, and the school produced several reports on new developments in camouflage design. Many of the students went on to serve in the camouflage units of the Army and the Navy.[89] American entry into the war clearly affected the work at the school, and like so many emigrés in various professional fields, the Bauhaus designers were eager to aid the fight against Germany, which provided them opportunities to employ their expertise in cooperation with the American state. Touting their connections with CIAM, institute members also hoped they could play an advisory role when it came to the postwar reconstruction of war-torn Europe.

The school also offered night classes to working professionals, which were of most direct interest to its corporate sponsors. These classes were geared explicitly toward the commercial applicability of more general concepts. In his inaugural address, Moholy-Nagy had enthusiastically embraced what he called the "splendid . . . American" concept of the night class as a new way to combine work and hobby, leisure and professional advancement.[90] By the early 1940s the school offered several such courses, including ones in display (from window dressing to fair exhibitions), in product design, and in advertising arts. The latter course was intended to convey the "fundamentals of visual communication" complemented by lectures on "consumer psychology, marketing, and business procedures."[91] The night classes were a success with employees of numerous corporations, which sponsored at least part of the tuition to help their employees gain additional training and expertise. The school's records for the winter of 1945 show students from the Container Corporation, Bell Telephone, Corning Glass, and Marshall Fields, as well as many smaller companies in courses for visual basics, product design, or commercial display.[92] Overall, the night classes received positive reviews: the head of display at Marshall Fields, for example, lauded the program and particularly its "European workshop method."[93]

In early 1945, Moholy-Nagy's reports to the school's board were brimming with optimism about its outreach within the world of arts and commerce.[94] He proposed new courses on salesmanship and consumer education and discussed plans for a research institute for product development to enhance collaboration with industry. Moholy-Nagy had just given a talk at the J. W. Thompson advertising agency in New York and was preparing an exhibition display for the U.S. Gypsum company. A special public relations report detailed plans for a scholarship program for Chicago high school students funded by local corporations and approved by the Board of Education. At the same time, the school furthered its connections with other emigrés, hosting talks by Gropius, Bayer, Sigfried Giedion, and French ad designer Jean Carlu. In November of 1944, the institute was even featured in a program of a local television station owned by Balaban & Katz, and the report surmised hopefully: "The future will produce other occasions of a similar nature, for the institute is ideal for televising."[95]

Indeed, the school and its design work received a fair amount of attention in the press and in trade publications. The *Furniture Manufacturer*, for example, detailed Moholy-Nagy's design philosophy. The article echoed the emigré's critique of accelerating design changes as "novelty propaganda" but also featured designs by students and teachers at the Institute of Design.[96] An article in *Industrial Design* allowed Moholy-Nagy to highlight innovative bent-plywood and knock-down designs of the kind favored by Kramer.[97] Wartime restrictions on the production of metal bedsprings had led members of the school to develop patented wooden veneer V-springs, as reported in the trade press.[98] Chicago's social world also paid attention to the work of the emigrés. A 1941 exhibition at the Katherine Kuh Gallery on the "Advance Guard of Advertising" included posters and packaging designs by György Kepes, Herbert Bayer, Herbert Matter, and Moholy-Nagy.[99]

Chicago as well as national newspapers, finally, ran numerous stories on the institute, its Bauhaus background, and its current projects. The *Chicago Sun* highlighted the school's devotion to research in new designs and informed its readers in 1942: "Today the very fountain pen in your pocket can trace part of its design back through American studios like Norman Bel Geddes' to abstract painters in pre-war Europe."[100] Indeed, between 1944 and 1946 Moholy-Nagy worked regularly for the design department of the Parker Pen company, influencing the styling of its products.[101] From the late 1930s to the mid-1940s, then, Moholy-Nagy and his "American Bauhaus" represented an influential contribution to American discussions on modern commercial design. In its more successful moments, the Chicago program combined the teaching of Bauhaus fundamentals and functionalism with research-based designs in collaboration with industry that became characteristic for postwar design in the United States.

Artists and Educators: The American Legacy of European Design Reform

Most of the reform-minded emigrés connected to CIAM and the Bauhaus saw themselves as artists and teachers first. While they developed manifold connections to American businesses, their primary impact on the U.S. design profession was in these fields. They helped set new artistic standards by experimenting with fundamental conceptions of color and form and by further pushing functionalist modernism onto the American scene. As educators, they conveyed to their students an awareness of the social context of design and the importance of meeting minimum standards and basic needs, but also of responding to the physical and psychological demands of consumers. Design, they believed, needed to

CHAPTER SIX

create a material world that improved the lives of people based on standards of quality, basic comforts, and efficiency.

Their social vision of design became adapted to the American promise of democratic mass consumption at midcentury. The career of Hungarian-born designer György Kepes provides one final example for this transatlantic transition. As a student in Hungary during the 1920s, Kepes had been involved with the country's political peasant movement and, like Moholy-Nagy, became attracted to radical artistic constructivism. Kepes left Budapest to join "the Moholy" in Berlin around 1930, working primarily with "new media" such as photography and film. While their artistic interest lay with experimental applications of new technology such as "photograms," they paid their bills with commercial contracts for advertising posters and photography.[102] As the political situation in Germany got ever more tenuous, Kepes followed Moholy-Nagy into exile, first to London and then to Chicago. While he had never been a student at the original Bauhaus, Kepes became a central figure among the teaching staff of the American Bauhaus, heading the light and color department. Together with Ukrainian-born Alexander Archipenko, former Bauhaus student Hin Bredendieck, and Moholy-Nagy himself, Kepes was part of the school's emigré group. Interested in "the meaning of order in visual experience in its present social context," Kepes found the Chicago Institute of Design to be a congenial environment to develop his ideas about the social ramifications of design.[103]

Kepes, too, occasionally engaged in design for commercial purposes during his years in Chicago. He regularly taught evening courses for professionals in the advertising field at the Art Directors Club of Chicago. He also worked as a consultant for the Container Corporation of America (CCA), for which Kepes designed and illustrated the "Paperboard Goes to War" public relations campaign during World War II.[104] Unlike Herbert Bayer, however, for whom practical design consultancy was of primary interest, Kepes was focused on academic writing and teaching. In 1942, he left Chicago for a brief teaching engagement in Denton, Texas, before moving to Brooklyn College on the invitation of Serge Chermayeff. In 1947, finally, Kepes took an influential position at the Massachusetts Institute of Technology, where he founded the Center for Advanced Visual Studies during the 1960s and would teach as a professor of visual design until his retirement in 1974.[105]

In part because of his radical experimentalism, Kepes was especially sought after as an educator of design professionals during the 1940s. He turned down offers to teach from several schools, including Washington University at St. Louis, which approached him for a design program in the commercial and industrial arts.[106] A short biographical sketch from the early 1940s emphasized his contribution to commercial art: "In the field of advertising art, Kepes has effectively applied the new visual language of our times to the task of effective communication."[107] Indeed, Kepes was well acquainted with industrial designers like Henry Dreyfuss, who on occasion consulted with him about "color symbology" and his insights regarding the cultural and psychological interpretations of colors for "application for industrial and commercial use."[108] In 1950, the Cambridge, MA, School of Design, which primarily trained commercial designers, asked Kepes to join its advisory board of academics and prominent practitioners (which also included emigré designer William Pahlmann).[109]

What attracted commercial artists to Kepes's design philosophy was his emphasis on graphic design's potential for social engineering. The 1944 publication of his book *Language of Vision* added to his reputation both among theorists of art and among those engaged in professional design education. The book, concerned with the theoretical underpinnings of modern graphic design, drew on his years in Chicago both in its conceptions and in the network of people involved in its production (from the publisher Paul Theobald to support from CCA's Egbert Jacobson). Modern life was chaotic, Kepes argued, citing European modernists from Kurt Schwitters to Jean Helion. Artists, therefore, needed to meet the advances of modern science in the "socio-economic and psychological realms." He acknowledged his indebtedness to Gestalt psychologists such as Max Wertheimer and Kurt Koffka, whose work had influenced his thinking about visual organization. Visual design, Kepes explained, was a means to "reunite" and "re-form" man through visual representation and dynamic iconography.[110] In more practical terms, *Language of Vision* provides numerous examples of actual advertisements by Kepes himself as well as by modernist colleagues such as Herbert Bayer, Xanti Schawinsky, and Jean Carlu. This focus on advertising design was no coincidence, as the foreword by Sigfried Giedion made clear. The book's intended audience was a young generation of professionals in advertising and other communication fields who, according to Giedion, were shaping the public taste of today and tasked to "rebuild America." It was the responsibility of art directors in industry and advertising, Giedion and Kepes believed, to "educate" rather than "corrupt" consumers by appropriately shaping their everyday visual experience through modern design.[111]

In fundamental ways, Kepes and his colleagues helped transform design education in the United States. Much like Walter Gropius at

Harvard, art historian Anna Vallye has recently argued, Kepes introduced a Bauhaus-inspired curriculum to MIT. More than just a new pedagogical approach, she further suggests, emigrés like Kepes and Gropius brought a technocratic interest and a social-democratic vision to the design discipline at their respective institutions that resonated with a broader American interest in social engineering and "New Deal" politics at midcentury.[112] Gropius, Kepes, and Josef Albers, who in 1950 became head of the design department at Yale, were among the most prominent cases, but other emigrés of the reform tradition similarly influenced American design education. Hin Bredendieck led the industrial design program at Georgia Tech from 1952 to 1971. Frans Wildenhein, who had studied with Moholy-Nagy at Weimar, became a prominent design teacher at the Rochester Institute of Technology. Through their publications, too, the emigrés helped shape professional design education. Studies such as Kepes's *Language of Vision*, Moholy-Nagy's *Vision in Motion* (1947), or Josef Albers's *Interaction of Color* (1963) made significant contributions to design theory in the United States.[113]

The Chicago Bauhaus, too, made its most lasting contributions to design education. The 1946 Museum of Modern Art conference "Industrial Design: A New Profession" demonstrates that when it came to design education, his American colleagues and even some of the most ardent "consumer engineers" deferred to Moholy-Nagy.[114] Joseph Hudnut introduced him at the conference as "the most able and vigorous and successful pioneer in educational discipline based upon objective analysis of the modern scene. We imitate him at Harvard and he is imitated over the world."[115] Moholy's emphasis on practical teaching of fundamentals and abstract concepts of space, color, and material resonated with the assembled members of the profession. Its contemporary champions were convinced of the impact of the American Bauhaus on U.S. design. In 1955, George Keck boasted about the educational achievements of his school, whose alumni had been "responsible for significant advances in industrial design" and whose "philosophy has penetrated the art education world to such an extent that there is today scarcely any art school of any stature in this country which has not adopted many of its principles."[116] Keck, of course, was a biased observer, but alumni of the American Bauhaus did indeed make their way into numerous teaching institutions during the postwar decades, including McGill, Cranbrook, and the University of North Carolina.[117] The American Society of Industrial Designers listed the school as one of the twenty leading programs for professional education in industrial design already in 1946.[118] In 1950, Walter Gropius described the institution as "the decisive link between the designer and industry in this country." Noting the high American regard for European goods during the interwar period when he had first visited the United States, he found the postwar period to be one of "growing pride in things American." American design, Gropius believed, had matured from the days of "bigger and better," while quality and good form had taken on greater importance. In this, he claimed, the Bauhaus tradition had had a noticeable influence.[119]

The European emigrés who came to the United States from politically charged traditions of interwar design reform were unlikely candidates to succeed in the context of American consumer capitalism or in an American society marked by the politics of the emerging Cold War. It comes as little surprise that FBI informants regularly speculated about possible communist sympathies among Bauhaus members or that in 1941 police searched the house of the fifty-eight-year old Harvard professor Walter Gropius, because they had been (falsely) tipped off that his "cellar was stocked with ammunition, guns etc."[120] Looking back at the complicated history of the American Bauhaus's engagement with commercial design in the United States, we can ask how significant its impact ultimately was. Independent commercial designers, such as Loewy or Teague, vastly outmatched their Bauhaus contemporaries with regard to the practical design work they produced. Design historians like Regina Blaszczyk, furthermore, have cautioned against overestimating the mass market influence of elite designers while overlooking the legions of "fashion intermediaries," the vast number of unknown design professionals in companies, department stores, and ad agencies who were much more intimately tied to the production process and ultimately more responsive to consumer inputs.[121]

Still, the radical modernist and reform-minded emigrés were an important part of the story of midcentury American consumer design in several ways. They helped shape the professional discourse during a formative period for American industrial design. As teachers in Chicago and elsewhere they passed on their concepts and methodologies, now adapted to an American context, to a new generation of American industrial designers. Their own commercial and experimental design work set standards that were discussed by trade journals and displayed in exhibitions such as "Good Design." Their work inspired, if often in modified form, the work of other professionals. As we have seen, American art directors and other fashion intermediaries in many ways took their cues not only from consumers but also from their emigré teachers and colleagues.

In 1946, *Time* magazine reported on the Chicago Bauhaus and on the work of Moholy-Nagy, emphasizing the corporate support for the

Institute of Design coming from companies such as Marshall Fields, United Airlines, and Sears. Their support, the magazine noted, was supposed to "pay off in the form of broadly trained designers equipped to create new products for future markets."[122] Critics of the emigrés often underestimated the degree to which the Werkbund, Bauhaus, and other reform traditions of the 1920s had already kept consumer appeal and mass production in mind. Designers such as Kramer, Bayer, or Breuer did not see a contradiction between original artistic achievement and commercial success. More important, the social engineering ideas that influenced the design philosophy of groups such as CIAM and the Bauhaus struck a chord among more commercially oriented consumer engineers. Their emphasis on functional modernism, on rationality and efficiency resonated in the contexts of productivity efforts, New Deal planning, and wartime and postwar consumption.

As we move toward postwar affluence, however, some of the designs championed by the emigrés proved to be too sober and too cerebral. Kramer's knock-down furniture, Gropius's ready-made houses, or the everyday items designed at the New Bauhaus—much of it lacked the color, the frills, the fashion to create a continuous demand for postwar growth. Only in the high-end and corporate market did companies such as Knoll manage to convert functionalist modernism into a colorful consumer fad. The "totality" of the reform vision, however, which placed goods in entire life-worlds shaped by designers, was increasingly shared by commercial design consultants and corporate marketing departments. The attention paid by emigré designers to analyzing consumer needs, consumer perceptions, and psychological processes, furthermore, became a core concern for design consultants, who increasingly integrated consumer research into their creative work.

SEVEN

"Streamlining Everything": Design, Market Research, and the Postwar "American" World of Goods

"We must 'streamline' not only our automobiles, offices, factories, and homes, . . . but ourselves as well," emigré designer László Moholy-Nagy was quoted in a 1942 *St. Louis Star Tribune* article. The piece, titled "A Blueprint for the Post-War World," detailed his designs and the work of the American Bauhaus, forecasting a coming integration of man and machine—the adaptation of consumers to technological modernity and vice versa.[1] Design and marketing became ever more closely intertwined after World War II. Marketing professionals and designers promised an affluent and streamlined postwar era in which new products interacted in novel ways with refashioned consumers: packaging and logos with symbols that enhanced recognition, forms and color schemes that appealed to subconscious desires, and product branding that shaped buying and consumption behavior. In advancing the notion of "streamlining" consumers, Moholy-Nagy captured the essence of consumer engineering, linking the design of products and environments to the psychological molding of consumers themselves.

The connection between product design and consumer psychology gained widespread currency among commercial designers from the 1940s onward; indeed it shaped

CHAPTER SEVEN

the very world of goods that American consumers inhabited after the war. The design business increasingly went beyond the styling of individual products and became focused on placing goods within larger ensembles—into commercial environments or "worlds of goods" designed to appeal to consumers. In her work on the "world of goods," anthropologist Mary Douglas famously argues that goods (and especially consumer products) are part of a larger system of social references: "[G]oods in their assemblage present a set of meanings more or less coherent, more or less intentional. They are read by those who know the code."[2] Goods thus appear as a means of social and cultural communication, a notion that consumer engineers anticipated: they were eager to place products within larger ensembles and within interactive consumer environments. Informed by parallel developments in consumer and marketing research, they furthermore framed products within broader product lines while carefully developing new brand and corporate identities.[3]

Its very ensemble character—the embeddedness of individual products within a larger environment of other goods as seen in retail stores, magazine spreads, and even movie sets—arguably made the American world of goods so appealing, boosting consumerist aspirations. The "populuxe" designs that characterized much of the 1950s thrived on the systematic integration of furniture and new consumer durables in homes that communicated a new life-style of consumer modernity.[4] This integration made consumption even more dynamic as it enticed consumers to update not simply individual goods but entire ensembles. Historian Jan de Vries has called this the "Diderot effect," based on the observation of French philosophe Denis Diderot that a new robe he acquired had led him to refurbish his entire study so that it would correspond to the style and newness of the garment.[5] While this phenomenon of "ensemble updates" could be found in consumer societies since the eighteenth century, it gained new significance in mid-twentieth-century consumer capitalism.

What drove midcentury designers to "streamline everything" and to contextualize products within a larger world of goods? An intellectual history explanation would emphasize the effect of holistic design approaches. Moholy-Nagy's design philosophy was rooted in an interwar vision of reform with a totalizing aesthetic that aimed to make human life-worlds fit the machine age. Cultural historians, by contrast, have stressed a tradition of commercial display art in early twentieth-century America, highlighting, for example, the work of stage designer and Austrian immigrant Joseph Urban and his protégé Norman Bel Geddes.[6] They initially designed scene settings for theater stages, but by the 1920s they were creating consumption displays for shop windows and department stores.[7] Bel Geddes's designs always retained a sense of theatrical composition. Nowhere was this more pronounced than in his famed "Futurama" display for General Motors at the 1939 World's Fair. The display did not simply present automobiles but placed them in a futuristic context that imagined a world of consumer prosperity to come. Such fair displays exerted a profound influence on the visual aesthetics of American consumer capitalism.[8]

Beyond design philosophies and the cultural tradition of commercial displays, business history opens up a third explanation: the increasing integration of marketing and commercial design. Historian Regina Blaszczyk has argued that leading midcentury designers such as Raymond Loewy and Norman Bel Geddes saw themselves primarily as elite "tastemakers" who were hesitant to accept "ordinary people's preferences." "Put simply," she writes, "*tastemakers* sought to reform consumer taste, [rather than] to mediate the relationship between producers and consumers and help companies better understand demand."[9] It is indeed important to question the myth of designers as creative geniuses who single-handedly alter the aesthetics of mass consumption. However, by the 1950s leading design firms, and especially the very elite "tastemakers," began to systematically incorporate market and consumer research into their design process. Consumer feedback became institutionalized within the design process, partially replacing the more implicit consumer knowledge of traditional "fashion intermediaries."

Efforts to create new "worlds of goods" and systems of symbolic reference rested on a systematic exploration of consumer behavior, motives, and desires. "The starting place for good design," designer Gerald Johnson explained to marketing professionals in 1948, "is the customer," and to understand what this customer needs, "we conduct a tremendous amount of market, consumer and material research before we start designing."[10] In their quest for perpetual innovation, postwar design professionals took an engineering approach to their creative work, mirroring the notion of "creative research" invoked by research consultants such as Dichter and Politz. In fact, design consultants came to play a similar function in postwar consumer marketing as their counterparts in market research, spurring on competitive innovation.

This chapter will trace the emergence of increasingly integrative conceptions of consumer design focusing on the examples of two prominent design consultancies headed by European immigrants. Raymond Loewy's design studio was one of the leading consulting firms of the mid-twentieth century; its service portfolio signified like few others the pervasiveness of industrial design efforts after the war. Based on the West Coast, the design firm of Walter Landor was both smaller and initially more focused on the food and beverage industry, making it an ideal case study for the postwar development of package design and the rise of the supermarket as the premier retail environment. Both Landor and Loewy shared a focus on research-driven design, and both championed more integrative design approaches centered on brand and corporate identities. In the language of consumer engineers such as Egmont Arens, this was applied "humaneering," and this chapter interweaves the two threads of professionalization in consumer research and professionalization in industrial design.

Design as "Human Engineering": Consumer Research at Raymond Loewy Associates

"Large corporations," Raymond Loewy claimed, "place their full confidence in our judgement as to what the public will buy [and] how far the Joneses will go design-wise."[11] Good design, he argued, should target the "MAYA" stage (i.e., "most advanced, yet acceptable"), bearing the marks of a modern, simple aesthetic while still retaining a mass market appeal. The precondition for MAYA design, Loewy argued, was a thorough understanding of the product and its merchandising potential.[12] Loewy's design firm represents an exemplary case study for the way in which American commercial design during the 1940s and 1950s became both more comprehensive in the scope of design services offered and simultaneously more research-driven. As marketing consultant to some of the largest consumer goods corporations in the United States, Raymond Loewy Associates (or Loewy/Snaith by the 1960s) had sixty-five associate designers in four divisions by the mid-1940s. At the core of the operation was still the product design division. Since the late 1930s, however, two additional areas of activity had emerged: design for transportation for companies including the Pennsylvania Railroad and the Greyhound Bus Company and architectural design, which prominently included retail stores and shopping centers. In the context of the war, a fourth division of the firm was concerned with designs for the War Department, including uniforms, menus, and the development of USO centers.[13] This expansion of services offered by Loewy Associates was typical for design studios in the postwar years.

Industrial design had evolved from a simple styling function into a complex industry. What had begun with the improvement of the appearance of already engineered products during the 1930s was now integral to both product development and corporate marketing. Loewy cast the designer's role as that of the advocate of continuous innovation, to "create and maintain awareness among his clients . . . of the necessity to improve . . . sales and market position."[14] When it came to product innovation, designers were now part of a larger task force that included engineers and sales managers. Design, Loewy's firm claimed, had become "a necessary sales tool," and it boasted successes such as sales increases after package redesigns for products like Suchard chocolate or Canada Dry soda.[15] In each case, Loewy's design consultants strove to be involved in the entire marketing process from product development to design and advertising all the way to retailing strategies.

As in all areas of postwar marketing, this increasingly entailed market and consumer research. The key to success, Loewy explained in 1953, was "product sales research." As a first step this simply meant gathering information about the specific industry in question and conducting extensive product research and testing. A second step focused on the retail level and involved store surveys and interviews with dealers as well as the study of shopping behavior. The third step of sales research, finally, focused on consumers and included consumer research and interviews. This was "extensive [and, one should note, expensive] research," as Loewy admitted, but, he claimed, it was "a service that American industry is willing to adopt."[16] Particularly within the type of large corporation that typically retained the services of Loewy's or other design firms, market research had become a widespread phenomenon. Now this was to inform the design process as well.

Research-driven design was controversial, however, and many designers had long resisted the notion that consumer surveys should inform their work. General Motors presents one of the earliest in-house efforts to integrate consumer feedback systematically into the production process. Its customer research department, headed by Henry Weaver, understood itself as the "missing link" between the consumer and the company's various departments.[17] Weaver believed that information gleaned from customer surveys regarding body and style preferences could help to improve the design of new GM models. By the late 1930s, he had come into conflict with lead designer Harley Earl, the influential head of the Art and Color Section. Sales managers wanted cars built with less chrome and frills based on survey results showing "practical consumers" who prized safety and dependability over extravagant styling. Earl scoffed at alleged consumer preferences for "more headroom" and "easy access," which defied the styling trends of the time, but gave in to sales managers' demands for the 1939 model year. In the designer's narration of the story, the results were boxy and misshapen cars that quickly convinced management not to let sales surveys infringe upon the authority of the design team.[18] However, consumer research did become more influential for design processes during the 1940s. Episodes such as the conflict at GM held two lessons for designers that Loewy and others in the profession quickly embraced. First, designers did well in cooperating not only with engineers but also with company sales managers and research directors from the outset. Second, while consumer research could be invaluable for design, designs based directly on stated consumer preferences did not usually work well. Surveys of consumer wants and behavior could guide designers in optimizing their work and increasingly came to be seen as indispensable, but no direct questions would be asked of consumers regarding their design ideas; the designer would remain the ultimate arbiter of good taste and form.

The 1946 Museum of Modern Art conference on industrial design as a profession devoted an entire section to the place of consumer research within design processes.[19] Many were skeptical with regard to the value that consumer surveys could add to their work and discussed survey research as a potential threat to the creative agency of designers. Industrial designer George Nelson warned that surveys could have a "positively destructive, degenerating influence on design." He felt they encouraged design for the lowest common denominator, "the result being mush. . . . The survey, as I have seen it used, is actually an instrument for evasion of responsibility by the manufacturer and the designer."[20] Nelson saw the "responsible" designer as someone who fully stood behind his or her work, invoking the tradition of the artist with a "distinct creative function."[21] László Moholy-Nagy honed in on this very point; "the designer's task is a creative task," he stated, with cultural aspects and a need for vision. To Moholy-Nagy, "the so-called market survey and consumer survey [are] mainly an unnecessary evil, mainly an excuse for negative design."[22]

While not all industrial designers embraced survey research with open arms, the discussion at the conference made clear that professional attitudes were shifting during the 1940s. Ray Patten, designer at General Electric, noted that his company had lots of experience with "all sorts of surveys" and employed "consumer research groups at plants throughout the country."[23] Even an ardent critic such as Moholy-Nagy conceded that the study of consumer needs and product functions was very much worthwhile for the creative process. By the early 1950s, the new edition of Harold Van Doren's *Industrial Design: A Practical Guide* featured an entire chapter on "research tactics." The textbook touted consumer research as a "new tool" for the designer and listed possible sources of available data helping to "define problems," including existing surveys, national publications (e.g., Curtis Publishing Co.), trade associations, and the Department of Commerce. If designers were interested in conducting original research for their client, the handbook advised that they could enlist the help of specialized research organizations, or—for "smaller jobs"—they could hire "intelligent" interviewers of their own.[24] To most of its leading proponents, consumer research had become a part of the design profession.

Thus, Loewy's firm was spearheading a broader movement toward research-driven design. The "task force" that Loewy often invoked when he discussed his firm's approach to new design jobs had initially included only the designer and the product engineer. By the 1940s, Loewy recounted in speeches, the task force was enlarged to include sales and advertising, and by the 1950s a "new member entered the TASK FORCE team: the market survey specialist," because increasingly "more accurate information about the market, its potential, the state of the competition, were needed."[25] Loewy Associates both obtained its market research data from outside sources (including clients and advertising agencies) and began to generate survey data of its own. To analyze the merely descriptive survey materials and to extrapolate problems and solutions, the company established its own research department around 1950. His was, Loewy claimed, the "first industrial design organization to establish such a division of market analysis."[26]

By the early 1950s, market analyses aimed at adapting products to changing markets. A trade journal article on Loewy's redesign of brands for Reardon paints, for example, emphasized the postwar growth of the do-it-yourself market. A "huge middle market" of new suburban homeowners emerged as a new customer base for the paint producer that had previously primarily sold to professional painters and construction companies. Loewy's Chicago office had done a thorough analysis of product packaging requirements, new packaging developments, and shipping and storage requirements for Reardon. "This type of study," the article stated, was "a Loewy specialty, [and] serves as a basis for every decision the company must make regarding its products in the competitive markets they will enter." The study not only identified DIY amateurs rather than professionals as a new target group but also explored their preferences and new ways to appeal to this consumer segment. The outcome was a sweeping redesign of Reardon's product ensemble: its containers, point of sales displays, shipping cartons, and merchandising program.

In addition, Loewy designed a new trademark for letterheads, envelopes, forms, and calling cards. By January of 1956, the changeover was complete and amounted to a full-scale "reorientation of the company in relation to the buying world, where every man or woman . . . is a potential customer."[27] Consumer research helped Loewy go beyond product styling to think more broadly in terms of brands and markets.

For some design projects, Loewy's firm did a considerable part of the consumer surveys on its own. A mid-1950s consumer survey for Armour meat products provides some insights into how such research was conducted. Loewy's task was to redesign the trademark logo and the packaging for Armour's line of canned meat products, and to that end the firm began with an evaluation of the existing designs. Loewy Associates surveyed a randomly selected sample of 2,600 "housewives" in seven major U.S. cities. The surveys asked respondents to describe the trademarks of Armour and its major competitors including Swift, Wilson, and Oscar Mayer, probing for details such as forms and color schemes and asking for customer assessments. The interviews were done by telephone by fourteen part-time interviewers (mostly women homemakers), who were to follow a detailed set of instructions in conducting the survey. We know that Loewy's staff checked the results, because one of the interviewers who apparently failed to conform to instructions had her interview results omitted from the final count of surveys. Based on the survey, Loewy recommended an overhaul of the Armour product line that made the brand name more prominent, eliminating the word "star" and replacing it with a simple graphic representation.[28]

Loewy's research capacities grew over the course of the 1950s. While their surveys did not reach the sophistication of those of specialized research firms such as A. Politz Research, they were professionally done and required the expertise of in-house researchers. By the early 1960s, an organizational chart of Loewy/Snaith Inc. listed four creative departments: (1) "Product and Equipment Design," (2) "Graphic Planning and Design" (for design of corporate logos or packaging), and (3) "Architectural and Interior Planning and Design." On the same organizational level as these departments, the chart now also showed a department for (4) "Research, Development, Analysis." The portfolio of this department included market research, economic analyses, and the assessment of market potentials, among other aspects.[29] The firm's division for marketing and market planning dated back to the early 1950s, and the company even briefly reorganized it as an independent subsidiary named Market Concepts Inc. in 1965.[30]

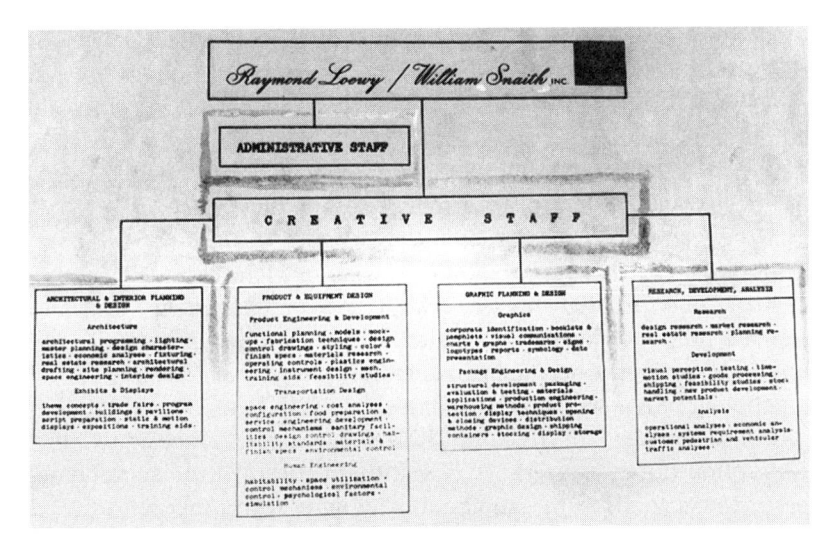

7.1　Organizational chart, Loewy/Snaith Inc., featuring the "Research, Development, Analysis" department (right), early 1960s. Source: Library of Congress, Prints & Photographs Division. Visual Materials from the Raymond Loewy Papers.

Besides consumer surveys, the research department offered other forms of experimental studies including visual testing with tachistoscopic equipment to measure impact, identification, and communication of products and trademarks, a practical application of the cognitive principles laid out by Gestalt psychology theorists.[31] It also offered product testing and concept and trend research, as well as corporate identity studies. For specialized tasks, the firm enlisted outside expertise, however. A late 1950s study on color combinations for British Petroleum, for example, was advised by Theodore Kawosky, a Dartmouth psychologist specializing in synesthesia.[32]

For the design firm, taking a product to market included several stages of research and analysis. Thorough consultations with the clients were followed by extensive research including the study of sales reports and "field research," such as a survey of shops and stores. Designs of prototypes were similarly field-tested with an eye to consumer reactions and possible marketing strategies regarding place and price. Product names were tested for their semantic properties and for consumer identifications, as was the packaging. By the end of the 1960s, Loewy's firm had done over 150 research projects for various clients including numerous Fortune 500 corporations. Besides the trademark analysis and a study of the sausage market for Armour & Co. discussed above and a brand name analysis for the National Biscuit Co., the firm conducted consumer product research for DuPont, studied new forms of store retailing for General Electric, and did facial tissue market studies for Kimberly Clark and a merchandising analysis for Gimbel Bros., among many others.[33] Loewy's design studio had developed considerable consumer research capabilities.

More than simply an additional service offered to improve sales, marketing research as portrayed by Loewy was part of a larger effort in "human engineering" in design. In *Consumer Engineering*, "humaneering" had referred primarily to the application of engineering principles to the interaction of consumers and goods, emphasizing the importance of new psychological research in connecting goods with human emotions and subconscious needs.[34] There was, however, a tactile dimension to this conception of "humaneering" as well that concerned the appropriateness of new goods to the human touch (along with other senses). Did a device, for example, fit snugly in the palm of one's hand? This essentially ergonomic aspect of human engineering greatly concerned mid-century designers, as architectural historian John Harwood has shown, because it represented the crucial element of man-machine interfaces, which became a common challenge as new (and often electronic) consumer goods proliferated after the war.[35]

Modern design research, Loewy and his colleagues claimed, would make new forms of "systems engineering" possible, which organized the complex relationship between man and machine systems. As the "great grandchild of modern psychological studies for the industrial world," Loewy predicted in speeches, "human engineering" would contribute to narrowing the gap between the senses and technology. Doing away with all forms of nuisances, he proposed, design was to maximize human efficiency while minimizing physical or mental discomfort. Loewy and his firm thought closely about the psychological effect material objects had on their environment. Ultimately, they strove to control and manipulate these environments to have a "stimulating" effect on consumer senses and shopping habits.[36]

Emphasizing "scientific" research activities became an important way of selling the services of a design consultancy. Loewy's research staff kept a close eye on psychological innovation in the field of consumer studies more broadly and even contributed to the development of professional knowledge. Building on experiments by psychologist Abraham Maslow, for example, Loewy/Snaith Vice President of Research Herbert Krugman conducted experimental studies on changing attitudes toward packages over time, investigating when a design reached its peak in appeal.[37] Others studied consumer reactions to product presentation prior to the act of purchase, employing the "latest learning theories and behavior mechanisms to find ways to expose underlying triggers to consumer action."[38] At a 1967 professional luncheon, Loewy summarized the work of his research department:

> In our marketing research and development division we apply psychological principles, knowledge and technology to marketing investigation. . . . A leading psychologist from a major university advises and consults with our product design panel when knowledge of basic human behavior is needed. [Successful product applications] result from the mass-market interpretation of some highly sophisticated aspects translated into commercially understandable selling symbology and form. This, at least, is the design-in-depth technique that our clients seem to require.[39]

Their "design-in-depth" and "human engineering" approach applied not simply to individual products but increasingly to ensembles of goods from product lines and corporate identities and entire shopping interiors.

Consumer research heavily informed design for retail interiors as well. As the firm moved more systematically into retail planning during the 1940s, Loewy Associates was early in employing research programs in this field. In 1947, for example, they studied metropolitan logistics in the Los Angeles area for Broadway Department Stores, analyzing optimal storage and sales distribution between central stores and suburban branches.[40] Around the same time, the Associated Merchandising Corporation commissioned a study that detailed developments in postwar retailing more generally.[41] In a retail landscape defined by self-service, Loewy noted in the early 1950s, sales clerks were increasingly "menial," untrained, and often eliminated entirely. Not only did sales displays have to take on the role of the sales clerk, but goods needed to be engineered to simply sell themselves.[42] The design firm thus conceptualized both the products and the environments in which they would be displayed and sold.

This integrative approach similarly applied to the emerging field of corporate identity design, which became central for Loewy's "Graphic Planning and Design" department. Again, product designs were embedded in a much broader design context based on consumer research results. A 1953 company memo notes that for numerous corporations the firm had designed or redesigned

CHAPTER SEVEN

practically everything that pertains, even remotely, to their operations. [One successful product leads to] the designing of every package and shipping carton, of the retail outlet stores, the delivery trucks, driver's uniforms, point of sales displays, all business forms and printed matter, a new company trademark, the firm's stationary, product testing equipment, experimental laboratories, new plants, cafeterias, workmen recreation areas.[43]

By designing recurring logos and color schemes for leading consumer goods producers, Loewy Associates helped to do much more than simply design and style products ranging from cigarette packages to automobiles. Instead, the firm was able to embed these products within broader contexts—within sales displays and larger retail settings as well as within product lines and encompassing corporate identities.

In 1950, *Cosmopolitan* magazine made the sweeping suggestion that Loewy had "probably affected the daily life of more Americans than any other man of his time."[44] This was an exaggeration that Loewy and his fellow design consultants happily embraced. In his autobiography he recounted a piece written by a journalist about how "Jack Smith," the average American citizen of 1950, encountered industrial design throughout the day: one cold winter morning Jack Smith wakes up on a mattress designed by Raymond Loewy, startled by an alarm clock designed by Walter Teague. He stumbles out of bed to find the light switch (Loewy design) and to turn on the heater (by Egmont Arens). In the bathroom, he showers in a Dreyfuss bathtub and shaves with a Loewy razor while standing on a Joseph Platt carpet. In the kitchen, he listens to the radio (Loewy) and uses a toaster designed by Harold Van Doren as well as a coffee maker styled, again, by Loewy. And so begins another regular day of a midcentury American consumer surrounded by a world of goods created by sophisticated consumer engineers.[45]

Such depictions, of course, have to be taken with more than a grain of salt as they were part of the self-promotion of professional designers. The all-encompassing design jobs, expanding from the product all the way to corporate identities, were good business for design studios such as Loewy's. In the same vein, the emphasis on research in Loewy's speeches and in the company's communications also served self-promotional ends. Systematic research and consumer studies, as we already saw with regard to the self-promotion of market research consultancies in chapter 4, increasingly became part of American corporate marketing at midcentury, and few firms in the business of marketing consultation could afford to stand on the sidelines of this trend. The observation that surveys and notions of psychological "human engineering" became ever more prevalent does not tell us very much about their effectiveness. It does suggest, however, that design and marketing research became ever more closely intertwined. Loewy's firm stood prominently at the forefront of this development, but it was only one among many design firms on this exact trajectory.

The Psychology of Packaging in the Supermarket Era: Walter Landor Associates

Walter Landor Associates similarly prided itself on being a pioneer of research-driven designs during the postwar decades.[46] The firm of German emigré designer Walter Landor thus presents a useful complementary case study to Raymond Loewy Associates. Landor's firm was a good bit younger, initially significantly smaller, and in contrast to the New York–based studios of Loewy and other design stars, Landor's office in San Francisco got its start on the West Coast. At the beginning, his client base was much more regionally focused and limited to the food and beverage industry. Over the course of the 1950s, however, Landor Associates broadened its business portfolio, developed its own approach to consumer research and integrated design, and ultimately emerged as one of the leading American design firms in brand systems and corporate identity design.[47]

While Walter Landor became a quintessential "West Coast" designer, he was rooted in the tradition of interwar European design modernism. This made him part of a sizeable group of modernist European designers and architects including Kem Weber, Richard Neutra, and Victor Gruen who had settled in California since the 1920s and helped shape the midcentury design of that region.[48] Landor had been born as Walter Landauer into a German Jewish family in Munich in 1913. Early on, he was influenced by his father, the architect Fritz Landauer, who introduced his son to the work of the Werkbund and Bauhaus movements. Landor later recalled his affinity for modern art during his school years in Munich and his enthusiasm for the Bauhaus during the 1920s as the "first real awakening of the people to the form and shape of consumer goods."[49] Artists such as Josef Albers and Walter Gropius, he suggested, had managed to push design beyond the simple formula of "form follows function." To some degree, he exaggerated his identification with the Bauhaus tradition as part of his later self-representation, presumably to underscore his design pedigree.[50] It is true, however, that already as an adolescent Landor was acquainted with modern artists such as Henri

CHAPTER SEVEN

7.2 Walter Landor (1913–95) discussing design ideas with his staff aboard the *Klamath*. Courtesy of Landor Design Collection, Archives Center, National Museum of American History, Smithsonian Institution.

Kay Henrion (born Heinrich Fritz Kohn), who became one of Britain's leading graphic designers.

It was London where Landor received his first significant professional experiences. Intrigued by an advertising course he took during his last year of secondary schooling, he participated in a six months' exchange in the early 1930s that placed him in the London-based advertising agency W. S. Crawford. Here Landor had his initial training in both advertising design and market research, and he went on to study graphic design at Goldsmiths College.[51] At Goldsmiths, Landor became acquainted with Milner Gray and Misha Black, who had established a new design studio of which Walter Landor, too, became a part as both a designer and a marketing specialist. Renamed "Industrial Design Partnership" in 1935, the firm pioneered modernist designs during the 1930s, early on pursued "corporate identification schemes," and engaged in research for design, ultimately opening a "design research unit" during the 1940s.[52] Landor himself became the firm's specialist for new plastic materials.

When the Industrial Design Partnership was commissioned to plan the Persia pavilion for the British Commonwealth exhibit at the 1939 New York World's Fair, Landor followed Misha Black across the Atlantic. Carrying letters from Milner Gray, Landor used his time in New York to acquaint himself with leading American designers, many of whom had their own pavilion projects at the fair. He met with Viennese emigré Victor Gruen as well as with Walter Teague, Gilbert Rhode, and, of course, Raymond Loewy. Most lasting, however, was the connection established with designer Henry Dreyfuss to whom he was introduced by emigré socialite and patron of the arts Anita Warburg. Landor had the chance to study Dreyfuss's studios, and he acquired a basic understanding of the business of industrial design as practiced in the United States.[53]

In late 1939, with Europe already embroiled in war, Walter Landor decided to settle in San Francisco. Initially he found employment as an associate professor for industrial design and interior architecture at the California College of Arts and Crafts in Oakland. Within little more than a year, Landor would marry his "star student" (as he liked to tell the story) and set up home in Russian Hill. Even on the West Coast, he retained his connections to European modernism, however. Landor, still calling himself Walter Landauer at the time, was part of Telesis, a group of planners, architects, and designers that aimed to bring modern architecture and planning to San Francisco. The group had connections to CIAM, Walter Gropius, and others of the Bauhaus tradition as well as to the London-based MARS group (which included Landor's former colleague Misha Black). Landor attended regular meetings of the group between 1939 and 1941 in preparation for an exhibition at the San Francisco Museum of Art that explored a new balance between the needs of the individual and of society.[54] Inspired by transatlantic visions of social engineering, the "technicians" of Telesis planned for a postwar California with a more integrated environment of social and technological components—functional homes, neighborhoods, and cities at low cost. These "technicians" included architects, landscape architects, urban planners, and, finally, industrial designers, with Landor representing the latter group.[55]

His involvement with Telesis connects Landor directly to interwar reform modernism with its holistic notions of design as part of a broader reconstitution of modern society. The group's "master plan" for San Francisco gave special consideration to consumer needs: "The industrial designer should not only be equipped to create merchandise that is more salable and simpler to produce, but also merchandise that is designed to fit social needs, to bring higher standards of living to an increasing number of people. He functions at a strategic point in the profit system, between the production-conscious manufacturer and the sales-conscious merchandiser. But his eventual concern must be for the use-conscious consumer."[56] The Telesis group saw the social use of design primarily as raising standards of efficiency and lowering the cost of everyday items. Landor, however, turned this broader social engineering approach to his profession quickly toward more commercial solutions.

With his wife as partner, Landor Associates opened as an independent design office in 1941. During the war years, Walter Landor held a position at Hewlett Packard as well, but after 1945 his firm became a full-time enterprise and the staff expanded. San Francisco's dearth of industrial production led Landor to focus primarily on packaging and labeling for canned or bottled food and beverages. Early clients included "Golden V" milk and the locally based but nationally active S&W food distributor, which had been forced to switch from cans to glass jars owing to wartime shortages. Work for Spice Islands and for Sick's Beer of Seattle's Rainier Brewing Company not only won Landor design awards and brought recognition to the new company but also marked the beginning of his use of design as a form of strategic marketing and of brand placement within consumer markets.[57]

By the 1950s, Landor Associates had emerged as one of the leading industrial design studios on the West Coast. The firm's offices at Pier 5 adjacent to the San Francisco harbor now employed dozens of designers, and the client portfolio expanded with regard to region and branches of industry. They designed for Safeway supermarkets, tobacco giant Philip Morris, and even steamship lines and commercial airlines (Commodore). Mexicali Beer in Mexico and the Japanese Sapporo Beer count among the first international clients, long before the firm opened offices abroad in Rome and Tokyo during the late 1960s and 1970s. By the second postwar decade, Landor Associates had gained national recognition with clients all over the United States and employing "about 300 artists at any given time around the world," according to Landor.[58]

The economic success of his firm was—as in the case of Raymond Loewy—derived in part from a professional self-understanding that fused creative design with a marketing perspective. From the outset, Landor Associates built on its founder's dual background in design and marketing. Industrial designers, Landor explained in 1947, should have a "broad perspective" to recognize long-term trends in merchandising and consumer preferences. For creatives this meant to extend their horizons into color psychology, architecture, and many other fields. Just like his London employer Industrial Design Partnership, commercial designers would engage in designing products and spaces (from showrooms to corporate offices) and all manner of packages (from labels to delivery trucks). "Designing for mass production is designing for mass appeal," Landor noted, and designers needed to balance creative leadership and consumer preferences: "It's the [designer's] job to lead, but he must also know how far the public is willing to follow him."[59] Anticipating Loewy's concept of MAYA, Landor, too, saw himself as a design consultant with a clear grasp of markets and consumers.

Accordingly, his design firm increased its consumer research capabilities throughout the 1950s and early 1960s. Initially, market research at Landor entailed more or less unsystematic surveys of consumers at supermarkets conducted by the designers themselves, interviewing patrons about product preferences or specific package designs.[60] By the end of the 1950s, however, the firm had its own research department, headed first by Herbert Kay and later by Hugh Schwartz as research directors. The new department conducted surveys and consumer panels, commissioned outside studies, and during the early 1960s began collaboration with the International Marketing Institute. Landor Associates employed a mix of quantitative and qualitative methodologies but put a special emphasis on focus groups and motivation research. Overall, Landor stressed consumer psychology and the notion of "designing for the consumer." A comprehensive introduction to the firm's research approach in the 1958 yearbook edition of *Good Packaging* underlined the importance of new "fool-proof research techniques" and "practical and trustworthy methodology." To that end, Herbert Kay as the resident research consultant called on external research organizations and devised new research techniques and consumer panels. To underline Landor's use of state-of-the-art methodologies, the article even documented a research symposium at its headquarters with nationally recognized consumer researchers such as the behavioral scientist Maurice Rappaport of Stanford and emigré consumption expert Ernest Dichter of the Institute for Motivational Research.[61]

Dichter's motivation research held particular appeal for Landor, who thought a product had a "personality" that was reinforced by its appeal to the customer's senses and who stressed the importance of distinction and individuality in modern consumer lifestyles.[62] On several occasions, Landor collaborated with Dichter's institute in the research process. In 1958, for example, Landor used Dichter's research in the redesign of coffee cans for S&W Fine Foods. A Dichter study on the U.S. coffee market had called for more imaginative promotion, suggesting that to consumers "coffee is more than *just* coffee. Coffee is a symbol of warmth and comfort and a way of life." Landor now attempted to use the new package as a sales promotional tool that would appeal to those "fundamental social and psychological drives" that made coffee popular according to the insights of Dichter's motivation research. They measured consumer responses to various can designs on display and eventually narrowed the field to six designs, which were independently tested by Landor and Dichter's Institute for Motivational Research. The favored design was then again tested for consumer reactions in different color treatments.[63] Landor touted this type of continuous "consumer response" input throughout the design phase as a cost-cutting mechanism (as it supposedly minimized marketing failures) and as an ongoing process in which even existing designs were subjected to consumer testing at regular intervals. With coffee cans and many other products, the guiding question became: what appeal does it have for the consumer?

Landor Associates referred to this integration of consumer research into various steps of the design process as "consumer response design." In their own work, the design studio used a whole set of motivation research tools to elicit and record consumer responses. Especially for products such as candy where the purchase decision was deemed to be "emotional" or impulse guided, researchers would conduct consumer observations of supermarket behavior, panel tests, and individual depth interviews with consumers both at home and at retail outlets.[64] Landor occasionally also employed quantitative market research, and psychological experiments increasingly rounded out the design research. These experiments included the analysis of a given product's visual properties and measured the duration of a consumer's glace at visual displays: the shorter the necessary exposure, the greater the "attention-catching properties" of product or package. Eye cameras were used to determine how quickly a product could be recognized and categorized by consumers.[65] Ultimately, the combination of lab experiments and motivational research was believed to reveal which design had the "maximum visual impact" and provided the appropriate "connotative symbols," stimulating responses from well-defined demographic segments of the buying public.[66] Landor cooperated with advertising agencies like McCann-Erickson, integrating packaging and advertising to build larger brand images.

Supermarkets were the premier environment in which a product had to prevail in the postwar decades. In new self-service stores, the product

"STREAMLINING EVERYTHING"

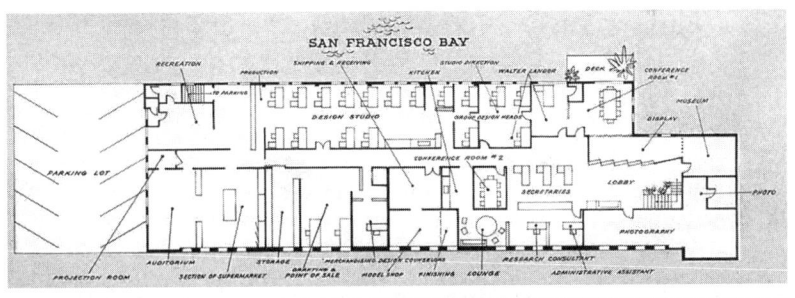

7.3   Landor Associates Research Facilities, c. 1958. Discussion panel with Landor, Ernest Dichter, Landor research director Herbert Kay, and other market research experts. Floor plan of design studio with mock supermarket. Source: *Good Packaging Yearbook*, 1958.

had to sell itself. Landor Associates research facilities thus included a small mock supermarket in which researchers could study consumers under "laboratory conditions." To the designer as consumer engineer, stores became "machines for selling" whose design depended on consumer psychology and effective merchandising. Their primary task was "aiding and abetting the package."[67] Merchandising Design Counselors Inc., a Landor subsidiary, studied merchandising conditions and helped develop the retail environments for Landor-designed products. They created display settings, point-of-purchase materials such as store refrigerator signs, and entire supermarket sections. With Safeway Supermarkets as a major client, Landor (like Loewy) came to regard retail design as a natural extension of the portfolio of the industrial designer.

The notion that "the package sells" and that packaging enabled manufacturers to address consumers directly became prominent in American consumer culture with the rise of self-service stores following World War I. Journals such as *Modern Packaging* (first published in 1927) and

211

organizations such as the Packaging Machinery Manufacturers Institute (established 1933) helped promote new packaging technologies and materials. By the end of World War II, "packaging management" had become an integral part of marketing and supply chain management.[68] In the age of the supermarket, the spirit of consumer engineering filled the consumer laboratory at Landor Associates, where designers worked on "psychologically keyed" designs with an "emotional appeal" able to promote "new buying habits." A *Good Packaging* feature on Landor's retail packages noted: "a rare product succeeds today without being dressed right; dressed to make the right kind of consumer impression." Modern and "visual minded" consumers demanded continuous redesign, and companies would be more competitive through frequent updates (without, of course, destroying familiar brand identities). Successful package design, Landor explained, needed "impulse purchase appeal," inspiring confidence and conveying a "feeling" of the product, a "package personality" that worked well in visual advertising and was "styled for instant recognition." In short, packaging that successfully moved products off shelves required not simply aesthetic design but "a research mentality; an analytical mind, an understanding of marketing and a deep interest in merchandising; a sympathy for people and their needs, a keen interest in finding out what goes on in their minds and their subconscious, and a humble attitude towards satisfying their preferences instead of one's own."[69]

Packaging became primarily a marketing device. Package design could provide products with a "personality" appealing to emotions and desires or simply to a desire for novelty. Embracing Marshall McLuhan's notion that the medium was the message, Landor advertised his packaging solutions as the "magical tool to communicate the newness of a product," a newness, he believed, that consumers increasingly came to expect.[70] By the end of the 1950s Landor predicted not only that package changes would accelerate even further in the decades to come but that the research orientation of package design would become ever more pronounced to reduce the risk in package investments.[71] Research, finally, focused not simply on individual products but on larger product environments. "Effective packaging presentation is a reflection of a total environment," Landor told the American Management Association in 1966. The product and its package design had to relate to the "kind of consumer" it sought to appeal to and to the "specific sales arena" in which it would be sold. The product furthermore had to relate to competitor products as well as, finally, to the "manufacturer's image" more broadly.[72]

Brand Images and Corporate Identities: Designing Corporate Worlds

The concepts of brand image and corporate identity gained tremendously in importance during the postwar decades. Companies increasingly coordinated their public appearance and paid attention to the visual presentation of product lines and brands. Much as the notion of product personalities had opened up consultation opportunities for firms specializing in motivation research, corporate identity contracts offered new revenue for design consultancies. By the 1960s, projects in graphic design involving the design or redesign of brand and company logos made up a significant portion of the overall business both at Loewy/Snaith and at Landor Associates. Both firms had acquired prestigious corporate image accounts since the 1950s, and corporate identity programs, which involved every aspect of a company's internal and public communication, were reflective of their increasingly comprehensive design philosophies.

While corporate and brand identities dovetailed nicely with the rise of commercial design and consumer psychology, they had a longer prehistory. Consumer brands, which were often closely identified with the companies behind them, date back well into the nineteenth century.[73] By no means limited to the American market, branding and the use of brands in advertising and sales strategies was a transnational phenomenon. In Germany, marketing specialist Hans Domizlaff developed early branding strategies for cigarette manufacturers already during the 1920s, and even before World War I, designer Peter Behrens had developed an integrated corporate design for the electricity corporation AEG that included advertising, product design, and even the corporate architecture.[74] In the United States, too, corporations became increasingly concerned with presenting a favorable public image during the first half of the twentieth century. Public relations became professionalized as a field and integrated into corporate management structures.[75] At the same time, midcentury trademark design (including work by emigrés such as Herbert Bayer and Bernard Rudofsky) began to systematically exploit the insights of Gestalt psychology as applied to design by György Kepes and others.[76] By the 1940s, the idea that the "company image" could be improved and controlled by means of professional design was systematically pushed by a growing number of design studios including Lippincott & Margulies.[77]

CHAPTER SEVEN

After the importance of a visually identifiable "corporate image" became widely recognized during the 1950s, subsequent decades saw even greater concern with less tangible aspects of corporate communication including notions of corporate "personality" and "identity."[78] U.S. corporations strove to appear personable and relatable, and postwar designers such as Loewy and Landor were able to capitalize on these developments. American companies, Loewy/Snaith claimed at the end of the 1960s, were facing an "identity crisis," with large corporations appearing bland and abstract. His company offered them "positive corporate identity programs," because "[t]he names, symbols, colors used by a corporation or institution; its décor and architectural façade, its very uniforms and stationary, make the initial impact on those who buy its products or use its services."[79] This kind of comprehensive service had been part of Loewy's portfolio since about 1945 with early corporate design work for companies such as International Harvester. For the airline TWA, Loewy's firm designed everything from aircraft interiors to uniforms, menus, silverware, and even airport installations and exterior ramps. Already in the 1940s, this work, too, was undergirded by research and done in cooperation with university research centers, public opinion pollsters, and institutes for applied psychology.[80] Over the subsequent decades, Loewy clients for corporate identity design included institutions as diverse as oil and gas companies, soft drink producers, and the U.S. Postal Service.[81]

Landor, too, recognized the growing importance of corporate identity design starting in the 1950s.[82] Beginning with work for Pabco, a maker of building products, Landor employed an approach that later came to be known as "brand systems design." As early as the late 1940s, his studio created a coordinated image program across Pabco's entire product line from paints to roofing, wallboard, and linoleum, and he devised the company's new trademark for domestic and overseas markets. In 1955, Landor similarly developed a "brand identity" for the Lone Star Brewing Company, consciously covering every step at which the public came into contact with Lone Star as a company and with its products.[83] Brands, to Landor, had a "personality"—again a notion that was just gaining traction in marketing circles during the late 1940s and 1950s—and the "effectiveness" of this personality could be measured. Landor's consumer response design thus promised to create brand personalities that would "strike a responsive chord in the consumer."[84]

Beyond the brands they offered, companies themselves needed to tend to their corporate face and personality if they wanted to attract consumers. In 1947, Landor offered tips to California companies concerning the "personality of the firm" and its "various identifying symbols," stressing the importance of trademarks, tags, letterheads, and packaging for identification, among other things. "Complete coordination of design from letter head, business card, into showroom design," Landor advised even smaller firms in industries such as fashion, "is necessary to reflect a fashion sense, making your message to the retailer look as if it came from a well-established well-run business." Consumers, of course, also would be swayed by the careful integration of products into a corporate identity, and Landor promised to deliver the "sales stimulus of an effectively designed setting."[85] By the end of the 1950s, Landor boasted that he had performed "successful design surgery on products, packages and corporate identities" for numerous companies including Safeway, Standard Oil, and U.S. Steel.[86]

Loewy and Landor were neither the first nor the only designers utilizing brand and corporate design approaches, but their firms were early movers in this expanding field. As case studies, the histories of their design businesses attest to the degree to which consumer psychology and market research systematically shaped product development and marketing during the postwar period. Products, brands, and companies were endowed with "personalities" that were geared to specific consumer preferences and market segments. Design had become an integral aspect of consumer marketing, and they succeeded in persuading leading American corporations to integrate industrial design consultation into their overall public relations and marketing approach. Loewy's firm could claim during the 1950s that for numerous companies they had designed "practically everything" that pertained to these companies' operations. The designer as "consumer engineer" saw himself as intimately involved in all aspects of marketing, advertising, and PR.

Industrial design had moved a long way from simple "styling" by the 1960s. Research oriented and increasingly comprehensive in its outlook, the profession had begun to consciously create "worlds of goods," from product lines with shared brand identities and standardized symbolic languages of corporate identities, to retail displays and showrooms that placed individual goods within a larger context. Such product ensembles were designed to fit consumers and their needs and desires, but they were also created to "engineer consumption" by stimulating purchases and by "streamlining" consumer behavior within carefully crafted environments. Even if they designed for sustained demand and introduced a dynamic of obsolescence through continuous design innovations and adaptations, midcentury design professionals such as Loewy and Landor did not see themselves as "waste makers." In an autobiographical sketch, Landor saw his biggest contribution to design as improving consumer environments for commercial and aesthetic purposes:

> I felt strongly all my life that products, packages, corporate images are part of our daily environment, and to the extent that I can I wanted to . . . improve the looks of things. It's not only a matter of pushing, marketing, making people sell and buy more, it's a matter of recognizing that the visual sensibility and sensitivity of people should be trained upward.[87]

Since the 1930s, Landor felt, such an improvement had been noticeable in the United States and elsewhere, and he saw his firm as contributing its share. The notion of "improving" aesthetic environments connects Landor to the transnational design reform movement coming out of interwar Europe. At the same time, Landor, Loewy, and other design consultants contributed to the emergence of a postwar world of goods considered uniquely "American" and with tremendous international appeal. Thus, the postwar translation of American consumer capitalism with its carefully designed aesthetics and its elaborate consumer research methodologies to the emigrés' countries of origin will be at the center of the final section of this study.

SECTION THREE

Transatlantic Return Voyages

Bridging Transatlantic Divides: Bringing Consumer Modernity "Back" to Europe

Postwar Europeans were fascinated with America's seemingly "irresistible empire" of goods,[1] but they frequently rejected American consumer capitalism as materialistic and foreign. European elites prided themselves on their culture and refinement, emphasizing the ethics of production and quality over sales appeal and the latest style coming from across the Atlantic. They attacked the crass commercialism of consumer engineering, the inherent inequalities of market segmentation, and the wastefulness of recurring style changes, which ran counter to the rationalism of interwar social engineering. As Western Europe entered a prolonged era of prosperity during the 1950s and 1960s, however, marketing practices were transformed there as well, often in close exchange with developments in the United States. Along with innovations in advertising and retailing, both consumer research and commercial design witnessed professionalization processes in postwar Europe. In the context of the Cold War, furthermore, consumer affluence became increasingly politicized as a marker of Western market economies. Consumer engineers now saw themselves on the front lines of an economic competition between East and West in which modern design and psychological appeals seemingly attained a new political meaning.

Many of the immigrants and emigrés in market research and design now returned to Europe or reconnected with their countries of origin. They participated in economic and cultural transfers that now crossed the Atlantic primarily from west to east, again playing the role of cultural translators in the professional spread and adaptation of marketing practices. Closely analyzing these transfers helps us to qualify persistent notions of "Americanization," however. Instead, we find processes of reconnection and adaptations to existing traditions of design and research across the caesura of World War II. The careers of the emigrés belie the perception of a backward Europe in need of "modernization" from the United States. They suggest a more mutual and multilateral process of "Westernization" and cultural globalization during the midcentury decades. The emigrés' prominent role, finally, helped European elites to accept American consumer culture with its motivation research and colorful design as part of a shared Western cultural tradition.

Thus, the final section of this book underlines the reciprocal nature of midcentury transatlantic exchanges in consumer marketing. Chapter 8 uses the furniture firm Knoll Associates as an example of the vast transatlantic networks in modern design that helped to shape elite consumption in both the United States and Western Europe. The aesthetic language of international modernism, which became a signature characteristic of postwar corporate America and its colorful world of goods, represented the most tangible legacy of the transatlantic transfers in midcentury design marketing. The example of Knoll introduces the significant role of European emigrés in presenting American mass consumption to postwar Europe. As they "returned" in different ways, chapter 9 demonstrates more systematically, they again performed the roles of knowledge brokers, cultural translators, and "transatlantic intermediaries." As returning scholars and artists, as guest lecturers and visiting professionals, as businessmen, or simply as correspondence partners, the former emigrés informed the postwar professionalization of market research and commercial design in Europe. Their presence made the adaptation of elements of American consumer capitalism more palatable to their European colleagues.

This acceptance had its limits, however, and notions of consumer engineering were met with growing criticism on both sides of the Atlantic as we move into the 1960s. As the conclusion notes, the transatlantic critique of consumer capitalism drew heavily on European emigrés as well, including most prominently thinkers of the so-called Frankfurt School such as Theodor Adorno. Intimately familiar with the work of their fellow emigré consumer engineers, these critics focused specifically on the social psychological implications and the physiological shortcomings of a world engineered by consumer experts. Their critiques dovetailed with broader grass roots revolts against technocratic social engineering in general as well as with changes in marketing management and mass production regimes. Together with a generational change, they brought the midcentury era of the consumer engineers to a close.

EIGHT

Corporate America and the International Style: The Transnational Network of Knoll Associates between Europe and the United States

"A European style emigrated, conquered America and then returned to a burned-out Europe, to succeed most rapidly in the very country where once . . . it had generated so much enmity as to be condemned and exiled." Thus begins a 1961 article in the architectural journal *Deutsche Bauzeitung* devoted to the ten-year anniversary of the design furniture firm Knoll International in the German market. "The return to Germany of the ideas and forms of the 'neue Sachlichkeit' (neo-Utilitarianism) of the twenties and thirties came about after the Second World War," the piece in the West German architecture journal continues. "Since the beginning of the forties, Knoll International in America has continued to a great degree to advance the work of European avant-garde architects in liberating design from the false and the showy."[1] Around 1960, the firm was at the pinnacle of its influence, defining "modern" furniture design for corporations and high-end consumers on both sides of the Atlantic. Knoll was steeped both in traditions of interwar European modernism and in American consumer engineering as it sold "modern" consumer design to postwar Europeans.

Knoll's success in midcentury America rested on the commercial adaptation of a modernist design style developed to a large degree by European emigrés and by the transnational network of modernist designers that we have traced in previous chapters. Knoll designers and Knoll customers consisted of a small elite of "tastemakers." When Knoll opened its first retail outlet to "average consumers" (rather than decorators and architects) in 1971, the *New York Times* reported with excitement that the "top producer of prestigious modern" was now accessible to a broader customer base. Knoll, the paper noted, had brought to the American market not only "Bauhaus standbys" and modern classics from Marcel Breuer and Mies van der Rohe, but also work by Scandinavian and Italian-born designers.[2] While its furniture was too expensive for most consumers, the firm's designs still profoundly informed modern commercial design in the United States and beyond, especially as the "Knoll look" came to define the public image of numerous American companies. Knoll did more than simply sell furniture, but rather it designed the spaces that surrounded the furniture to become part of a modern corporate identity.[3]

Knoll's furniture signaled an association with an international world of design and with the design stars that were part of it. Especially with regard to fashion and furniture, postwar Americans still looked to designers and designs that were associated with "Europe."[4] The company thus provides an ideal case study for the reciprocal transnational design transfers in midcentury consumer marketing between the United States and Europe. Knoll's approach was different from the mass market design work of industrial designers such as Raymond Loewy and Walter Landor with regard to the consumer segments they targeted. Market research and consumer surveys do not feature prominently in Knoll's designs, where the artist's vision rather than anticipated consumer reactions still set the tone. Yet with corporate and institutional clients making up a sizeable share of its sales, Knoll, too, incorporated professional research and a thorough understanding of client needs into its design jobs, as the emergence of its "planning unit" headed by Florence Knoll shows. Her designs drew on the same comprehensive vision of design that placed individual goods (or pieces of furniture) within larger ensembles. The "Knoll look" thus represents another version of the midcentury world of goods represented in showrooms, magazine spreads of lavish residences, or corporate boardrooms. As companies expanded their corporate image programs, they looked to architects to redesign their office buildings.

Company building programs, art historian Alexandra Lange has suggested, were part of broader "corporate images" strategies that typically included logos, product design, and advertising. Taken together, they represented "commercial applications of the sort of total design taught at the Bauhaus and by its emigres."[5] The "international style" of Mies van der Rohe's 1957 Seagram building in New York or Eero Saarinen's technical center for General Motors was complemented by a modernist interior design style that Knoll and its competitors developed at midcentury.

The history of Knoll Associates not only highlights the transnational character of midcentury commercial design but provides a powerful example of immigrant entrepreneurship as well.[6] The firm, built by German American Hans Knoll and his second wife, designer Florence Knoll, illuminates the close network of European born and trained architects, designers, and illustrators that made up the transatlantic design elite of the interwar and postwar decades. Knoll gave their work a prominent commercial outlet and furthered their influence on a small but affluent segment of the American consumer market. As the above article from the *Deutsche Bauzeitung* indicates, however, Knoll's case also shines a light on the return of commercial modernism to Europe after World War II. Here, Knoll acted as an American company, promoting "American" or "international" design forms with obvious "European" roots.

Marketing "European" Modernism: Knoll Associates in the United States

From its inception, Knoll Associates thrived on a business model that relied on transatlantic networks and on bringing interwar modernism into the commercial furniture business. Drawing on its founder's immigrant background, the company helped pioneer a design-driven marketing approach to the U.S. furniture industry, which was undergoing dramatic changes during the interwar years. Hans Knoll came from a family that had already been in the furniture business for several generations back in Germany. In 1865, his grandfather Wilhelm Knoll had opened an outlet for leather-upholstered furniture in Stuttgart and soon began manufacturing as well. In 1890, the company became an official supplier to the Württemberg royal court. The family company—which by that time had expanded to include a second manufacturing site near Stuttgart and a factory in Vienna and also had a licensing agreement with a British manufacturer—was taken over by Wilhelm's sons Wilhelm and Walter in 1907. Hans's father, Walter Knoll, remained involved in the family venture for a while, but by 1925 he had founded his own company, Walter Knoll & Co. This new firm produced decidedly modern furniture that drew on design innovations by Werkbund and Bauhaus artists of the period. These designers strove to create quality designs for mass production, emphasized functional forms, and experimented with new materials and production forms, such as bent wood and tubular steel designs. Accordingly, Walter Knoll & Co.'s successful "Prodomo" system, introduced in 1929, featured modernist furniture based on steel tubes.[7]

The midcentury rise of modernist design in furniture thus had already begun during the interwar period for the Knoll family. Hans Knoll himself was born in 1914 in Stuttgart and learned the furniture trade as an apprentice in the family business. Already as a youth, he spent time abroad in Switzerland and in Great Britain as part of his practical education. While his brother Robert was in the United States during the early 1930s, Hans worked with Jantzen Knitting Mills in Brentford, England (from 1933 to 1935), as well as for Plan Ltd. (1935 to 1938), companies with business ties to his father's firm.[8] Plan Ltd. distributed the Prodomo furniture for the British market, and the firm provided Hans Knoll with his first exposure to a company that exclusively focused on marketing modern designer furniture. The firm specialized in metal-tube designs and was headed by Serge Chermayeff, the Russian emigré who later became the second director of the Institute of Design (or American Bauhaus) in Chicago.[9] When Plan Ltd. ran into financial trouble in 1938, Hans Knoll briefly returned to Germany, only to set out for the United States soon afterward. According to the recollections of a later business partner, the relationship between Hans Knoll and his father, Walter, was strained. The father's overbearing character and his enthusiastic support for the National Socialist regime had contributed to Hans's initial move to England in 1936 and his ultimate decision to form his own business in the United States.[10] However, Knoll appears to have been contemplating the move from Britain across the Atlantic for some time, following the path of many European modernist designers including Bauhaus emigré László Moholy-Nagy. While setting up his Chicago school in 1937, Moholy-Nagy received a letter from his wife, Sibyl, who was then still in England. She told her husband of Hans Knoll, a young man dealing in patent furniture, who planned to visit Moholy in the United States and who was bringing "quite new and interesting ideas" with him.[11]

For Hans Knoll, more than for many of the other emigrés previously discussed, the transatlantic transfer of design ideas was part of a conscious business plan. Knoll's goal was to bring "modern furniture" to the United States at a time when the strict and functionalist forms of the Bauhaus designers were beginning to gain traction among Americans but were nearly entirely absent from the American furniture industry. "It was my whole idea," he later claimed, "to develop new products working with well-known designers and to encourage their particular talents."[12] Upon arrival in New York City in 1939 he initially intended to distribute his father's company's designs (especially the Vostra line of furniture), but then he incorporated an independent firm, Hans G. Knoll Furniture Inc., with a small office at 444 Madison Avenue.[13] Within two decades, Knoll Associates would leave a mark on the industry and contribute to a growing emphasis on design and designers themselves within the business.

The rise of Knoll Furniture Inc. occupies a transition period for American furniture makers. Since the late nineteenth century, the American furniture industry had been dominated by a network of corporations that produced residential and office furnishings in large quantities. Chicago, Illinois, and especially Grand Rapids, Michigan, formed the center of furniture production early in the century, producing predominantly reproductions of earlier period styles as well as arts-and-crafts style pieces.[14] The industry had long been a bastion of specialized batch production, with companies focused on frequent model changes and shifting consumer demand.[15] By the interwar period, however, furniture production had begun to shift to the American South, where labor costs were lower and raw materials were more easily accessible. North Carolina emerged as a serious competitor for the traditional industry, particularly in the field of residential furniture. The Great Depression posed another severe challenge to the industry, and by the 1930s Grand Rapids corporations were searching for alternative markets for their products. Many shifted entirely to commercial and office furniture and to contract work for public facilities. World War II and the postwar economic boom led to a dramatic expansion in government and corporate office space that would make nonresidential furniture a viable alternative for the coming decades.

As in other industries at the time, industrial design became professionalized and increasingly integrated into marketing strategies in the furniture industry. The Michigan-based Herman Miller Corporation, Knoll's principal competitor in the postwar era and best known today for its Aeron chair, pioneered the industry's foray into the market for high-end commercial and residential furniture.[16] The Depression had forced Herman Miller's management to search for solutions to frequent design changes and the demand of department store buyers for continuous nov-

8.1 Hans Knoll (1914–1955) and Florence Knoll (1917–2019), 1946. Courtesy of Knoll, Inc.

elty. From 1930 onward the company was advised by industrial designer Gilbert Rhode, who emphasized the growing importance of design (and especially of modern design) for the company and for the industry in general. Rhode applied the expanding notion of professionalized design to the furniture industry: designing furniture meant systematic planning and designing entire spaces for consumers. Blunting the power of retailers and their buyers, Herman Miller opened a showroom in New York in 1941 and began to market their furniture directly to consumers and interior decorators through a catalogue. By the 1940s, designers such as George Nelson and Charles Eames were helping to make Herman Miller an explicitly design-driven corporation and playing a central role in the firm's marketing and advertising.[17] This business model set Herman Miller apart in a still largely traditional industry, and its approach made it a direct rival to Knoll with his own design-driven vision.

CHAPTER EIGHT

 Hans Knoll came onto the American scene as an outsider, yet he could draw on his training in the family firm back in Germany and on his already substantial international experience. His father's company had had connections to modernist architects and designers, supplying furniture, for example, to the Werkbund's Weissenhof model development near Stuttgart, built in 1927. Weissenhof, built by numerous avant-garde architects as part of an international exhibition, foreshadowed later International Style architecture and was conceived as a model for modern workers' housing.[18] Knoll's association with Plan Ltd. in London had similarly attuned him to the possibilities and challenges involved in the serial production and mass marketing of modernist furniture that had been part of the Werkbund's interest as well. Knoll's business benefited from a growing interest in the United States in functionalist European design, and the growth of Knoll Associates paralleled the rise in influence of Bauhaus emigrés in American arts, academia, and professional schools.[19] Although business records for the early years are lacking, Hans Knoll presumably brought some capital and business connections with him to the United States. Thus, he arrived by no means empty-handed, and his business plan rested on a relatively solid foundation.

 Still, the company struggled in its early years. Initially, Hans G. Knoll Inc. lacked designers who could create furniture pieces. An early associate was Danish-born architect Jens Risom, whom Knoll knew from London. Risom designed the first line of furniture for the company and teamed up with Knoll to promote the firm on a nationwide tour in 1941.[20] Knoll made use of German American acquaintances and visited people with links to his father's company on the West Coast and elsewhere. On the production side, too, immigrant ties came into play. Knoll Associates moved its production facilities in 1945 from New York City to East Greenville, Pennsylvania, an area with a strong German American community where it is still headquartered today. In addition to inherited German business connections and Hans Knoll's reputed business savvy, it was government contracts during World War II that kept the company afloat in the beginning. Major contracts included offices for the United States Information Service (USIS), for Secretary of War Henry Stimson, and for the United Service Organization (USO) in Times Square.[21] The company's eventual impact on transatlantic design and marketing history, however, was largely due to two factors more closely connected to Hans's wife, Florence Knoll: (1) its vast network of modernist designers and (2) its new and systematic approach to interior design.

Transatlantic Design Modernity: The Use of Emigré Networks

Transatlantic networks with prominent roles played by immigrants and emigrés were particularly important in the design development of the company. Knoll's second wife, Florence, herself the daughter of Swiss immigrants to the United States, played a crucial role in this regard. The need for a trained interior designer, a position often filled by female architects at the time, brought Knoll in contact with his future wife, Florence Schust.[22] By 1943, Florence had begun to work regularly with the company, and over the course of the 1940s, the firm's emphasis shifted increasingly from residential to commercial work as well as from the Scandinavian wood designs of Risom to more radically modernist forms. Shortly after his arrival in the United States, Knoll had married Barbara Southwick, a student at the National Academy of Design in New York City, with whom he had two children, Peter and Maia.[23] This first marriage was short-lived, however, and Knoll married Florence, who would play a significant role in the success of Knoll Associates. In 1946, Florence Schust (now Knoll) became a partner in the firm, which changed its name to Knoll Associates.

Celebrated as the "First Lady of the modern office," Florence Knoll stands out as one of the few prominent women among the midcentury consumer engineers discussed in this book. The emerging design profession of that era was still widely dominated by men, with notable clusters of prominent women only among the designers of fashion, textiles, and home interiors.[24] Yet it was Florence Knoll rather than Hans who was well connected within leading design circles in the United States.[25] Born in 1917 in Saginaw, Michigan, Florence Knoll became one of the most renowned interior designers of the twentieth century. She was trained as an architect at the Cranbrook Academy of Art outside of Detroit, under the tutelage of Finnish-born architect Eliel Saarinen, who was president of Cranbrook during the 1930s. Not unlike the German Bauhaus in some respects, Cranbrook was an artist colony and school that by that time had become one of the centers of modernism in design and architecture in the United States and a nodal point for transatlantic exchanges. The school helped train a generation of leading designers including Eero Saarinen, Charles and Ray Eames, Harry Bertoia, and Ralph Rapson, a group of which Knoll was an integral part.[26]

Florence Knoll's design education was heavily embedded in the networks and institutions of transatlantic modernism. Following her years at Cranbrook, she studied at the Architectural Association in London between 1938 and 1939 and subsequently began to apprentice with the architectural offices of Bauhaus emigrés Marcel Breuer and Walter Gropius in Boston. In 1941, she finished her architectural education at the Illinois Institute of Technology in Chicago under the supervision of Bauhaus emigré Ludwig Mies van der Rohe. Before coming to Knoll, she had also briefly worked for architect Wallace Harrison in New York. Florence Knoll's connections to Cranbrook-trained designers as well as to the European modernists who had fled to the United States from the National Socialists would prove to be a crucial asset to the young company that Hans Knoll had founded.

Following the war, the company expanded its lines of designer furniture, making use of Florence Knoll's numerous connections in the design world. Over the years, Knoll Associates built up a network of modernist designers they could draw on, including a substantial number of immigrants and emigrés. The group included several Italian-born designers such as Harry Bertoia, who designed some of the most successful pieces of the Knoll collection, including his lattice steel-wire Diamond chairs. Florence Knoll had known Bertoia from his work at the Cranbrook metal workshop.[27] Milan-trained Franco Albini contributed glass-topped desks and wire and glass shelving systems. Pierre Jeanneret, the Swiss-born cousin and erstwhile collaborator of Le Corbusier, had his "scissor" chair sold through Knoll. The Bauhaus tradition also featured prominently in the Knoll repertoire, including designs from Hans Bellman, a student of Mies van der Rohe's at the Bauhaus who later worked at Zurich, Basel, and Harvard.[28]

Knoll reproduced numerous earlier Bauhaus designs from the interwar period under licensing agreements, including most notably Marcel Breuer's "Wassily" chair made of bent metal tubes and Mies van der Rohe's "Barcelona" chair, both of which would come to be considered as modern classics.[29] Danish architect Abel Sorenson also counted among the early designers with Knoll along with many former Cranbrook students and collaborators. Ralph Rapson, for example, who had studied at Cranbrook and then worked with Moholy-Nagy in Chicago on interiors for the B&O Railroad, designed metal outdoor furniture for Knoll. Isamu Noguchi and George Nakashima designed for Knoll, and Eero Saarinen contributed several chair designs to their collection including the "Womb chair" in 1946 and his iconic molded "tulip" chairs in 1956.[30] Besides a group of designers from the Pratt Institute in New York, the network also included artists from countries such as Argentina and Japan, making it a truly global cast. Still, in their training and overall design outlook, most of the prominent Knoll contributors were influenced by the type of functionalist design modernism that had been especially prevalent in interwar Europe.

Knoll's challenge was to assemble these individual designs into a coherent "look" calculated to appeal to upscale clients and consumers. Showrooms allowed the company to assemble the various pieces of modernist furniture on offer into carefully crafted ensembles of goods. Expanding far beyond its original New York office, the firm opened showrooms for its furniture in several American cities. The showrooms helped in "normalizing the strange shapes and European imports" for customers across the United States, as did prominent magazine spreads about Knoll's home and social life, historian Andrea Lange has argued.[31] Indeed, by 1950 the company had established additional showrooms in Chicago and Atlanta, and further locations in Boston, Dallas, San Francisco, and elsewhere soon followed.[32] The showrooms provided Knoll with an opportunity not only to showcase the interior design strengths of the company but also to market directly to clients, corporate customers, and interior decorators.[33] As noted above, this form of direct marketing along with catalogue sales had been introduced by Herman Miller already in 1941, and Knoll's first catalogue quickly followed in 1942.

When it came to creating a modernist "Knoll look," connections to European emigrés were central to the firm's business strategy in other ways as well. For its graphic design work, for example, Knoll Associates turned to Swiss-born Herbert Matter, a personal friend of Hans Knoll's. Matter had studied art in Paris, where he worked with A. M. Cassandre and Le Corbusier in creating photography and poster art for advertising and marketing. Like so many emigrés in the design world, he then came to the United States for the 1939 World's Fair in New York to help prepare the Swiss pavilion and the Corning Glass pavilion. Matter, who later became a professor of photography at Yale, designed the company's iconic "K" trademark logo in the late 1940s. He was also involved in several prominent advertising campaigns for the firm. Matter's 1955 ads featuring Saarinen's womb chair and other acclaimed pieces represented Knoll's first concerted national advertising effort, and the campaign made a splash among Madison Avenue advertisers because of its unusual modernist motifs, its humorous approach, and its sparse use of type copy.[34]

Textile design was another central element of the "Knoll look." When Florence Knoll decided to design and produce textiles in-house through a new subsidiary, Knoll Textiles, she also turned to emigré connections (particularly women) such as textile designer Anni Albers, who had been affiliated with the Bauhaus prior to her emigration to the United States and taught at Black Mountain College after the war. Cranbrook's Loja Saarinen, who had a Finnish background, also created influential textile designs for Knoll. In the textile field in particular, European connections remained vital, and the company retained suppliers and artists in Scandinavia, Italy, France, and West Germany among other places.[35] Hungarian-born immigrant designer Eszter Haraszty, finally, was the company's color specialist and headed the textile division from 1949 to 1955. Haraszty, who was trained in Budapest and also worked as a consultant to emigré architect Victor Gruen, was partially responsible for the vivid modernist color schemes that characterized Knoll's signature look.

The networks Hans and Florence Knoll had been able to build within the design community provided a comparative advantage to the young company. While it was Florence more than Hans who had the direct links to the design world, his German background came into play as well. Pennsylvania was chosen as a site for their first large production facility, Florence Knoll later recalled, in part because the area's high number of German American immigrants with traditional ties to furniture and cabinet making provided an ideal workforce for the company.[36] According to furniture designer Richard Schulz, many of the workers even spoke Pennsylvania Dutch, a dialect not unrelated to that of Hans Knoll's Southwest German origins.[37] More than simply drawing on such connections, however, Knoll Associates succeeded by assembling high profile designs from their network into a coherent "look" that appealed to upscale American consumers as a well as to a growing corporate clientele in the postwar years. The company effectively marketed interwar "European" functionalism as a new transatlantic "international style" for the modern American interior.

The Planning Unit: Marketing Interior Design as Corporate PR

The "Knoll look" and its proliferation throughout the offices of corporate America were the result of professional design marketing and planning. Already during the war years, Knoll created the so-called "planning unit," a designated group of employees headed by Florence that played a central role in the company's success. While Hans remained in charge of the business end of the firm, she increasingly shaped the design vision of the company and facilitated a shift toward creating entire environments and product ensembles. The planning unit was founded in 1943 to prepare Knoll's New York showroom, and initially it had only a small staff and few active projects. Still, the planning unit was an innovation in that the company began to work very closely with clients to create interior designs that met their needs; they designed office spaces rather than simply the office furniture to fill them. The unit's expansion and growing importance in the firm corresponded closely to the postwar corporate building boom and to the inclusion of built environments in corporate public relations strategies.

The "planning unit" brought a new degree of professional research into the interior design process, and Florence Knoll insisted that she was much more than merely an interior decorator. Instead, as art historian Bobbye Tigerman has argued, Knoll understood interior design as a "systematic and efficient process" closely modeled on the work of architects.[38] In preparing designs for corporate clients, Knoll and her team would conduct extensive interviews with company executives and staff to determine needs and anticipate design solutions.[39] When Knoll was awarded the 1961 annual "gold medal" by the American Institute of Architects (a recognition she shared that year with Bauhaus emigré Anni Albers), the press release praised her planning unit as an "experimental laboratory for new furniture."[40] A *New York Times* commentator later observed that Knoll's planning unit helped pioneer the use of design studies "to determine how work procedures affect office furniture and lay-out." Knoll used an "architectural approach in designing interiors" that led to the integration of lighting, ducts, and outlets as well as of furniture, colors, and fabrics into what ultimately became a "total design." Her Cranbrook training and the Bauhaus influences likely contributed to an encompassing design approach and to her attempts at creating offices as "total-design showrooms."[41]

While the work of the planning unit was individually suited to each client, a trademark "Knoll look" emerged that entailed a strict functional modernism and bright, vivid color schemes, as well as an almost residential furniture arrangement for offices.[42] During the mid-1950s, Knoll furnished the suburban office headquarters of the Connecticut General Life Insurance Company outside of Hartford. The company's 1956 annual report proudly described the horizontal buildings that "our studies showed best accommodated the flow of work through our offices." Knoll Associates along with architectural firms had worked closely with the company for five years, beginning with "studies of our business processes in relation to design and construction."[43] The result was a custom-designed interior from the office plans and furniture to the fabrics and colors. Knoll's color palette typically included "hot violent shades" of pink, violet, fuchsia, yellow, and peacock blue, which made their room

CHAPTER EIGHT

8.2   The "Knoll Look": Office of CBS president Frank Stanton as designed by the Knoll "Planning Unit," 1954. Archives of American Art.

designs instantly recognizable and provided an image of vibrant modernity. Combining a consistent color scheme with iconic furniture designs, the "Knoll look" has been likened by art historians to the Bauhaus notion of a "Gesamtkunstwerk."[44]

During the 1950s, this "Knoll look" became a staple in corporate offices across the United States. It promised to meet the needs of modern office organization while at the same time adding to corporate public relations efforts by signifying aesthetic sophistication. Notable clients during the late 1940s and 1950s included Alcoa, the Carnegie Endowment for International Peace, the Dow Chemical Company, the Federal Reserve Bank in Detroit, the Heinz Corporation, the Center for Advanced Study at Palo Alto, and the Universities of Rochester and Michigan. An influential early client was Nelson Rockefeller, a prominent businessman and philanthropist art collector whose offices Knoll redesigned in 1945 as Rockefeller returned to private life after serving as assistant secretary of state.[45] For his office, Florence Knoll first surveyed office staff with regard to work routines and space requirements—the kind of workflow analysis that would come to define planning unit work. Of particular importance for Knoll Associates' reputation and business development was the interior design work done for the Columbia Broadcasting System (CBS) during the early 1950s. In 1954, Knoll redesigned the office

234

of CBS president Frank Stanton.[46] This office much like Knoll's work in general received extensive coverage both in the national press and in leading trade journals such as *Interiors* and *Architectural Forum*, thereby amplifying the cultural impact of the "Knoll look."

At the height of their American success during the mid-1950s, the company underwent a dramatic organizational transformation. Hans Knoll died in 1955 as the result of a car accident on the island of Cuba, where the company not only had a showroom but also redecorated the U.S. embassy.[47] A young woman, Hans Knoll's girlfriend, also died in the crash—while Florence and Hans were still running the company jointly, they had begun to go their separate ways privately and were preparing to divorce.[48] The executive management of the Knoll Group, which included Knoll Associates, Knoll International, and Knoll Textiles, now passed on to Florence, who remained its guiding force until her retirement in 1965. In 1958 she married the Florida banker Harry Hood Bassett, and in 1959 she sold Knoll Enterprises to Art Metal Construction Company for an undisclosed sum, while retaining the presidency of all three Knoll companies. Art Metal, a New York–based office furniture company with the General Services Administration (GSA) as its major client, was hoping for creative input through this acquisition. Reflecting her pioneering design approach with the "planning unit," Knoll joined the board of the new parent company as "design and research director."[49] After her retirement in 1965, the company continued producing designer office furniture, changing ownership several times over the subsequent decades.[50]

Knoll Associates never rivaled the large furniture manufacturers in Grand Rapids, North Carolina, and elsewhere, and most American consumers would never have considered—or been able to afford—Knoll pieces. Early on the firm briefly considered broadening its customer base, decreasing prices by some 20 percent as volume increased during the second half of the 1940s, selling chairs for as low as $20 and loveseats for $50 (roughly $250 and $630 in 2019 value).[51] Generally, however, the company did not compete on price and remained on the higher end of the furniture market. The American mass market of the 1950s, described as the era of "populuxe" by historian Thomas Hine, was also less open to the company's modernist style than the corporations and an avant-garde elite of consumers who ranked among Knoll's clients.[52] Still, Knoll left its mark on the American furniture industry, grossing about $3 million annually in the early 1950s (or $28.42 million in 2019 value), and its products were sold in over 250 stores in the United States alone.[53] The Knoll look and designs were frequently copied

CHAPTER EIGHT

and imitated. The company's business strategy of emphasizing strictly functional modern forms and materials and assigning a prominent role to design and designers gained wider acceptance within the industry. The interior design work of the planning unit similarly set standards for corporate remodeling in subsequent decades.[54]

The company's most enduring legacy, however, lies less with its commercial impact than with its influence on the world of mid-twentieth-century design art. Like its two founders, Hans and Florence Knoll, Knoll Associates as a firm straddled the line between art and commerce. The Museum of Modern Art (MoMA) in New York, which had played a singularly important role in popularizing Bauhaus modernism in the United States, features centrally in Knoll's success as well. MoMA exhibitions and competitions for modern and "good" design were part of a larger effort by museum officials Alfred Barr and Edgar Kaufmann Jr. to popularize modern design, frequently promoting European-born artists. During the 1950s, designs marketed by the company and Florence Knoll's work were repeatedly part of these "Good Design" exhibitions. Today, numerous Knoll pieces are part of the permanent design collection of the museum.[55] The connection to the art world also played out in the company's design work for several American museums, including the Virginia Museum of Fine Arts in Richmond and the Dallas Museum of Fine Arts in 1953.[56]

Both socially and as entrepreneurs, the Knolls played a role in bringing together American and European modernists from the Cranbrook and Bauhaus circles with businessmen and corporations on both sides of the Atlantic.[57] By the early 1950s Hans and Florence Knoll lived in a spacious Manhattan apartment overlooking the East River, and Hans had become a member of New York's exclusive River Club, which had been founded in 1930 by Kermit Roosevelt and counted such leading families as the Vanderbilts among its members.[58] He also belonged to the Arts Club of Chicago, which had a long history of supporting modern art in the United States dating back to the interwar period. For the Knolls, personal connections to business clients such as Frank Stanton and Nelson Rockefeller mixed with those with designers such as Harry Bertoia, the Eameses, and the Saarinens.[59] They were part of a larger symbiosis between modern art and architecture and corporate America that was characteristic of the postwar boom years and that was fostered by a host of institutions, including museums, universities, and foundations.[60] As especially large multinational corporations on both sides of the Atlantic increasingly adopted modern designs for their products and other elements of their corporate identity formation, Herman Miller and, prominently, Knoll Associates provided the interior design and office furniture to go with them.[61] As the "Knoll look" became characteristic of corporate modernity in the United States, it not only helped place interwar "European" design in company boardrooms but became part of what some refer to as the postwar "Americanization" of Western Europe.

Return to Europe: Exporting "American" Design as "International" Style

Soon after the war, Knoll Associates "returned" to Europe, opening branch offices in several European countries. In some ways, this amounted to more than simply an internationalization strategy of a successful furniture firm. In bringing (what came to be known as) the modernist "international style" to postwar European consumers as a form of modern "American" design, Knoll helped pave the way for a postwar "Westernization" of design styles and a shared transatlantic modernity in the postwar years. In a wider sense, the company became part of Cold War efforts on the part of the United States to utilize "good design" as a tool of cultural diplomacy, representing American goods abroad.[62]

Knoll International was first set up as a subsidiary in 1948 with a showroom in Paris to facilitate the internationalization of Knoll's business. Soon after the war was over, both Knolls had traveled back to Europe frequently, and Florence in particular enjoyed spending time in France for professional inspiration and recreation.[63] An office in Stuttgart followed in 1950, where ties to the firm of Hans Knoll's father were rebuilt. Both the Walter and the Wilhelm Knoll companies had shifted much of their production to war needs after 1939, and both had been badly damaged by Allied bombing campaigns. Jens Risom began to design furniture for Walter Knoll while that company was producing furniture for Knoll International. During the mid-1950s, however, soon after Hans's death, the cooperation between Walter Knoll and Knoll International came to an end.[64] Still, this connection back to Germany helped Knoll Associates to expand internationally in the immediate postwar years. In Italy, Knoll International also used local suppliers, such as well-established design furniture firms Cassina and Fantoni. A manager from competitor Herman Miller later observed that their European networks allowed Knoll to "have the jump on us" in Europe with respect to suppliers and markets.[65]

Knoll's marketing strategy in Europe matched the American model: the company built up a reputation particularly among architects and interior designers by maintaining high design standards and producing small batches of furniture of high quality. The individual branches and subsidiaries of Knoll International were managed fairly independently, while Florence Knoll at the New York headquarters focused on the relationship with designers—new designs and designers (which frequently came from new European markets) had to be personally approved. New York defined the overall design palette, while regional managers could decide which particular pieces would be marketed regionally. This provided flexibility to adapt to differences in the European markets, where, for example, less volume restricted possibilities for diversity in textile patterns. Knoll International introduced wardrobes—never part of the company's American repertoire—in the European market, relying on European designers to suit the tastes of European households and to accommodate typical European building plans without built-in closet space.[66] Over the course of the 1950s and early 1960s, they opened showrooms in several European cities such as Brussels, Milan, Stockholm, and Zurich, as well as many more offices globally in thirty countries, including Canada and Cuba. Knoll became the face of modern American furniture abroad—and the company projected the signature "Knoll look" globally while allowing for localized product modifications.

The openings of new European showrooms were typically staged as public events with local design elites. In Germany, Knoll regional manager Toby Rodes marketed Knoll as cutting-edge international furniture design with modern forms that were integral rather than foreign to European design tradition. He converted the company's showrooms into true sales areas with all displayed items on sale (including vases, ashtrays, and other accessories designed by German designers including Wilhelm Wagenfeld). Rodes used the showrooms for lectures and PR events with prominent European architects such as Egon Eiermann and Gio Ponti as part of a broader effort to promote explicitly modern design in home and office furnishings (which initially had a market share of only around 1 percent during the 1950s). Rodes also represented Knoll in the "Verbund-Kreis," an informal German trade group that included managers of companies interested in promoting modern design such as Braun electronics, Pfaff sewing machines, Rosenthal porcelain, WMF silverware, Bremer Tauwerke (producer of Sisal carpets), and Rasch wallpaper. The group conceptualized a traveling collection of furnishings for the "modern apartment" to persuade Germans (and entire classes of high school students) in various German cities of the advantages of modern design in the era of West Germany's "economic miracle" (*Wirtschafts-*

wunder). For the 1957 International Building Exhibition in Berlin, Knoll furnished an exhibition apartment planned by Bauhaus emigré Walter Gropius. Here, just as in their showrooms which were decorated in a minimalist fashion with monochromatic ingrain wallpaper, Rodes and Knoll pushed for strictly modernist design in postwar Germany.[67]

Knoll International very consciously emphasized the exclusivity of its designs (as if marketing a Rolls Royce, a *Spiegel* report observed) and its uncompromising, comprehensive modern vision. About 60 percent of Knoll's business in Germany involved companies and state agencies, with only a few private customers, but Knoll aimed at the "taste leaders." "It is not our goal to penetrate every apartment through mass production," the Stuttgart office chief stated, even if that meant turning down business from wealthy industrialists of the "economic miracle" who desired a Knoll chair or couch for their living rooms. Knoll International aspired to design and furnish representative office spaces, and Toby Rodes, who had become European director, was confident that "the right people will come to us."[68] The broader mass market did not remain unaffected by Knoll designs, however, as numerous German furniture companies including the Christian Holzäpfel KG and Tecta Works launched more affordable, Knoll-inspired models. According to contemporary German observers, Knoll served as a kind of experimental laboratory for the furniture industry in general. When the German Design Council honored Knoll in 1960 for its design work, it noted the influence that this American company had had on postwar German design developments more broadly.[69]

By 1960, Knoll International had become the interior designer for many European corporations and worked on government offices in Germany and elsewhere on the continent. Prominent customers included AEG, Hoechst, Krupp, MAN, Daimler Benz, Rosenthal, and Volkswagen.[70] Clients often strove to project an image of modernity that put them visually on a par with leading American corporations in their field. Especially in West Germany, the embrace of design modernity, as Paul Betts has observed, went well beyond the corporate world and became a shared marker of cultural and political elites.[71] Accordingly, the German ministry of the interior was a Knoll client, as were the state governments of Hessia and Hamburg. Knoll furnished the great hall of the Baden-Württemberg Landtag (state parliament) building in Stuttgart and the Hessian state library in Wiesbaden.[72]

As in the United States, the company's marketing of its "planning unit" projected an image of utmost professionalism and a near scientific planning procedure when it came to high profile European design projects. Florence Knoll surveyed the personnel and analyzed the workflow, the *Deutsche Bauzeitung* observed, coordinating "with great care her design to the given measurements of things to be accommodated—files, file-cards, typewriters, adding machines and dictaphones, telephones, steno-blocks, pencils, whiskey bottles and containers for paper clips."[73] The German trade journal was not merely impressed with Knoll's systematic approach to interior design, which it claimed would lead to a revolution in office design. More broadly, the *Deutsche Bauzeitung* commentator proposed, Knoll's designs would help to change "human relationships in the work area," which would prove beneficial for Germany as a country in need of more "democratic teamwork on all levels." For all its elitist appeal, the American "international style" of Knoll's furniture and design also retained democratic connotations to some contemporary observers especially in postwar West Germany. In this, Knoll was part of a more general discourse about "Americanization" through mass consumption in Cold War Western Europe.

Indeed, Knoll International had profited from the Cold War competition over consumer product design in several ways, including through large orders placed by American administrations in Europe. In 1951, the U.S. State Department ordered furnishings for ninety houses of American civil servants in Germany, and Knoll also provided the interiors for several of the so-called "Amerikahäuser," American cultural missions across West Germany.[74] Knoll International in Stuttgart also refurnished the U.S. consulates in Frankfurt, Düsseldorf, and Stuttgart as well as several embassies abroad, including those in Stockholm and Copenhagen. The Stuttgart office, which also housed the European "planning unit," won a large General Services Administration (GSA) bid in 1956 to furnish several U.S. embassies in Africa.[75] These state contracts not only provided important revenue for Knoll's overseas subsidiaries but also were part of a "cultural Cold War" by which the American government tried to woo postwar Europeans.[76]

The international style represented by Knoll signified not only American consumer modernity but also a form of refined taste compatible with existing European design traditions because of a shared aesthetic vocabulary. In 1972, for example, the Musée des Arts Décoratifs in Paris devoted a major retrospective to Knoll designs.[77] At the same time, designers at Braun, one of the most innovative companies in postwar German product design, named Knoll International along with the work of Charles Eames and others as significant inspirations for their work.[78] The company was thus part of a wider transatlantic exchange and with its employment of European and American designers functioned as a translator and intermediary in a postwar transnational discourse over modern consumer design.

In contrast to Landor and Loewy's studios with their mass market designs, Knoll Associates focused on the "high-end" consumer segment. Yet the furniture manufacturer, too, designed worlds of goods—their product ensembles displayed in showrooms and offices would, like few others, come to define the "look" of corporate America in the postwar decades. Hans and especially Florence contributed to the professionalization of furniture and interior design and were part of the growing array of external experts that helped shape the marketing and PR efforts of American corporations at midcentury. While they clearly belonged to a transatlantic elite of "tastemakers," they were far from aloof from the commercial world. As design brokers they helped popularize a specific form of modernist design that was frequently copied and adapted, and it became emblematic for the work of midcentury consumer engineers.

Knoll communicated a language of "modern" consumer design that heavily drew on the networks of European emigré designers discussed in previous chapters and turned their work into commercial success in the United States. European design traditions did retain a significant sway in the United States, even at the height of the postwar era of American mass consumption. In areas such as furniture design Americans continued to take cues from designers that could leverage a European pedigree. Immigrants such as Hans Knoll along with the design network the company (and especially his wife) maintained played a central role in this vibrant transatlantic exchange and helped to shape U.S. commercial culture by repackaging interwar European design innovations by Bauhaus artists and others as "American" design modernism after World War II.

At the same time, Knoll prominently promoted what contemporaries saw as an "American" international style in postwar Europe. The 1961 *Spiegel* cover story on Florence Knoll and her company presented an opportunity for the magazine to rehash the story of the Bauhaus and its functionalist style for a wide postwar readership, a story that included their persecution by the Nazi government and their subsequent emigration to the United States. Now the circle seemingly came to a close, and Mies, Breuer, and Gropius, who had influenced Knoll, returned to Europe through a company that, the *Spiegel* noted, referencing *Life* magazine, had become as iconic as Tiffany's in the United States. Such

CHAPTER EIGHT

connections to the Bauhaus tradition helped make Knoll accessible and acceptable to European elites. At the same time, Knoll designs came to exemplify the look of "American" consumer modernity that held such tremendous appeal after the war. As the following chapter will show, numerous emigré consumer engineers contributed to exactly this process of making "American" marketing and consumer capitalism accessible and relatable to postwar Europeans.

NINE

The "Return" to Europe: Emigrés as Cultural Translators and the Transformation of Postwar European Marketing

World War II brought the United States into a position of unparalleled global economic and political power. To consumer engineers, American consumer capitalism offered a unique selling point in the emerging Cold War: "The whole world admires and envies American products, American appearance, American quality. We should and I believe we will take advantage of this receptive attitude," Raymond Loewy told students at Harvard University in 1950. "Democracy," he argued, "is hard to sell. No one has yet been able to make its high spiritual values of freedom, liberty and self-respect a 'packaged' item to be sold to the rest of the world.... Democracy hasn't been merchandized. Until then, we must use substitute solutions. Foremost among them is the American product. The citizens of Lower Slobovia may not give a hoot for freedom of speech, but how they fall for a gleaming Frigidaire, a streamlined bus or a coffee percolator. So let's use this extensively."[1]

The citizens of Lower Slobovia, a fictional poverty-stricken and backward country featured in the popular comic strip *Li'l Abner*, served as a stand-in for the millions of postwar Europeans whose hearts and minds had to be

243

won over to the "American way of life." American military power, diplomatic alliances, and political reconstruction efforts on the war-torn continent were augmented by material support in the form of aid shipments, technical assistance, and, most important, the so-called "Marshall Plan." The promise of consumer affluence, historians have suggested, played a vital role in adding a kind of soft power to American political efforts. To be sure, American corporations had brought their consumer goods and marketing practices to European markets already during the early decades of the twentieth century. After World War II, however, such efforts became much more systematic, and by the late 1940s, the State Department was partnering with the industry's Advertising Council to "sell" Europe on free enterprise and consumer capitalism.[2] American foundations and government organizations sponsored exhibitions across the continent that touted the American standard of living and extolled the virtue of American-made consumer goods. Symbolizing American "modernity" and promising broad-based affluence, consumer marketing and commercial design became weapons in the Cold War conflict.[3]

Students of contemporary European history have long debated an alleged "Americanization" of European culture and economies in the decades following World War II.[4] European companies, some have argued, restructured to adopt new American marketing and management practices and became increasingly oriented toward the mass production and mass marketing of consumer goods.[5] These transformations, however, also sparked conflict and resistance. In France, for example, the growing influence of U.S. corporations and of new consumption patterns perceived as "American" was hotly contested during the 1950s and 1960s.[6] While many younger Europeans embraced new goods and consumption styles as a welcome challenge to existing social orders, social elites tended to reject "American" consumption forms as debased "mass culture."[7] Especially in West Germany and Austria, subject to fundamental political reconstruction after the war and the demise of Nazi rule, mass consumption was tied up in a "cultural Cold War," as social and political elites had to be wooed to the Western cause.[8] American influence on European business and consumption practices, however, did not result in a complete transatlantic convergence of mass consumerism as differences in social and economic structure as well as in the cultural meaning of consumption practices prevailed.[9]

Moreover, "Americanization" in marketing entailed not simply outright imports but rather a careful and selective adaptation of specific elements.[10] Such transnational corporate learning processes, negotiating differences between American and European consumption and marketing cultures, opened new career opportunities for the emigré consumer engineers. This chapter will trace their "return" to Europe after the war as well as their impact on Western European consumer marketing between the late 1940s and the early 1960s. The focus will be on West Germany, which had a special role in this transatlantic exchange. Positioned on the front lines of the East-West conflict, it was most directly affected by American cultural and reeducation policies. Many of the emigrés in design and consumer research also came from German-speaking countries, where their postwar networks were consequently most strongly developed. Still, many aspects of transfer and adaptation discussed below for West Germany applied to other European countries, as examples will illustrate.

Postwar Europe witnessed a period of sustained economic boom with rising standards of living. As European societies slowly crossed the threshold to becoming "mass consumption societies," the erstwhile emigrés served as influential networkers, knowledge brokers, and cultural translators in this transformative process. They advised the American efforts to expand the "irresistible empire," and they familiarized Europeans with "modern" American marketing methods. Much as they had done earlier in the United States, the emigrés brought a transnational dimension to professionalization processes in market research and commercial design in postwar Europe. At conferences, in publications, and in their practical work, they helped to promote core elements of consumer engineering. Lazarsfeld, Katona, and Dichter, for example, advocated new forms of consumer research in Western Europe, while Raymond Loewy and others pushed for marketing-oriented design among their European colleagues. As design and research consultants, they set out to bring American-style marketing to European companies and helped American firms succeed with European consumers.

The design and marketing knowledge they brought with them to postwar Europe was already transatlantic rather than purely "American" in character, as we have seen. By focusing on the emigrés as cultural translators and on the modified "reimports" they facilitated, this chapter challenges conventional narratives of "Americanization" after World War II. Instead, it presents a story of circular or reciprocal exchange processes in the field of consumer marketing and design akin to the concept of "Westernization."[11] The returning emigrés played a crucial part in that story, because their very presence helped to normalize "American" consumer culture to postwar European elites, making it somewhat less foreign and more palatable. The returning emigrés did not simply act as cheerleaders of American consumer modernity, however. Through their cultural translations they were able to engage skeptical colleagues and consumers in Europe and, in some instances, to "Europeanize" modern marketing practice.

(R)emigrés as Transatlantic Mediators: Return Voyages and Postwar Networks

Historians have stressed the importance of individual "transatlantic mediators" in developing a shared political culture for the postwar West.[12] In addition to journalists, diplomats, and government experts traveling back and forth across the Atlantic, former emigrés and exiles were particularly influential as transatlantic facilitators during postwar reconstruction in the early years of the Cold War. West German society in particular was transformed in the decade following World War II with the help of emigrés whose work was frequently tied to larger American reconstruction and reeducation efforts sponsored by the U.S. State Department or other organizations such as the Ford Foundation.[13] Returning political exiles, for example, helped to "westernize" the German labor unions and to give the Social Democratic Party an Atlantic orientation.[14] Elsewhere on the continent, emigrés such as Max Ascoli and Mario Einaudi in Italy also served as influential voices from American exile and contributed to the postwar transformation of their country.[15] The "return" of emigrés took a wide variety of forms, including temporary stays and recurring visits.[16] Taking positions as visiting professors, for example, was among the most important ways in which returning emigrés contributed to transnational knowledge flows back to postwar Europe.[17] Among the consumer research and design experts, the temporary return was most common. While some did return permanently, most influenced European developments as sojourners through their personal, academic, and professional contacts and networks.

Few of the emigrés discussed in this book embodied the figure of the transatlantic sojourner better than Paul Lazarsfeld. He was paradigmatic for a new type of transatlantic networker continuously on the move between the two continents, helping to "export" new social science methodology back to Europe. After the war, his Bureau of Applied Social Research conducted media research in Europe for the government-sponsored Voice of America, and Lazarsfeld became an active member of Cold War social science networks maintained by the Ford Foundation. As early as 1948, he returned to Europe as a visiting professor at the University of Oslo. During the 1950s, Lazarsfeld traveled back on various occasions on government business, conducting research in France and Germany, visiting London and Paris on behalf of UNESCO, or giving lectures in Austria, Poland, and Yugoslavia.[18] A July 1959 letter home to New York, provides an impression of the nature of these return trips, which were funded in part by the Ford Foundation:

> The Ford part of my summer is just over and it kept me on my toes every minute. . . . The seven weeks were divided about as follows: One week Paris, three weeks Austria and three weeks Yugoslavia. . . . It is worth mentioning, however, how confusing the rapid shifts between countries can be. . . . One day, e.g., I talked in the morning in German on personal influence (Vienna) and in the evening in French on mathematics and sociology (Paris). . . . I was at three major conferences. One organized by UNESCO on mathematical social sciences, one by the European Economic Community on market research in the European Economic Community (both in Paris) and one on problems in the university (held in Dubrovnik).[19]

Marketing remained part of his agenda, and he maintained ties to his old Viennese networks. With support from the Ford Foundation, he and emigré economist Oscar Morgenstern initiated the Vienna Institute for Advanced Study in 1963, which in turn provided a venue for American social scientists as well as for other emigrés such as Friedrich Hayek and Charlotte Bühler to reconnect across the Atlantic.[20]

Only an estimated 10 percent of the German-speaking emigrés returned to Europe from the United States for good.[21] Many had simply built a new life for themselves in America, and a few, such as Herta Herzog-Massing, returned only in retirement. Those who did want to return faced many obstacles. Returning emigré academics frequently had difficulties finding suitable positions within the still very conservative German universities. The University of Frankfurt presented an exception as it reinstated Max Horkheimer in his old position as professor in 1949. Together with Theodor Adorno, who permanently returned to Germany in 1953, he reestablished the Frankfurt Institute for Social Research, a central institution for transatlantic exchanges in social research. The designer Ferdinand Kramer also permanently returned to Frankfurt; his exile contacts to Adorno and Horkheimer helped to secure him a position as architect of the university.[22] Despite their more distant attitude toward American commercial culture, the Frankfurt group still played their part as transatlantic mediators. Kramer, for example, gave numerous lectures on American department stores and industrial design across postwar Germany and Switzerland during the 1950s.[23]

Emigrés provided prominent voices in postwar debates on mass consumption; the number of lectures they delivered at universities, professional congresses, and before popular audiences is vast. Some had recurring stints as guest professors such as Josef Albers, who repeatedly taught at the newly founded design school in Ulm. Others, like Loewy or Dichter, addressed professional associations, hoping to influence professional norms and developments in their respective fields. Quite frequently, their work and appearances were covered by European newspapers and professional magazines. Artists left their imprint on the postwar European scene through exhibitions of their work, such as Herbert Bayer or Leo Lionni, who ultimately returned to Italy in 1962. Repeatedly the emigrés expressed their desire to pass on their knowledge about "America." In 1954, for example, Sybil Moholy-Nagy, by now a professor at the Pratt Institute in Brooklyn, wrote to her friend "Teddy" Adorno about her wish to offer lectures on American architecture during her upcoming visit to Germany.[24] Even if emigrés returned only for a short time, as teachers their impact on the European academy could be significant.

Most emigrés regarded their activities as cultural mediators as serving the economic and political interests of both their home countries and the United States during the Cold War. Several of them even involved themselves directly in U.S. foreign policy and cultural diplomacy. During World War II, many European exiles had offered their expertise on European societies and economies, serving as intelligence analysts and advisors to agencies such as the Office of Strategic Services (OSS).[25] During postwar reconstruction and the Cold War, emigrés could again be found in advisory roles, utilizing their expertise and personal networks and producing government reports. Emigré economist Richard Musgrave, for example, co-authored a report for the Economic Cooperation Administration (ECA) on postwar Germany's fiscal challenges that advised the Germans not to neglect their consumer goods sector and to follow the American example of a consumption-driven growth policy.[26]

George Katona returned to Europe several times during the early 1950s as a liaison in consumer research. He surveyed research institutes at the behest of the American government to assess the state of market and opinion polling in postwar Germany. Because of his prior work in Germany, Katona observed in one report, German people, "usually, talked to me without considering me a representative of the Occupation Authority or a foreigner."[27] Emigré designers, too, were in the employ of postwar American foreign policy. They were among those whose work was exhibited as "American design" in Cold War exhibitions sponsored by the ECA and other agencies. Designer Will Burtin, who had fled Germany in 1938 at age thirty, had designed training manuals for the OSS and after the war worked for the State Department. By the 1950s, Burtin was organizing exhibits on modern design that were contracted by the government-funded Traveling Exhibition Service (TES) as part of its cultural diplomacy in Europe.[28]

From the United States, too, emigré consumer experts contributed to postwar exchanges. Those in academia occasionally trained students coming to study in the United States. Theodor Adorno, for example, during the 1950s sent students from Frankfurt to study with George Katona in Michigan and with Paul Lazarsfeld in New York.[29] Traveling delegations of European professionals to the United States sought out the emigré marketing experts, and there was extensive correspondence between emigré scholars at American universities and their colleagues back in Western Europe.[30] Their publications also helped them to reach European academic, professional, and popular audiences. Loewy's memoirs on the emergence of the industrial design profession, for example, were translated into numerous European languages including French, Spanish, and Italian. The widely reviewed German version of Loewy's book appeared with publisher Econ-Verlag, which was particularly active in making American consumer modernity accessible to German readers, translating as well works by Ernest Dichter and George Katona.[31] In numerous ways, then, emigré experts were involved in postwar knowledge flows across the Atlantic and reciprocal processes of "Westernization."

"American Style" Marketing? Consumer Research in Postwar Europe

As they returned to postwar Europe, emigré consumer researchers felt they were coming back with new insights to share, touting the achievements of American scientific marketing, which—with their active involvement—had blossomed over the course of the 1930s and 1940s. "I hope, you don't perceive me as a stranger," George Katona opened his keynote speech (in German) at a 1950 Frankfurt conference on opinion research. "I have studied in Germany and, until 1933, I have lived in Germany." Referencing his emigration experience following the Nazi seizure of power in the first sentence, Katona conveyed to his audience not only his familiarity with the country, but a sense of distance. He then continued to discuss what "we" as Americans had achieved in the intervening seventeen years in the field of opinion research.[32] Not surprisingly to those familiar with his work, Katona stressed the importance of behavioral and attitude research in American consumer studies. Citing Kurt Lewin's insights into group dynamics along with wartime studies on motivation and attitudes, Katona offered his German audience the prospect of democratic consumer research contributing to economic growth and a coming age of mass consumption.

To be sure, no straightforward "Americanization" of German marketing and consumer research occurred during the 1950s and 1960s. The English term "marketing" indeed increasingly appeared in the German literature, indicating a new, more comprehensive understanding that combined sales policy and advertising with product development.[33] This did not yet amount to a full-fledged turn toward marketing management during the 1950s, but marketing practices in postwar Europe did change, and American influences on advertising, public relations, and market studies were clearly discernible. American companies in Europe often served as innovators for new forms of advertising and market research.[34] However, European economies already had well-developed marketing cultures, which tempered the impact of American transfers. In Germany, for example, advertising and market research had seen an early professionalization since at least the interwar years, which continued under the National Socialist regime.[35] Indeed, the German opinion researchers Katona encountered were prone to resist new methodologies perceived as "American" in the social sciences, as he observed in a report to the Allied High Commission. "Germany is, of course, different from the United States," was one objection Katona would frequently hear from his German colleagues, who tended to prefer qualitative studies to the more empirical "American" approach he offered.[36]

Still, European marketing practice professionalized and became more "scientific" after the war, especially in advertising and retailing.[37] Across Europe new store formats provided access to a growing variety of brand goods.[38] In Germany and elsewhere, small stores and consumer cooperatives continued to dominate the retail landscape into the 1960s, even as American-style self-service "supermarkets" began to gain in importance.[39] While credit use and credit marketing remained controversial and much more limited than in the United States, installment selling and small loans similarly saw an expansion with banks and retailers across Europe.[40] A look at the advertising industry similarly demonstrates the continued European ambivalence toward modern marketing formats after the war. In Germany, the ad business saw renewed competition as American firms returned to the market and larger German agencies adopted the full-service model.[41] Advertising spending rose significantly as consumer spending took off during the 1950s, and advertisements touted a growing array of brand goods in new media including television.[42] By 1960, J. Walter Thompson, McCann, Young & Rubicam, and several other prominent U.S. advertising agencies were present in the German market, representing primarily American firms but also dozens of German companies including Reemtsma cigarettes, Henkel household products, and the electrical manufacturer Siemens.[43] At the same time, however, many in the German advertising profession continued to harbor resentments against what they portrayed as the crass materialism, manipulative psychology, or uninspired "nose-counting" of Madison Avenue. Postwar admen such as Hanns Brose still saw themselves as traditional self-made men and viewed "scientific" marketing with profound skepticism.[44]

Such skeptical attitudes toward American marketing were not restricted to Germany but could be found across Europe. In Britain, too, J. Walter Thompson was among the leading advertising agencies of the 1950s, but JWT London still "sought to soften its American-ness" by "adapting, revising and rejecting" the policies of its American parent company.[45] Yet European corporations paid close attention to new developments across the Atlantic. Through their subsidiaries in the United States and through study trips organized by industry organizations, German companies kept abreast of new marketing developments internationally.[46] The government-sponsored Society for the Promotion of German-American Trade also organized study tours in cooperation with American technical assistance programs.[47] In 1954, for example, German marketing experts visited the United States to learn about PR organization and new advertising methods.[48]

Within market research, we find the same mix of domestic traditions and American influences. Postwar European consumer research drew on interwar antecedents and at the same time was frequently intertwined with political reconstruction efforts during the Cold War. Early survey researchers in Great Britain had their roots in social reform and interwar planning efforts and were later linked to government research for Britain's nascent welfare state. French market surveys similarly grew in the context of state planning, while in West Germany, the well-known Allensbach Institute highlights the intersection of commercial surveys with political attempts to shape a new consuming public.[49] When Katona studied German survey research organizations in 1950, he found their number to be "surprisingly large." Most were, however, small, underfunded, and methodologically not quite up-to-date by American standards. Even the largest public polling institutions such as Elisabeth Noelle-Neumann's Institut für Demoskopie in Allensbach and the

Emnid Institut in Bielefeld, Katona observed, relied heavily on commercial market surveys as a primary source of income.[50]

By the end of the 1950s, the number of commercial research organizations and their activities in the German market had expanded significantly.[51] In addition to Allensbach and Emnid, the DIVO society in Frankfurt, Infratest in Munich, and several smaller agencies offered a mix of opinion and market surveys. The Gesellschaft für Konsumforschung (GfK) in Nuremberg and the Hamburg-based Gesellschaft für Marktforschung (GfM) led the group of institutions focused solely on commercial research. A few organizations also offered psychological studies, including Bernt Spiegel's Institut für Marktpsychologie in Mannheim as well as European subsidiaries of Social Research Inc. (SRI) in Hamburg and Dichter's Institute for Motivational Research in Zurich.[52] A 1960 report by JWT's Frankfurt office counted "over 50 nationally known economic research organizations" that had been set up in Germany mostly after 1945. JWT emphasized the innovative impulses coming from European subsidiaries of American research firms such as Nielsen, including new methods in panel and motivation research.[53] In addition, McCann and other American agencies that advised German companies such as Henkel and Daimler Benz on their marketing strategies now demanded the same research procedures they were accustomed to back home on the American market.[54]

As a result, German market research firms were keen to keep up with new methodological developments in the United States.[55] In an extensive 1954 report for the industry leader GfK, advertising consultant Hanns Kropff surveyed new American techniques in media research, emphasizing advances in areas such as random sampling and psychological interviewing methods.[56] Another report detailed the state of American advertising research, illuminating new approaches to estimating the impact of radio and billboard advertising as well as new efforts in psychological consumer research. Both reports noted the work of the emigré professionals in the United States, most prominently Paul Lazarsfeld, Alfred Politz, and Ernest Dichter.[57] Other private research organizations including Emnid and Allensbach similarly looked across the Atlantic to improve their survey work. Pollster Elisabeth Noelle-Neumann of the Allensbach Institute first studied American survey methodology at the University of Missouri in 1937/38, which serves as an important reminder that besides the emigrés there had been other transatlantic exchanges in market research during National Socialism. After the war, Noelle-Neumann established contacts with Lazarsfeld and with the Frankfurt Institute for Social Research.[58] The German Marshall

Plan ministry also devised a study tour of market research organizations in the United States.[59] Both government and industry actors regarded consumer research as an important area for economic "modernization."

Transatlantic transfers helped to bring the new psychological and sociological approaches of consumer engineering to postwar Germany. Historian Nepomuk Gasteiger finds that emigrés such as Lazarsfeld, Dichter, and Katona were instrumental in disseminating an understanding of consumers as members of discrete sociological market segments whose behavior was influenced by psychological desires and motivations to be explored and exploited by marketing experts.[60] Europeans, too, now set out to master the "art of asking why." Representatives of German market research such as the GfK's Georg Bergler claimed that their organizations had long paid attention to the psychological aspects of consumer behavior, but Bergler acknowledged the impact of American trends in the field of psychological consumer research on Europe and especially Ernest Dichter's role as a "prophet" of motivational research during in the 1950s.[61]

Among European market research professionals a similar divide between qualitative and quantitative approaches emerged as between "Tiefenboys" and "nose counters" in the United States.[62] In the mid-1950s, motivation research had begun to make its first inroads into European consumer research.[63] The 1957 meeting of the European Society for Opinion and Market Research (ESOMAR) in Gothenburg saw a direct confrontation between proponents of traditional quantitative methods and those favoring new psychological approaches.[64] In professional journals, European marketing experts touted recent achievements in qualitative research, noting, for example, research on the symbolism of goods by Herta Herzog and Pierre Martineau of the Chicago School, the communication theories developed in the context of the BASR at Columbia, or the concepts of dynamic social theory as discussed by Kurt Lewin.[65] Their thinking about the role of attitudes and motivations and about the psychological and sociological constructedness of consumer behavior now became part of the European professional debate.[66]

Advocates of a statistical and data-driven approach to market research, to be sure, also found reception in the European discourse. Hanns Kropff, the doyen of German advertising research, who had been studying American research methods since the 1920s, was one of the leading critics of Dichter during the 1950s. Citing Politz's attacks on motivation research methodology, Kropff felt the new approach was too commercial and at best offered only hypotheses.[67] Because many European research organizations were already engaged in quantitative research, Politz's own firm was comparatively slow to enter the European market. Only by the late 1950s did Politz Research Inc. set up a German office through its media advertising research subsidiary, Universal Marketing Research.[68] Despite its many critics, however, it was the psychological approach that intrigued European marketing experts the most. Market researcher Hans Schad observed in 1958: "Sigmund Freud is not less well-known in Germany than in the United States and especially in this area of advertising research we should not have needed a nudge from the U.S."[69] As they began to embrace new, "American" psychological methods in market research, Europeans recognized the unusual "detour" these methods had taken across the Atlantic. By 1960, psychological studies focusing on brand image and product personality could be found in Western Europe as well.[70] On both sides of the Atlantic, market researchers such as Ernest Dichter or Bernt Spiegel in Germany now filled the role of knowledge entrepreneurs.[71]

Translating Consumer Engineering: Ernest Dichter as Transatlantic Mediator

"American" marketing implied more than mere methodological issues to European observers. It invoked concerns about new and unchecked forms of material consumption as well as the threat of consumer manipulation. In the spring of 1960, German America correspondent Peter von Zahn dedicated an entire episode of his show "Images from the New World" to investigate the role of advertising in the United States. Taking his cues from critics such as Vance Packard, von Zahn not only emphasized innovations in advertising arts and in research but also portrayed the large agencies of Madison Avenue with "cynical disparagement," as a JWT representative noted. Besides agencies such as JWT, McCann-Erickson, and Benson & Mather, the program featured Ernest Dichter, who presented his motivation research as a key to understanding the dynamics of American consumer culture.[72] To the discomfort of German elites like von Zahn, Dichter unabashedly advocated the transfer of new consumer engineering methods in an attempt to translate American consumer modernity to Western Europe.

Dichter presents a particularly interesting case study of cultural translation because of his dual role as an advisor to both European and American organizations, for which he heavily drew on his personal experience as an emigré. In advertising his studies to American companies, he took care to point out that he would personally supervise the research his company proposed to conduct abroad. Dichter proclaimed himself an expert in marketing because of his degree in psychology and an expert on Europe because of his migration experience: "[Europeans] accept me . . . ; in my particular case because I can combine European answers with American know-how. . . . I do speak their languages and they will sort of hug me and say, 'Well, you're really one of us.'"[73] His immigrant background served as a prominent asset for his role as advocate of American interests in postwar Europe.

Dichter had already been active in Europe as a conduit for American marketing thought during the 1930s. Addressing French marketing experts while still in Paris in 1937, he related not only his experiences in Vienna but also his insights into American market research gained from an affiliation with the Psychological Corporation. Reporting on Lazarsfeld's research on consumer motivations, Dichter suggested that France, too, should establish an institute for psychological market research and adopt the "American" practice of maintaining brand barometers.[74] After the war, Dichter returned to Europe in multiple capacities, delivering talks as an itinerant professional expert and advising clients as a research consultant based in New York and Zurich. Dichter and his employees promoted their brand of "modern" motivation research across the continent.[75] His firm did consulting work for companies such as Schwarzkopf, Braun, C&A, and Renault in order to help them position themselves in both the American and the domestic markets.[76] During the 1960s, he also held professional seminars and consulted for advertising agencies such as Troost in Düsseldorf. By the end of that decade, Dichter's subsidiary had a second office in Frankfurt and counted air carrier Lufthansa, consumer goods manufacturers like Knorr, and retailers such as Quelle among its clients.[77] Dichter also conducted studies for government agencies, analyzing for example the popular image of Germany in the United States for the German Information Service.[78] With his European clients, Dichter benefited from his expertise as an "American" researcher versed in modern marketing methods, albeit one with European roots.

As European economies recovered, Dichter offered insights on European markets to American consumer goods producers as well. Despite fears of "Americanization" among some Europeans, many U.S. firms actually found themselves forced to adapt to the European market in subtle ways, leading some historians to speak of a "Europeanization" of American companies engaged abroad.[79] European consumers, Dichter noted, were different from their American counterparts, and selling on European markets demanded a special kind of expertise and a sense for "localizing" advertising. Not surprisingly, Dichter called for psychological studies as a key to understanding foreign markets: "Counting how many bars of soap are purchased in England or how many people buy Durban dentrifice in Italy in a year provides us with valuable information, but it is not research. It is census taking. . . . When I want to know how to get more Italian women to buy fur coats, or to use more margarine, I have to understand first the cultural background of fur coats or margarine in Italy."[80] Dichter compared his motivational research to an anthropologist's study of a tribe—the "tribe" in this case being Italian, English, American, or German consumers. Proclaiming the need for "a psychological bridge over the ocean," he offered his market services on either side of the Atlantic.[81]

As a consultant to his American clients, Dichter most directly positioned himself as a cultural broker. In numerous studies, speeches, and essays, he gave advice to American businesses seeking to enter European markets. Playing heavily on his transatlantic expertise, he alerted Americans to the pitfalls of doing business in Europe and to the necessity of considering local conditions. Writing for trade publications as a means to advertise his own consulting services, Dichter surveyed the "unsatisfied needs of Europe today" and discussed lingering transatlantic differences in consumer habits. Europeans "would *like* to have as substantial a breakfast as Americans," he observed for example in an essay for the trade journal *Business Abroad*, but they had not yet been stimulated enough to spend on cereals, pancakes, or waffle mixes. While the principal meal for many was still lunch, he noted a trend away from this custom and predicted opportunities for "conveniently-packaged" lunch snacks. He made similar observations regarding furniture and other goods but stressed that Europeans still tended to hold on to goods for too long. The crucial lesson to be learned from such observations was that European consumers were only now on the "verge of accepting the concept of obsolescence." Through psychologically informed marketing, Dichter believed, U.S. firms could overcome the traditional "immorality of waste" and sell Europeans on "the idea of rejuvenation and replacement—to sell the American concepts of modernization, disposable products, and 'trade-ins.' "[82]

The spirit of consumer capitalism had not yet fully reached Western European markets, and Dichter's studies advised American firms on how to adapt to this fact while promoting a change in attitude. In Europeanizing their marketing, paying attention to the cultural dynamics of a national market was vital even in an age of emerging global brands, and a comparison between European and American consumer behavior, he urged, really "involves a comparison between their different cultures."[83] There was no universal "world customer," and Dichter grouped countries into various categories depending on their level of affluence, their degree of social stratification, and their internal political dynamics, among other factors. The United States stood out in many respects, but even within Western Europe Dichter found vast variance in consumer attitudes and behavior.[84] While some brands, such as Coke, Gillette, or Volkswagen, succeeded by not changing their "look" (creating something of a "commercial United Nations"), Dichter warned that generally countries retained basic differences in packaging style, for example. The "rather brassy looking American can" of one U.S. hairspray producer, he related, had been "lost in translation" on European consumers accustomed to package designs with greater solidity and elegance.[85]

Some differences arose simply from divergent levels of material affluence, but transatlantic gaps in cultural attitudes could be revealed through the careful study of language. In a proposal for a motivational research study of the German market for prefabricated homes for Hodgson Homes, for example, Dichter cautioned that the word "prefabricated" carried different connotations in Germany and in the United States. A desire for "permanence," his psychological research suggested, was still strong with German consumers, and the term "prefabricated" might conjure notions of temporary, low-quality housing. German interest in additional "superficial" gadgets and add-ons, by contrast, was lower than in the United States, where "prefabricated" often suggested a fixed set of integrated modern amenities. To translate the American product successfully for the German market, Dichter thus proposed to study various "words and phrases, to come up with the best possible term for this type of home."[86]

Local and national knowledge was crucial for marketing translations. Finding the proper language to approach consumers was a recurring trope in Dichter's writings. "Sales to the women of Europe can be killed with kindness," he warned in an article that speculated about transatlantic gender differences. "Unlike most Americans, European women have mixed feelings about gaining equality with men," Dichter asserted bluntly, and American companies would lose in sales if they failed to understand this. Lacking the "throw-away" mentality of her American counterpart, Dichter claimed, the "European housewife" still "hears the voice of the devil" in advertisements for "one-step" cleaners or "instant foods." She feels that "the righteous thing to do is make the soup and bake the cakes herself." Thus, he suggested, "if you are selling a labor saving device to the French or German or Italian housewife, stress the point that the device will permit her to please her husband even more than before."[87] Regardless of the dubious merits of such advice from a contemporary perspective, it demonstrates his awareness that translation in marketing required profound cultural knowledge as well as finding the proper language to adapt to different consumer expectations, for example with regard to gender or homeownership.

Mixed European attitudes toward the United States presented a particular challenge for American companies, in Dichter's view. "Americans can sell everything except America," Dichter commented in the late 1950s, a time when America's prestige had declined somewhat among its Western allies. Americans, Dichter felt, needed to "humanize the European concept of the American businessman," to move away from the cold, cash-loving, size- and efficiency-obsessed Yankee, toward an affable personality who could discuss politics and culture along with business. Apart from the larger political and social contexts, there were specific elements of American consumer capitalism that concerned the marketing expert. "Europeans are aware—perhaps almost too much aware—of our techniques of mass-production and mass-distribution."[88] What Europeans had to understand, Dichter felt, was that mass consumption did not pose a threat to individualism but rather allowed for a resurgence of individual styles and tastes.

One way to mollify anti-American sentiments in Western Europe, Dichter argued, was to downplay the "Americanness" of products. This could be achieved by couching advertisements in a language of international modernism rather than invoking American consumer modernity per se. In the housing study, for example, he advised that instead of making prefabricated houses "typically German or typically American, it may be most desirable to make this an international design."[89] A "Made in U.S.A." product was "slightly damaged by the over-all image of anything American," Dichter noted, as Europeans had picked up on the warnings of consumer critics like Packard, considering American products designed for obsolescence, brash and gaudy.[90] Another way to navigate such European resentments was to properly adapt products and their marketing to the various national contexts. "In the most primitive sense," Dichter wrote, "translate properly your instructions." Again, language was key, but advertisements also had to point out that a product, while American, was really coupled with, for example, French taste. "Leave off the flag. . . . You have to put 'Made in USA' as small as possible." Pillsbury's cake mixes flopped in England, he suggested, because the company did not study local conditions deeply enough and "because they presented it as an American idea."[91]

The most effective strategy in dealing with European markets, according to Dichter, was in effect a strategy of cross-cultural synthesis. The basic idea was to emphasize the European "roots" of American consumer culture or its particular elements while highlighting transatlantic adaptation processes: "we really got it from you originally and we copied it, we developed it a little further."[92] General Motors, he pointed out, had used this approach with regard to its European subsidiary Opel, claiming that original "German" design innovations were brought to the United States and then re-exported back to Europe. Such a creative translation, synthesizing European tradition with American consumer modernity, was, as he called it, a "subtle" but effective "form of flattery."

Dichter was aware that the same phenomenon applied to marketing and consumer research more generally: "isn't it funny that most of the marketing people that really made good in the States, quite a number of them happen to be people from Vienna or Germany / went to college there . . . ; Alfred Politz, Professor Lazarsfeld, Ernest Dichter. [They] have come back to you as American experts but whom are you kidding?"[93] While their careers could only have happened within the context of American consumer capitalism, Dichter suggested, without the influence of emigré experts, postwar American marketing might have been quite different and perhaps less palatable to Western Europeans. Of course, much of Dichter's writing was conceived as a way of self-promotion for himself and his consulting business. However, he was hardly alone among the returning emigrés in promoting American-style mass consumption in postwar Europe.

Dichter and his fellow emigrés found themselves ideally positioned to broker between European and American consumer culture. George Katona struck a very similar tone in his speeches and publications on the continent.[94] Katona was instrumental in bringing behavioral approaches to European marketing, familiarizing colleagues on the continent with the marketing applications of Lewinian social psychology.[95] The basic methodology of his consumer confidence measurements was eventually adapted to European economies as well. More broadly, Katona saw it as his mission to persuade Europeans to adopt the frame of reference of affluent societies. Changing European attitudes toward consumer credit was a crucial element in this, as he explained at a 1962 Eurofinas credit conference in Vienna. Credit financing, he believed, helped consumers to better budget their finances and to "raise their level of human aspirations" to ensure future prosperity through consumption-driven growth.[96] In contrast to optimistic Americans striving for ever more goods, Katona found Europeans still maladapted to postwar affluence and modern mass consumption. Psychologically they had an "apprehensive" and "defensive" posture toward consumption. The "American" emigrés therefore collaborated with European colleagues to analyze what they saw as a "gap between the reality of a rapidly developing mass-consumption society and its perception by the people."[97] Making the American way of life more palatable, consumer engineers believed, required refined psychological marketing along with modern product designs that Europeans could embrace as their own. Here, too, returning emigrés would play a pivotal role.

Mass Market Aesthetics: Commercial Design as a Transatlantic Transfer

When Raymond Loewy addressed German design professionals at the Industrieform e.V. in Essen in 1955, he came to them both as a representative of the American Society of Industrial Designers and, more broadly, as an advocate of American-style consumer engineering abroad. Loewy also came to Germany via Paris where in 1953 he had established the Compagnie de l'Esthétique Industrielle as a European subsidiary of his design consultancy. An ongoing cooperation with the porcelain manufacturer Rosenthal had familiarized him with the German design scene and some of its leading representatives. As Loewy promoted his professional ideal of the industrial designer as a marketing expert, he knew that some in the audience were skeptical of his openly commercial agenda. To woo them, Loewy called for not only an "aesthetic reeducation" of consumers but a democratization of "good taste" which for too long had been the privilege of elites. He was careful, moreover, to contextualize his mission and modern American design in general within a longer European tradition.[98]

As in the areas of advertising and marketing research, one should be careful not to conjure up the notion of a postwar "Americanization" of European commercial design. Here, too, American impulses met with ingrained domestic traditions in industrial and commercial modernism that dated back to the nineteenth century and included the British Arts and Crafts movement, French poster art, and the Werkbund and Bauhaus in Germany. European style, fashion and design retained a certain cachet in the United States even at the height of the postwar "American century." Among American consumers—and especially among those with higher incomes—European-designed goods remained highly sought after, and many American department store buyers still looked across the Atlantic.[99] Scandinavian modernism and the handcrafted appeal of Danish furniture fascinated U.S. decorators during the 1950s, and Italian goods from leatherware to automobiles were in high demand.[100] Thus, while American designers, institutions, and firms wielded a new degree of influence over the development of the international design profession, the field remained genuinely transatlantic during the postwar decades.

In Germany, the emigration of Bauhaus luminaries and the professed antimodernism of the National Socialist regime did not bring about a complete break with design modernism, but the country found itself in need of greater reorientation than its European neighbors did.[101] The Werkbund reconstituted itself as an influential institutional link between arts and industry in 1947. The organization had long counted firms such as electrical giants AEG and Siemens among its members and now added new firms with a strong profile in design such as Braun electronics, Pfaff sewing machines, WMF tableware, Knoll International's Stuttgart office, and porcelain manufacturer Rosenthal. Leading Werkbund designers such as Wilhelm Wagenfeld looked to reconnect to Weimar era modernism and notions of "good form." In 1949, a first Werkbund exhibition on housing in Cologne recalled the interwar interest in minimum standards and in its austerity acknowledged prevailing conditions of postwar scarcity and an overriding need for speedy reconstruction in West Germany.[102] Any notion of styling consumer goods for accelerated consumption clashed not only with the sober rationalism of the German design tradition but also with far from affluent domestic conditions.

Initial debates over the state of postwar design were sparked by concerns over the international competitiveness of German consumer products as articles of export. A 1949 trade show of German consumer goods in New York's Rockefeller Center had ended in disaster from the perspective of design professionals. Reviewing the German designs on display, Loewy employee Carl Otto complained about a lack of modern styling and their "outdated" 1930s look. Germany, Otto felt, had forgotten the lessons of the Werkbund and Bauhaus which had been since picked up by designers in other countries.[103] MoMA assistant curator and German emigré Herwin Schaefer spoke even more harshly of the German exhibition as "a hoard of high-produced kitsch and rubbish."[104] The exhibition fueled a critical debate in West Germany that ultimately led to the creation of the Rat für Formgebung (Design Council) in 1951, comprising representatives from the government along with industry associations and advertising and design professionals. Modeled after the British Council on Industrial Design, the Design Council organized international exhibitions and domestic design competitions. It also provided a coordinating function between various other design organizations such as the more commercially oriented Industrieform e.V. in Essen and served as a forum for the development of modern design in postwar Germany.[105]

Functionalist modernism, historian Paul Betts has argued, became the "aesthetic signature" of educated, well-to-do West German elites during the 1950s.[106] Simple, functional designs in the Bauhaus and Werkbund tradition signaled a break with the politicized Nazi past and a form of cultured internationalism. Such modern forms also provided a means of social distinction in contrast to more popular and gaudy designs such as organic furniture styles. Both the consuming elites and design professionals professed their disdain for American "styling" and its chrome-adorned populuxe style. Thus, when Loewy's memoirs first appeared in German translation in 1953, they caused quite a stir.[107] His call for a marketing orientation of design provoked an intensive debate about the direction of the design profession in Germany. Critics attacked American design trends as commercially corrupted design and as a debasement of the traditional ideal of "good form." Cologne design professor Walter Kersting, for example, rejected the notion that Germany's international standing could be improved by adopting a new culture of design from the United States.[108] Many critics associated the notion of styling with irrational wastefulness, which appeared especially egregious in light of postwar scarcity. At best, they argued, a commercial orientation would lead to mediocre and unimaginative designs.[109]

Proponents of American design, by contrast, countered that while designers such as Loewy focused on profit, their design was equally informed by the ideal of "good form" and by the simple shapes that had motivated the Bauhaus.[110] Herwin Schaefer, now officially a representative of U.S. cultural diplomacy in the employ of the State Department, stressed the role of the Museum of Modern Art in bringing European modernism to contemporary American design at a 1952 design conference in Darmstadt with representatives from Great Britain, France, Germany, and the Netherlands.[111] Voices within the German economics ministry, furthermore, raised concerns about German competitiveness on the U.S. market if design did not keep up with international developments. Consequently, the ministry organized a 1953 study tour to the United States for designers and industry representatives.[112] Industrialists such as Philip Rosenthal, finally, called on German design "purists" to embrace the commercial spirit of the United States where not only had designers become a major force in industry but variations in form and color also fostered individual consumer choice and provided a boost to sales numbers. "Creating the desire to buy," Rosenthal argued in the spirit of consumer engineering, would serve as "the engine of a free economy."[113]

Design forms of American provenance did make inroads into postwar Europe by the mid-1950s.[114] A number of firms especially in the electronics industry, in furniture, and in housewares began to emphasize industrial design in their production process.[115] German design journals and professionals, moreover, expressed keen interest in the American design profession and in a more integral role of designers in industry. Again, former emigrés played a significant role in this exchange. Several study tours to the United States featured visits with European emigrés working for American companies or as independent consultants. One group of German professionals, for example, visited the offices of Loewy and Peter Muller-Munk and met with emigré architects such as Walter Gropius, Ludwig Hilbersheimer, and Ludwig Mies van der Rohe. The group visited design schools including the Harvard School of Design and the Institute of Design in Chicago, and among the U.S. corporations they toured, Knoll Associates was included along with such industrial giants as IBM, General Motors, and General Electric. When they visited Corning Glass in western New York, they were welcomed by German-born design curator Axel von Saldern.[116] The group also met with Egmont Arens and other leading consumer engineers to learn about integrating designers into corporate marketing strategies.

American design education was widely discussed in 1950s Germany. The design and advertising journal *Graphik* ran a long story on practice-oriented American design education in 1956, highlighting the impact of emigré design educators on U.S. institutions such as the Czech-born designer Antonin Heythum at the University of Syracuse and, of course, all the European designers at the former American Bauhaus in Chicago.[117] A Stuttgart exhibition on industrial design education in the United States similarly noted the influence of Bauhaus conceptions on contemporary American design.[118] The biggest institutional innovation in postwar German design education, the 1955 foundation of the Ulm Institute of Design (Hochschule für Gestaltung), was heavily influenced by the American Bauhaus tradition.[119] The school sought to reestablish the tradition of Bauhaus education in Germany, and the prospect of substantial sponsorship from American reeducation funds shifted the emphasis toward industrial design early on in the conception of the school. Led by former Bauhaus student Max Bill, the Ulm Institute of

Design became influential within the German design profession during the postwar decades. It promoted a rational functionalist style that became influential for West German industry and corporate design. Josef Albers, who repeatedly came to the school as a guest professor, upheld the connection to the Bauhaus. Its impact on the aesthetic language of postwar American goods appeared obvious to German observers during the 1950s. Emigrés such as Gropius and Marcel Breuer were proudly portrayed as successful protagonists not only of a European but also of a thriving American design culture.[120] A group of German designers visiting Albers at Yale noted his influence on American education and on the design of U.S. consumer goods. Seeing the designs of his students, they reported: "we could finally understand where . . . the extraordinary use of color came from which we had encountered everywhere in American furniture and department stores."[121]

This colorful nature of American consumer capitalism appeared tied to new forms of professional product marketing. Professional journals discussed the centrality of continuous design change and of a lively and changing color palette to consumer goods design in the United States. Much as American advertising journals had once portrayed European commercial design innovations in the interwar period, postwar West German design journals now looked abroad to scout new trends. Examples include a portrait of Walter Allner's graphic design work at *Fortune* magazine and a report on the prominent advertising designs of Saul Bass, whose style, again, had been influenced by György Kepes and László Moholy-Nagy of the American Bauhaus.[122] The growing importance of packaging design to consumer marketing was another recurring topic of transatlantic design reception. Italian-born designer Frank Gianninoto, who had founded the U.S. Package Designer Council, featured prominently in several stories on new American packaging methods. As new packaging materials became available and self-service stores began to spread in Western Europe, concepts of brand product packaging or corporate and brand image design garnered the attention of professionals on the continent as well.[123]

New forms of packaging design came to postwar Europe directly through the work of American consultants as well. Walter Landor's firm was active on various Western European markets already at the end of the 1950s. He corresponded, for example, with an institute for Dutch-American trade relations about the possibility of his firm's branching out to Europe and possibly engaging in the market for package design in the Netherlands for breweries and other clients. Dutch firms, Landor was told by his overseas partners, already had their own designers, but many were still "keenly interested in American packaging designs" to make their products more competitive.[124] Landor's firm was similarly active in consulting for clients in Ireland, Great Britain, and elsewhere on the continent. In Germany, Landor was in touch with Rosenthal as well as with several large brewing companies by the early 1960s, and he collaborated with advertising agencies. In a letter to a representative of the Reemtsma tobacco company, Landor stressed the significant "interest that West Germany is taking in our work."[125] While Europe would remain a small part of his business, the former emigré still helped to transfer new practices in packaging design across the Atlantic.

No industrial designer filled the role of transatlantic intermediary more prominently than Raymond Loewy. Translations of his books and his speeches before professional designers in Great Britain, West Germany, and elsewhere assured him wide press coverage across Europe by the mid-1950s. Marketing specialists eagerly received his design philosophy, noting his emphasis on consumer psychology.[126] His firm had established an office in London as early as 1934 and after the war opened another short-lived venture there in 1947, but his Paris subsidiary proved more successful with clients such as retailer Monoprix and, eventually, the airline Air France.[127] His Compagnie de l'Esthétique Industrielle, Loewy told design professionals in Paris, helped European companies benefit from American advances in the development and application of those corporate image strategies for which Raymond Loewy Associates had been a pioneering force since the 1940s.[128] Some of the more prominent industrial designers of postwar Europe, finally, also learned the trade as employees of Loewy's design studios. Albrecht Graf Goertz, for example, worked on Loewy's Studebaker designs before he returned to Europe to do design work for BMW and other corporations during the postwar decades.[129]

Industrial design became a significant part of postwar West German industry. To be sure, the number of independent design consultants remained small compared with the United States, and only a few companies such as Braun led the way in establishing modern industrial design departments. Yet design professionals increasingly played a role in industry, and product design became a recognized part of business management as an element in rationalizing production and in promoting sales.[130] Emigré design professionals emphasized the importance of "marketing" as a new concept to their European audiences. Merchandising and sales orientation pushed designs toward popular "streamlined" forms and toward disposable items following the logic of conspicuous consumption and planned obsolescence, Ferdinand Kramer observed in

1955. Despite this critical assessment, Kramer acknowledged that modern design necessarily depended on marketing, and he underscored the importance of understanding customer needs through consumer psychology and market research.[131] Marketing, emigré designer Peter Muller-Munk noted in a speech he gave to European colleagues in Vienna, required designers to collaborate with sales and technical staff. Crucially, he added, it also provided designers with the opportunity to become the voice of the consumer within the corporation.[132] Much like market researchers at the time, Loewy, Muller-Munk, Landor, and other returning emigrés envisioned professional designers who would act as creative intermediaries between consumers and industry, improving the lives of European consumers and ensuring continuous economic growth during the Cold War.

"Good Design" as Cold War Cultural Policy: The Role of the Emigrés

Historians of the Cold War routinely invoke the 1959 "kitchen debate" between U.S. vice president Richard Nixon and Soviet premier Nikita Khrushchev to illustrate the political importance of consumer goods in the East-West conflict. The two politicians debated the relative standards of living on either side of the Iron Curtain in a U.S. model house display set up with the latest consumer technology at the American National Exhibition in Moscow.[133] The Moscow exhibition was part of a series of exhibitions of consumer goods designed to show off the "soft power" of American mass consumption to Eastern and Western European audiences.[134]

The exhibitions were initially conceived during the Marshall Plan years to provide Europeans with a concrete vision of prosperity during postwar reconstruction. A first, small, exhibition called "How America Lives" in Frankfurt in 1949 was followed by "America at Home," which opened in Berlin in 1950. Featuring modern appliances, the exhibit garnered widespread attention. The traveling ECA show "Industry and Craft Create New Home Furnishings in the USA" traversed much of Western Europe from Munich and Milan to Paris and Amsterdam with numerous examples of modern commercial design. By the mid-1950s, exhibitions increasingly included entire supermarket sections along with domestic settings to promote consumer capitalism as an economic system beyond individual goods. The products, moreover, were chosen for their modern forms much like an international variant of the Museum of Modern

Art's "Good Design" exhibits. Indeed, Edgar Kaufmann Jr., the museum's design curator, became a leading advisor to the government's exhibitions in Europe, and, again, emigré designers were heavily involved in this Cold War cultural policy.[135]

Emphasizing design modernism was a calculated appeal to European elites.[136] The exhibits portrayed American consumer design as inspired by the culture of international modernism in order to allay European fears of importing a debased and commercialized materialism. In Paris, the Museum of Modern Art and its French American director, René d'Harnoncourt, fought French elite bias against American materialism by including select industrial design in a 1955 exhibition of American art.[137] For an exhibit of American consumer design in Stuttgart in 1951, both ECA administrator William Foster and MoMA curator Edgar Kaufmann emphasized the continuities between American and European design traditions in their opening addresses. Everyday life and its material objects in the United States, Foster assured his audience, had "their roots in European tradition," and while divergent economic developments had brought about significant differences in style, Americans often "still prefer goods which are informed by European design."[138] Kaufmann noted that progressive design in the United States had received a significant impetus from European countries, noting particularly the Bauhaus and its American legacy.[139] Such rhetorical gestures sought to integrate "American" consumer modernity within a broader history of a shared Western civilization. The notion of a transatlantic design modernism also fit more global Cold War narratives about economic and cultural "modernization," as similar exhibitions of consumer goods in countries such as India demonstrated.[140]

Both in Europe and globally, American displays of design internationalism heavily relied on emigré designers. Edgar Kaufmann, for example, composed his international exhibitions by featuring the modernist design styles that the emigrés had helped to popularize in American commercial design. The Stuttgart exhibition included designs from the offices of Raymond Loewy, Peter Muller-Munk, and Knoll Associates, and it featured Castleton China designed by Eva Zeisel and Peter Schlumbohm's Chemex coffeemaker. Marlie Ehrmann and Anni Albers both had textiles included in the exhibition, and the wallpaper display showed a creation by György Kepes. Many American-born designers of course were also included in the exhibition, but the share of immigrants and emigrés was noteworthy. Their group at Stuttgart further included German American graphic designer Tommi Parzinger, Swedish Americans Greta Grossman and Rolf Key-Oberg, Italian American Francesco

Collura, who worked for General Electric, and the Austrian American emigré designers Otto Natzler and Emil Lichtblau. As demonstrated in previous chapters, the prominent inclusion of European-born artists was reflective of the American commercial design profession at midcentury. Not all of American postwar designs, to be sure, adhered to the modernist tradition pushed by MoMA and on display in Stuttgart. This modernist slant in the exhibitions and the prominent inclusion of emigré designers provided European audiences with a somewhat biased view of American consumer goods design, but it likely appealed to the European elites who came to view them.

The organization of the consumer goods exhibitions also relied on emigrés as transatlantic intermediaries. Herwin Schaefer, for example, returned to West Germany as a cultural affairs officer for the U.S. government.[141] Emigré architect Peter Blake also advised the United States Information Agency in setting up exhibitions in 1950s Europe.[142] The most important emigré designer to support American foreign policy initiatives, however, was likely Cologne native Will Burtin. Burtin, who had been art director at *Fortune* during the late 1940s and worked for major American corporations including Union Carbide and Eastman Kodak, designed several exhibitions for the Smithsonian Institution's Traveling Exhibition Service (TES) during the 1950s.[143] One prominent example of his work was an exhibition of American packaging design on display at different America-Houses across Germany. The exhibit included package designs from graphic designers Saul Bass and Lester Beall along with work from Raymond Loewy Associates (his Armour meat packages). Numerous designs from the Container Corporation of America and works by Walter Landor were also included in the show, which highlighted American advances in modern packaging with new materials such as cellophane and strategies for supermarket selling and the development of corporate and brand identities.[144]

In their exhibits, Burtin and his collaborators did more than showcase American consumer capitalism. They sought to channel a larger vision of repairing a "shattered Western Culture" that had inspired Walter Paepcke's design conferences at the Aspen Institute in which Burtin was involved along with Herbert Bayer and other emigrés.[145] Industrial design had become a central element of American material culture by midcentury. Yet it was hardly a purely American phenomenon, as art historian James Plaut argued in 1954 in *Perspektiven*, the German-language periodical of the Ford Foundation, noting the emigré contributions to the field. Here, the strong influence of European immigrants and emigrés on American designs through transatlantic transfers becomes part of an official U.S. narrative during the Cold War.[146] The work of the emigrés became part of an effort to sell American design to Western Europeans both as consumer products and as artifacts of a shared cultural tradition.

As transatlantic intermediaries, emigrés aided in the cultural translation of modern consumer capitalism to postwar Europe. They contributed to the professional transformation of Western European marketing, and in both consumer research and commercial design they promoted new methods of systematic consumer engineering. Europeans, too, began to probe the psychology of consumers, to target appeals to specific segments of the consuming population, and to create a wide array of new consumer goods with a modernist appeal. European consumer research and design, however, were not completely Americanized in the decades following the war. The design internationalism presented at American exhibitions was at best a "blended" form of Americanization, as Greg Castillo has observed, that integrated design innovations coming from the United States into existing national consumer cultures.[147]

Business cultures and consumer tastes posed challenges to and set limits on the complete adoption of consumer engineering practices in postwar Europe. In both design and consumer research existing professional traditions tempered the influence of "modern" and "American" methods. European social scientists were often as skeptical of new empirical and psychological approaches as design artists were of the commercialized American vision of industrial design. As the U.S. government realized, the emigrés' prominence in international design modernism made American consumer culture more palatable to elite consumers in Germany, France, and elsewhere. Instead of crass materialism and planned obsolescence, furniture from Knoll and other emigré-inspired designs recalled the familiar ideal of "good form" and provided a vision of modern yet cultured mass consumption as a shared Western standard. Such an appeal to elites was a staple of American cultural foreign policy during the postwar decades, and many of the returning emigrés became influential figures in the transatlantic elite networks of the so-called cultural Cold War.[148]

Some contemporary observers even hoped that the postwar transnational exchanges might lead to a kind of "Europeanization" of consumer capitalism. In a 1953 radio interview, emigré architect Richard Neutra reflected on his life between the continents and his new role as a global emissary of American consumer modernity. In his many years abroad, Neutra told his European audience, he never forgot to love Europe. In spite of American wealth and power derived from technological innovation and despite its abundance of mass production, he believed that America still needed Europe. In Neutra's view, the United States with its emphasis on mass marketing did not yet have a recipe for a truly global consumer culture: "America with her dangerous technological overabundance needs the European interest in physiological sciences."[149] What Neutra meant was that greater attention should be paid to the psychological and physiological needs that consumer goods were designed to satisfy. Many emigré consumer researchers and designers, of course, had made exactly this point in their work over the past decade to adapt existing marketing practice to different social and cultural contexts. At the same time, Neutra's comments hint at a growing debate about the limits and problems of consumer engineering. Indeed, critiques of postwar mass consumption frequently focused on the very elements that consumer engineers had introduced to marketing since the 1930s. Emigré experts were at the center of this emerging transatlantic debate as well, which ultimately helped in bringing the era of consumer engineering to a close.

Consumer Engineering: Challenges and Legacies

By the early 1970s, erstwhile emigré and "father" of the suburban shopping mall Victor Gruen had turned into an ardent critic of American consumer capitalism. Approaching retirement age, he had moved back to his native Vienna in 1968 and subsequently set up the Victor Gruen Foundation for Environmental Planning with offices in Vienna and Los Angeles. In his writings as well as in public speeches and symposia that his foundation organized, Gruen not only critiqued American cities without a true core but also warned of the social and environmental costs of expanding automobility, suburban land use, and a sprawling consumer culture.[1] In doing so, he struck a chord similar to that of his fellow emigré Richard Neutra when he had warned of America's "technological overabundance" nearly two decades earlier. By now, however, Gruen's critique was part of a much broader critical discourse on modern mass consumption and the social, cultural, and environmental consequences of "wasteful" obsolescence and expert manipulation. Advertising and marketing efforts to engineer continuous growth in material consumption, in particular, became the focus of attacks on the postwar brand of consumer capitalism by social theorists, public intellectuals, and a growing consumer movement on both sides of the Atlantic.[2]

Ironically, this transnational revolt against postwar marketing also drew on intellectual foundations laid by members of the same cohort of emigrés who had championed the rise of consumer engineering. In fact, to most students of twentieth-century mass consumption the cultural critique of the Frankfurt School scholars will most readily come to mind when considering the relationship of European emigrés and American consumer capitalism. In exile, Theodor Adorno, Max Horkheimer, and other protagonists of the Frankfurt Institute for Social Research were closely connected to the group around Lazarsfeld and their work in media and consumer research beginning in the 1930s.[3] While working at McCann, Herta Herzog read the proofs for a study by Leo Löwenthal on political agitators and psychological manipulation through media use that appeared in the "Studies on Prejudice" series edited by Horkheimer.[4] By the 1950s, the Frankfurt School scholars were acting as transatlantic mediators in translating "American" empirical social research methods to their European colleagues.[5] The Frankfurt scholars had ties to modernist design circles, too, including Ferdinand Kramer both in New York and later at the University of Frankfurt. Adorno and his colleagues frequently gave surprisingly nuanced assessments of the consumer society they had experienced in American exile to audiences back in Germany.[6] Still, Adorno remains most remembered for his indictment of the commercial "culture industry" of mass market entertainment and advertising and of the passive consumption engineered by modern consumer capitalism. "The culture industry," Adorno declared, "fuses the old and familiar into a new quality. In all its branches, products which are tailored for consumption by masses, and which to a great extent determine the nature of that consumption, are more or less manufactured according to plan. . . . This is made possible by contemporary technical capabilities as well as by economic and administrative concentration. The culture industry intentionally integrates consumers from above."[7]

The notion of a corporate "integration" of consumers into a system of mass consumption specifically criticized aspects of consumer capitalism from psychological analysis to modern styling that their fellow emigrés had pioneered. In letters, Adorno had warned Lazarsfeld against psychological approaches to consumer research as early as 1937, because, he felt, an emphasis on the psychological factors of consumer decision making would overlook the social forces that structured consumer action.[8] Consumer psychology opened the door to manipulation, and Max Horkheimer similarly criticized Kurt Lewin's social psychology as an effort to "rapidly eliminate the subject" and a threat to individual agency.[9] Other emigré scholars affiliated with critical theory such as Erich Fromm and later Herbert Marcuse also voiced concern about the psychological appeals with which postwar marketing sought to stimulate sustained consumption.[10] In turn, American critics of postwar mass consumption such as sociologist David Riesman were inspired by the psychological analysis of emigré intellectuals like Fromm.[11] Meanwhile, critics of design in the technological age, including emigrés Bernard Rudofsky and Victor Papanek, not only attacked what they perceived as fashion-driven styling but condemned the technocratic and "dehumanizing" standardization efforts that permeated the "machine age."[12] In sum, these critics called into question the fundamental premises on which midcentury scientific marketing and commercial design rested.

More than merely an intellectual critique of consumer engineering, the 1960s and early 1970s witnessed the rise of a sustained transatlantic consumer movement.[13] Consumer advocates admonished producers to keep consumer needs in mind, and political consumerism called for educated and informed consumers to be in control of their own decisions. Going beyond indictments of deceptive advertising, psychological manipulation, and profit-centered rather than consumer-oriented designs, this movement against consumer engineering was part of a broader transatlantic rebellion against various forms of social engineering.[14] In both Europe and the United States protests targeted large-scale urban planning efforts and technocratic social policies, demanding individual agency in the face of expert control. The protests challenged the hegemony of Cold War liberalism and of a capitalist modernity to which, as we have seen, postwar marketing and consumer design had crucially contributed.

As the era of technocratic "high modernity" came to a close by the 1970s, so did the careers of many of the emigrés highlighted in this study. When they moved into retirement during the 1960s and 1970s, many of the emigrés sold their companies. Without their founders, few of the consulting firms managed to retain their influence on the marketing scene. Alfred Politz Research, for example, was sold in 1967 to the information technology provider Computer Services Corporation and was soon dissolved as a distinct entity. Dichter's Institute for Motivation Research was sold to an industrial conglomerate in 1971, where it was quickly jettisoned as unprofitable. Dichter regained control over the firm and continued business until the early 1980s, but the company never fully recovered. Loewy Associates, too, was sold and went bankrupt during the mid-1970s as Loewy returned to his native France in retirement. Knoll Associates, by contrast, managed to maintain its role as a design leader in furniture, but Florence Knoll had left the company in 1965, and the firm changed ownership several times over the subsequent decades. Among the academic institutions, we also find a mixed record with the emigré-founded organizations. The Institute of Design survives to this day, offering masters progams in design and design methods as part of the Illinois Institute of Technology. The University of Michigan's Institute for Social Research still publishes the Consumer Sentiment Index, although George Katona retired in 1972. Paul Lazarsfeld's Bureau of Applied Social Research at Columbia University, however, was closed in 1977, barely a year after Lazarsfeld had passed away. While some designers such as Eva Zeisel (who died in 2011, age 105) remained professionally active into the twenty-first century, the vast majority of the emigrés ended their marketing-related work between the late 1960s and the early 1980s.

The demise of consumer engineering, however, was more than the result of generational change and of a social and intellectual critique of technocratic social engineering. The consumer engineers also encountered a changing business environment that altered the demands for marketing knowledge and the structure of the consulting industry. As professionalization in marketing progressed, the opportunities for outsiders from arts and academia were increasingly closed off. Advertising agencies and corporate departments now hired artists and research specialists trained in the professional programs that emigrés had helped to build over the midcentury decades. By the 1960s, the advent of computers in marketing departments undercut the success of qualitative consumer research methods such as motivation research. The kind of psychological interpretation offered by Dichter and his colleagues now seemed less "scientific" than empirical results based on massive computing power. At the same time, however, the escalating costs of "full-service" marketing consultation offered by the traditional, large advertising agencies such as McCann and JWT drove more and more corporations to switch from a commission to a fee system for their advertising accounts. This tightened revenue from large accounts and opened new opportunities for smaller agencies with less research-driven and, they claimed, more creative marketing approaches.[15]

Market research and research-based design by no means disappear as we move into the 1970s. Yet consumer engineering's faith in individual experts and their capability to engineer products and consumers declined. Consumer goods corporations became increasingly conscious of the importance of a segmented public and of individualized consumer needs, which they attempted to serve with new forms of flexible production and marketing management.[16] Thus, as the Fordist mass production regime in the consumer goods industry declined, technocrats and expert "tastemakers" lost their importance, and flexible marketing management approaches could do without the conception of a malleable consuming mass.[17]

Still, it would be too simplistic to posit a clean break around 1970 with a neat transition from an era of "high modernist" consumer engineering to one of a postmodern pluralism of consumer lifestyles. To the contrary, the ideas of market research and design explored in this book spawned important legacies that connect the interwar emergence of professional marketing to this new phase of consumer capitalism. These legacies include the role of systematically engineered creativity, the continuous drive for scientific sophistication, and an increasingly global negotiation of marketing strategies. Whereas urban planners and other champions of high modernity may have been surprised by the grassroots backlash they encountered beginning in the 1960s, marketers were somewhat better prepared. Engaged with the "paradox of innovation" since the 1930s, consumer engineers had learned to take consumer tastes and desires seriously. While they believed in the power of experts to predict and mold consumer behavior, they also were aware of the limits of such efforts within complex and dynamic modern societies. In this sense, consumer engineering itself contributed to the demise of technocratic high modernism in industry because it stressed the need for creativity and increased flexibility, which would mark a new era of global consumer capitalism in the final decades of the twentieth century.

In their emphasis on consumer aesthetics and continuous innovation, the midcentury consumer engineers anticipated what historians have described as a "creative revolution" in advertising and marketing.[18] Advertisers in particular embraced a new, rebellious tone beginning in the 1960s and engaged in (small and generally well-calculated) transgressions of social norms that often picked up on themes of the countercultural youth movement.[19] Such new creative campaigns drew on the conceptions of brand image and the symbolism of consumer goods developed by psychological consumer research beginning in the 1930s. The campaigns often built upon hedonistic notions of unrestrained consumer desires that motivation researches such as Ernest Dichter had long promoted.[20] Similarly, the new creative campaigns contributed to a further aestheticization of consumption and consumer goods, a process that also traced its roots to the interwar culture of transatlantic modernism. If in modified form, the aesthetics of commercial modernism became ever more mainstream beginning in the 1970s in areas ranging from furniture to graphic design.[21]

While creativity and aesthetics gained in importance for comprehensive marketing management, the drive toward further scientific sophistication did not abate. As marketing departments grew and marketing knowledge became more specialized, concepts of commercial communication and consumer research became ever more complex. With regard to media messaging and reception, communications experts increasingly stressed the social embeddedness of mass communication. They followed the lead of Lazarsfeld and his colleagues in paying attention to the role of opinion leaders and of audience participation in shaping the meaning of advertising messages, product experiences, and consumer aspirations. The growing significance of market segments and the life-worlds of specific groups of consumers to which products could be related also drew on notions of varied consumer motivations developed during the midcentury decades. The 1960s shift back toward empirical models and computer-based analysis in marketing research did not lead to a declining interest in consumer psychology. Instead, cognitive and experimental consumer studies based in part on insights of Gestalt psychologists gained in currency, as did efforts to put psychological categorizations on a more solid empirical footing. On both sides of the Atlantic, researchers now attempted to operationalize brand image conceptions in new empirical ways, as historian Ingo Köhler has shown. German market researcher Bernt Spiegel, for example, developed market models and consumer typologies based on "emotional preferences," an approach that drew on Lewin's conception of social fields and the behavioral attitude research of George Katona and others.[22] Especially with the economic crises of the 1970s, "scientific" market research again promised to help companies steer clear of risks and miscalculations with regard to consumer markets in a seemingly new era of uncertainty.[23]

Historians today tend to discuss the 1970s as a caesura in Western societies, separating the postwar "boom years" with a Fordist labor and consumption regime from a new era characterized by unstable growth but also by a more pluralistic society and individualized forms of consumption.[24] This story about the transatlantic rise and decline of consumer engineering fits the notion of a particular form of structured high modernity coming to an end as we move into the final three decades of the twentieth century. However, we must not overemphasize the 1970s caesura with regard to mass consumption or embrace the contemporary diagnosis of a broad-based "value change" without qualification.[25] The midcentury buying public, as consumer engineers recognized early on, was never a homogenous consuming mass. Instead, consumer markets were increasingly conceptualized as segmented already at midcentury, and consumer researchers were engaged in uncovering a diverse spectrum of motivations and desires. For all their bureaucratic structures, midcentury "Fordist" consumer corporations were hardly dull or lacking a sense of innovation or aesthetics. The tension between the need for continuous creativity and the drive for systematic scientization of marketing dates back well into the interwar years. "Postmodern" consumer capitalism with segmented lifestyle marketing and individualized consumption aesthetics rested squarely on foundations laid by midcentury consumer researchers and commercial designers.

The transatlantic exchanges in marketing facilitated by the cohort of emigré consumer engineers did not come to an end during the final decades of the twentieth century either. The dynamics of these transatlantic exchanges changed, however, as Europe and the United States again became more evenly competitive starting in the 1960s and transatlantic flows were increasingly part of more global processes. As European companies expanded in the U.S. market, they adopted and adapted "American" methods of marketing and advertising.[26] European advertisers gained in strength as they competed with U.S. firms at home as well as abroad.[27] In fact, by the 1970s European advertising agencies such as the French Publicis network became global actors that challenged the midcentury dominance of the large U.S. firms like JWT and McCann. Madison Avenue even experienced a "British invasion" with some American firms being bought up while new "creative" agencies such as London's Saatchi & Saatchi (founded in 1970) succeeded with a new and challenging visual style. More than simply a copy of American "creative" pioneers such as Doyle Dane Bernbach (DDB), the European creatives were inspired by the graphic design work of earlier emigré artists such as Bayer and Moholy-Nagy and the visual traditions of interwar modernism.[28] In general, European design traditions—Scandinavian furniture, French fashion, German appliances, and Italian automobiles—were once again in high esteem in the United States. Much as in the interwar years, American commentators during the 1970s regarded the modern design of European (luxury) goods as a competitive advantage vis-à-vis an ailing consumer goods sector in the United States.[29] European exports, it appeared, had benefited from a postwar professionalization in commercial design to which returning emigrés such as Raymond Loewy had contributed.

A meaningful distinction between "European" and "American" design, however, became increasingly difficult to make. Much like transatlantic modernism in commercial design, an increasing number of multinational corporations and the globalization of marketing further muddied transatlantic distinctions. Marketing consultancies increasingly became part of global networks and holding companies. McCann-Erickson, for example, is today part of the Interpublic Group of Companies with subsidiaries and offices in over one hundred countries and of a structure of specialized agencies that includes direct marketing firms as well as market research and design consultancies.[30] Landor Associates provides a good example for this globalization process in marketing. The firm had heavily internationalized already under Landor's leadership during the 1960s and early 1970s with contracts and offices in Europe and Asia. Today Landor Associates maintains 26 offices in 20 countries, but soon after Landor's retirement in 1989 the firm had become a subsidiary of the advertising agency Young & Rubicam. The Young & Rubicam network in turn belongs to the London-based WPP group that also owns other ad agencies including Ogilvy & Mathers and the American PR firm Hill & Knowlton. Especially since the 1970s, the corporate image strategies that Landor and other design firms developed have increasingly aimed at the "global corporation."[31]

Global brands, however, did not necessarily entail a globally homogenous consumer capitalism. The ideas about cultural translation and local adaptations espoused by emigrés such as Dichter, Knoll, or Loewy did not fall on deaf ears with consumer multinationals. With the exception of certain iconic brands, globally active corporations increasingly "localized" their products and marketing.[32] In doing so, they continued a pattern of cultural globalization based on hybridization and a mélange of interrelated and locally specific consumption practices that the emigré consumer experts had promoted at midcentury.[33] Experts with immigrant backgrounds did not disappear as cultural translators in this new global marketing world either, but in an age of increased global mobility and instant communication, their role has become less conspicuous compared with the midcentury era. Still, individual marketing experts with transnational careers could be found even toward the end of the twentieth century. One need only look at German-born designer Hartmut Esslinger. Educated in design in southern Germany during the late 1960s in the tradition of Bauhaus modernism, he won the 1969 "Good Design" award of the German Design Council. After working for German IT companies and for Sony during the 1970s, Esslinger created the "Snow White" design language for Apple in 1982, which would set the aesthetic tone for the design of Apple products over the subsequent decade.[34] In some ways, then, the legacy of midcentury consumer engineering stretches even to Silicon Valley and its emerging digital economy.

Between the 1930s and 1960s, this book has argued, consumer marketing changed in profound ways. Emigré designers and consumer researchers have been instrumental in fostering some of the most appealing aspects of modern consumer capitalism, its variety and innovative flexibility as well as its ability to mass produce affordable goods that speak to consumer desires. Despite their backgrounds in social reform, however, the consumer capitalism they helped engineer did not completely fulfill the democratic aspirations of its midcentury advocates. Instead, the consumer engineers helped perpetuate a consumer culture and economy that remained riddled with inequalities along the lines of race, class, and gender. Exploiting individual desires and influencing consumer behavior for commercial gain, furthermore, they also exasperated the materialist and ultimately socially and ecologically unsustainable aspects of consumer capitalism that continue to pose significant challenges to this day.

Acknowledgments

This book has witnessed its own share of Atlantic crossings between Germany and the United States during the nearly ten years that it was being conceived and written. I am grateful to a long list of people for their input and support. Hartmut Berghoff helped me develop the project at the German Historical Institute (GHI) in Washington and saw it come to completion at Göttingen University. Similarly, Gary Cross and Uwe Spiekermann shared their insights during the early conceptual stages and later on the finished draft. Reinhild Kreis, too, got to know the project in Washington and then read and commented on countless drafts all the way to the final stages. The manuscript was accepted by the Philosophical Faculty of the Georg-August University Göttingen as part of my habilitation in 2018, and I am grateful to the reviewers in this process, Adelheid v. Saldern, Christian Kleinschmidt, Stefan Haas, Günter Silberer, Dirk Schumann, and especially Hartmut Berghoff, for their comments and feedback. Jeff Fear at Glasgow University read and commented on the entire manuscript, as did two other anonymous readers at the University of Chicago Press. They all helped me hone my arguments in different but important ways.

This book originated in a research project, "Transatlantic Perspectives: Europe in the Eyes of European Immigrants," that I directed between 2010 and 2014 at the GHI and that relied on generous funding by the German Ministry for Education and Research. Much of the early research was discussed with my wonderful project team, Andreas Joch, Barbara Louis, Corinna Ludwig, and Lauren Shaw. Hartmut

ACKNOWLEDGMENTS

Berghoff, Friedrich Lenger, Hartmut Kaelble, Sally Gregory Kohlstedt, and Jan Hesse provided feedback as advisory board members to this project. I also very much value the insights and comments from many other GHI colleagues including Richard Wetzell, David Lazar, Mark Stoneman, Christina Lubinski, Ines Prodoehl, Stefan Hördler, Miriam Rürup, Mario Daniels, Mischa Honeck, Britta Waldschmidt-Nelson, Laura Rischbieter, David Kuchenbuch, Sebastian Teupe, Stefan Link, Kevin Rick, and Joshua Davis, among many others. In discussions, Alexander Nützenadel, David Blackbourne, Ute Frevert, Dirk Hörder, Kathleen Neils Conzen, and other members of the GHI advisory board also gave the project helpful input during its early stages. At Göttingen, Robert Bernsee, Alexander Engel, Stina Barrenscheen, Gabriella Szalay, and especially Ingo Köhler were among those whose comments on presentations and drafts guided this project toward its ultimate shape.

The book further benefited from presentations at academic congresses over the years, including the American Historical Association, the American Studies Association, the Business History Conference, the European Business History Association, the Gesellschaft für Unternehmensgeschichte, and the German Studies Association. I also received valuable feedback on the project at numerous conferences and research seminars, including at Berlin (Humboldt), Copenhagen Business School, Düsseldorf, Frankfurt, Glasgow, Kassel, Konstanz, London (GHI), Minneapolis, Marburg, Regensburg, and Vienna. Colleagues such as Regina Blaszczyk, Peggy Re, Stefan Schwarzkopf, Volker Depkat, Simone Lässig, Daniel Wadhwani, and Nancy Green generously shared their insights on topics ranging from market research, consumer design, and immigrant entrepreneurship to elite migration, knowledge transfers, and biography as history. I am especially grateful to the participants at the 2015 conference "Consumer Engineering" in Göttingen as well as to Molly Nolan and the contributors to the 2014 volume we edited together on "transatlantic crossings" at the middle of the twentieth century.

There are many institutions to which I would like to express my gratitude because they have enabled me to conduct my research: the Archives of American Art, the Bauhaus Archiv, the Benjamin Archiv (Berlin), the Bentley Library (University of Michigan), the German Bundesarchiv (Koblenz and Berlin), the Daley Library (University of Illinois, Chicago), the German Exilarchiv, the German Design Council, the Gesellschaft für Konsumforschung, the Hartman Center for Sales and Advertising (Duke University), the Horkheimer Archive (Frankfurt), the Illinois Institute of Technology, the Library of Congress, the Hagley Museum and Library, the Museum of Modern Art, the National Archives and Records

Administration, the Paul Felix Lazarsfeld Archiv (Vienna), the Smithsonian Institution Archives Center, St. John's University Archives, the Staatsbibliothek (Berlin), and the Werkbund Archiv. At the University of Chicago Press, I would like to thank my editorial team including Douglas Mitchell, Kyle Wagner, and Susan Tarcov for their insights and support. Finally, my research has benefited from many tremendously helpful interns and student assistants at the GHI and here at Göttingen. Last, but not least, Christel Schikora's organizational skills helped me in countless ways. Thank you all!

As much as the finished work brings joy to the author and (hopefully) the reader, the process of writing can also take a heavy toll on friends and family. I am very happy that my kids followed me along on this transatlantic adventure, and I am deeply thankful to Reinhild for encouraging me over these past few years, for providing imaginative advice and gentle prodding, and for helping me get this project across the finish line.

Abbreviations for Archival Sources

AAA Kepes	György Kepes Papers, Archives of American Art, Washington, DC
AAA Knoll	Florence Knoll Papers, Archives of American Art, Washington, DC
AAA M-N	László and Sybil Moholy-Nagy Papers, Archives of American Art, Washington, DC
BeL Katona	George Katona Papers, Bentley Library, University of Michigan, Ann Arbor
BeL Likert	Rensis Likert Papers, Bentley Library, University of Michigan, Ann Arbor
BhA	Bauhaus Archiv, Museum für Gestaltung, Berlin
BhA Bayer	Herbert Bayer Folders, Bauhaus Archiv, Berlin
BhA Breuer	Marcel Breuer Folders, Bauhaus Archiv, Berlin
BhA Gropius	Walter Gropius Folders, Bauhaus Archiv, Berlin
BhA Kramer	Ferdinand Kramer Folders, Bauhaus Archiv, Museum für Gestaltung, Berlin
BhA Moholy-Nagy	László Moholy-Nagy Folders, Bauhaus Archiv, Berlin
BjA Adorno	Adorno Papers, Benjamin Archiv, Akademie der Künste, Berlin
BuArch	Bundesarchiv, Koblenz
DaW	Dichter Archiv, University of Vienna
GfK	Gesellschaft für Konsumforschung, Nürnberg
HaC Lanigan	Denis Lanigan Papers, Hartman Center for Sales and Advertising, Duke University, Durham, NC
HgA Dichter	Dichter Papers, Hagley Museum and Library, Wilmington, DE

ABBREVIATIONS FOR ARCHIVAL SOURCES

HgA Loewy	Loewy Papers, Hagley Museum and Library, Wilmington, DE
HkA	Horkheimer Archiv, University of Frankfurt
IIT	Institute of Design Records, Illinois Institute of Technology, Chicago
LaP	Walter Landor Papers, Landor Design Collection at Smithsonian Institution Archives Center
LoC	Library of Congress, Washington, DC
LoW	Raymond Loewy Papers, Library of Congress, Washington, DC
MoMA	Museum of Modern Art Library / Online Archives
Nara	National Archives and Records Administration, Washington, DC
NaX	Exilarchiv, Deutsche Nationalbibliothek, Frankfurt
PLA	Paul Lazarsfeld Archiv, University of Vienna
PP St. Johns	Politz Papers, St. John's University, New York
RfF	Rat für Formgebung / German Design Council, Frankfurt
UIC	Daley Library, University of Illinois, Chicago
WbA Kramer	Ferdinand Kramer Papers, Werkbund Archiv, Museum der Dinge, Berlin

Notes

INTRODUCTION

1. On Gruen, see Jeffrey Hardwick, *Mall Maker: Victor Gruen, Architect of an American Dream* (Philadelphia: University of Pennsylvania Press, 2004).
2. Lizabeth Cohen, *A Consumers' Republic: The Politics of Mass Consumption in Postwar America* (New York: Alfred A. Knopf, 2003). See also Meg Jacobs, *Pocketbook Politics: Economic Citizenship in Twentieth-Century America* (Princeton: Princeton University Press, 2007), and Robert Collins, *More: The Politics of Economic Growth in Postwar America* (New York: Oxford University Press, 2002).
3. Victor Gruen and Larry Smith, *Shopping Towns USA: The Planning of Shopping Centers* (New York: Reinhold, 1960).
4. Gruen and Smith, *Shopping Towns*, 140–71, quote p. 148.
5. Lizabeth Cohen, "From Town Center to Shopping Center: The Reconfiguration of Community Marketplaces in postwar America," *American Historical Review* 101 (1996): 1050–81. On Gruen's transatlantic career see also Joseph Malherek, "Shopping Malls and Social Democracy: Victor Gruen's Postwar Campaign for Conscientious Consumption in American Suburbia," in *Consumer Engineering: Marketing between Expert Planning and Consumer Responsiveness, 1920s–1970s*, ed. Jan Logemann, Gary Cross, Ingo Köhler, and Mark Stoneman (New York: Palgrave, 2019).
6. Victoria de Grazia, *Irresistible Empire: America's Advance through Twentieth-Century Europe* (Cambridge: Harvard University Press, 2006).
7. Victor Gruen, *The Heart of Our Cities: The Urban Crisis: Diagnosis and Cure* (New York: Simon and Schuster, 1967).

8. Jan Logemann, *Trams or Tailfins?* (Chicago: University of Chicago Press, 2012).
9. On ethnic consumption, see e.g. Donna Gabaccia, *We Are What We Eat: Ethnic Food and the Making of Americans* (Cambridge: Harvard University Press, 1998).
10. See Andrew Godley, *Jewish Immigrant Entrepreneurship in New York and London, 1880–1914: Enterprise and Culture* (Basingstoke, UK: Palgrave, 2001), and recently Hartmut Berghoff and Uwe Spiekermann, eds., *Immigrant Entrepreneurship: The German-American Experience since 1700* (Washington, DC: German Historical Institute 2016).
11. See, e.g., Jeffrey Fear and Paul Lerner, "Behind the Screens: Immigrants, Émigrés, and Exiles in Mid-Twentieth-Century Los Angeles," *Jewish Culture and History* 17 (2016): 1–21.
12. See, e.g., David Hounshell, *From the American System to Mass Production, 1800–1932: The Development of Manufacturing Technology in the United States* (Baltimore: Johns Hopkins University Press, 1985).
13. On the growth consensus of the 1940s, see Cohen, *Consumers' Republic*, and Alan Brinkley, *The End of Reform: New Deal Liberalism in Recession and War* (New York: Alfred A. Knopf, 1995).
14. On high modernity, see James C. Scott, *Seeing like a State: How Certain Schemes to Improve the Human Condition Have Failed* (New Haven: Yale University Press, 1998).
15. See Robert Gordon, *The Rise and Fall of American Growth: The U.S. Standard of Living since the Civil War* (Princeton: Princeton University Press, 2016). See also Alexander Field, "The Most Technologically Progressive Decade of the Century," *American Economic Review* 93 (2003): 1399–1430, and Field, *A Great Leap Forward: 1930s Depression and U.S. Economic Growth* (New Haven: Yale University Press, 2012).
16. Oliver Zunz, *Why the American Century?* (Chicago: University of Chicago Press, 1998).
17. See Mary Nolan, *The Transatlantic Century: Europe and America, 1890–2010* (New York: Cambridge University Press, 2014).
18. Joseph Schumpeter, "The Creative Response in Economic History," *Journal of Economic History* 7 (1947): 154.
19. Ibid., 151.
20. Joseph Schumpeter cited in Thomas McCraw, *Prophet of Innovation: Joseph Schumpeter and Creative Destruction* (Cambridge: Harvard University Press, 2007), 9.
21. Roy Sheldon and Egmont Arens, *Consumer Engineering: A New Technique for Prosperity* (New York: Harper, 1932).
22. Ibid., 1.
23. Ibid., 52. See also Bernard London, *Ending the Depression through Planned Obsolescence* (New York, 1932), and Giles Slade, *Made to Break: Technology and Obsolescence in America* (Cambridge: Harvard University Press, 2006), esp. 47–81.

24. Ibid., 54–55.
25. Ibid., 55.
26. See, e.g., D. G. Brian Jones and Mark Tadajewski, eds., *The Routledge Companion to Marketing History* (London: Routledge, 2016); Hartmut Berghoff, Philip Scranton, and Uwe Spiekermann, eds., *The Rise of Marketing and Market Research* (New York: Palgrave Macmillan, 2012); and Roy Church and Andrew Godley, eds., *The Emergence of Modern Marketing* (London: Frank Cass, 2003). See also Susan Strasser, *Satisfaction Guaranteed: The Making of the American Mass Market* (Washington, DC: Smithsonian Institution Press, 1995), and Pamela Laird, *Advertising Progress: American Business and the Rise of Consumer Marketing* (Baltimore: Johns Hopkins University Press, 2001), as well as Thomas Frank, *The Conquest of Cool: Business Culture, Counterculture, and the Rise of Hip Consumerism* (Chicago: University of Chicago Press, 1998), and Alexander Sedlmaier and Stefan Malinowski, "1968 als Katalysator der Konsumgesellschaft: Performative Regelverstöße, kommerzielle Adaptionen und ihre gegenseitige Durchdringung," *Geschichte und Gesellschaft* 32 (2006): 238–67.
27. Gerulf Hirt, *Verkannte Propheten? Zur "Expertenkultur" (west-)deutscher Werbekommunikatoren bis zur Rezession 1966/67* (Leipzig: Leipzig University Press, 2013).
28. See Hartmut Berghoff, ed., *Marketinggeschichte: Die Genese einer modernen Sozialtechnik* (Frankfurt am Main: Campus, 2007).
29. Dawn Spring, *Advertising in the Age of Persuasion: Building Brand America, 1941–1961* (New York: Palgrave Macmillan, 2013).
30. See Matthias Kipping and Lars Engwall, eds., *Management Consulting: Emergence and Dynamics of a Knowledge Industry* (Oxford: Oxford University Press, 2002). The term "knowledge industry" incidentally gained currency during the 1960s with two emigrés, economist Fritz Machlup und management consultant Peter Drucker. See Peter Drucker, *The Age of Discontinuity: Guidelines to Our Changing Society* (New York: Harper & Row, 1968), and Fritz Machlup, *The Production and Distribution of Knowledge in the United States* (Princeton: Princeton University Press, 1962).
31. See Christopher McKenna, *The World's Newest Profession: Management Consulting in the Twentieth Century* (Cambridge: Cambridge University Press, 2010).
32. Matthias Kipping, "American Management Consulting Companies in Western Europe, 1920 to 1990: Products, Reputation and Relationships," *Business History Review* 73 (1999): 190–220.
33. Andrew Godley and Mark Casson, "'Doctor, Doctor . . .': Entrepreneurial Diagnosis and Market Making," *Journal of Institutional Economics* 11 (2015): 601–21.
34. See Regina Blaszczyk, ed., *Producing Fashion: Commerce, Culture, and Consumers* (Philadelphia: University of Pennsylvania Press, 2008). For the furniture industry see Per Hansen, "Networks, Narratives and New Markets: The

Rise and Decline of Danish Modern Furniture Design, 1930–1970," *Business History Review* 80 (2006): 449–83.
35. Regina Blaszczyk, *Imagining Consumers: Design and Innovation from Wedgwood to Corning* (Baltimore: Johns Hopkins University Press, 2000).
36. Jeffrey Meikle, *Twentieth Century Limited: Industrial Design in America, 1925–1939*, 2nd ed. (Philadelphia: Temple University Press, 2010).
37. See David Gartman, *Auto Opium: A Social History of American Automobile Design* (London: Routledge, 1994), and John Harwood, *The Interface: IBM and the Transformation of Corporate Design, 1945–1976* (Minneapolis: University of Minnesota Press, 2011), on the prominent role of Elliot Noyes at IBM.
38. On advertising art and interwar aesthetic capitalism, see Terry Smith, *Making the Modern: Industry, Art, and Design in America* (Chicago: University of Chicago Press, 1993).
39. James Sloan Allen, *The Romance of Commerce and Culture: Capitalism, Modernism, and the Chicago-Aspen Crusade for Cultural Reform*. 2nd ed. Boulder: University Press of Colorado, 2002).
40. See Blaszczyk, *Producing Fashion*, and Thomas Hine, *Populuxe* (London: Bloomsbury, 1987).
41. See Stefan Schwartzkopf, "In Search of the Consumer: The History of Market Research from 1890 to 1960," in Jones and Tadajewski, *Routledge Companion to Marketing History*, 61–84.
42. Ingo Köhler, "Imagined Images, Surveyed Consumers: Market Research as a Means of Consumer Engineering, 1950s–1980s," in Logemann, Cross, and Köhler, *Consumer Engineering*.
43. For a European perspective, see, e.g., Kerstin Brückweh, ed., *The Voice of the Citizen Consumer: A History of Market Research, Consumer Movements, and the Public Sphere* (Oxford: Oxford University Press, 2011).
44. See, e.g., Ronald Fullerton, "The Birth of Consumer Behavior: Motivation Research in the 1940s and 1950s," *Journal of Historical Research in Marketing* 5 (2013): 212–22, and Ronald Fullerton, "The Art of Marketing Research: Selection from Paul F. Lazarsfeld's Shoe Buying in Zurich (1933)," *Journal of the Academy of Marketing Science* 18 (1990): 319–27.
45. See Lawrence Samuel, *Freud on Madison Avenue: Motivation Research and Subliminal Advertising in America* (Philadelphia: University of Pennsylvania Press, 2010), and Sarah Igo, *The Averaged American: Surveys, Citizens, and the Making of a Mass Public* (Cambridge: Harvard University Press, 2007). See also Nepomuk Gasteiger, *Der Konsument: Verbraucherbilder in Werbung, Konsumkritik und Verbraucherschutz* (Frankfurt: Campus, 2010).
46. Andreas Reckwitz, *Die Erfindung der Kreativität. Zum Prozess gesellschaftlicher Ästhetisierung* (Berlin: Suhrkamp, 2012), esp. 133–45.
47. See especially de Grazia, *Irresistible Empire*. On Americanization in commercial and youth culture, see also Uta Poiger, *Jazz, Rock, and Rebels: Cold War Politics and American Culture in a Divided Germany* (Berkeley: University of California Press, 2000), and Robert Rydell and Rob Kroes, *Buffalo Bill in*

Bologna: The Americanization of the World, 1869–1922 (Chicago: University of Chicago Press, 2005). For business and industry, see, e.g., Harm Schröter, *Americanization of the European Economy: A Compact Survey of American Economic Influence in Europe since the 1880s* (Dordrecht, The Netherlands: Springer, 2005).

48. Christian Kleinschmidt, *Der produktive Blick: Wahrnehmung amerikanischer und japanischer Management- und Produktionsmethoden durch deutsche Unternehmer 1950–1985* (Berlin: Akademie Verlag, 2002).
49. Logemann, *Trams or Tailfins?*
50. Still a model for studying transnational transfers in the Atlantic world during the first half of the twentieth century is Daniel Rodgers, *Atlantic Crossings: Social Politics in a Progressive Age* (Cambridge: Belknap Press, 1998). On transnational history more generally, see Ian Tyrell, "Reflections on the Transnational Turn in United States History: Theory and Practice," *Journal of Global History* 3 (2009): 453–74.
51. On the reciprocity of transatlantic transfers, see Nolan, *Transatlantic Century*, and Jan Logemann and Mary Nolan, eds., *More Atlantic Crossings? European Voices and the Postwar Atlantic Community* (Washington, DC: German Historical Institute, 2014).
52. See, e.g., Jan Nederveen Pieterse, *Globalization and Culture: Global Melange* (Oxford: Rowman and Littlefield, 2004).
53. Harold Kassarjian, "Scholarly Traditions and European Roots of American Consumer Research," in Giles Laurent, Gary Lilien, and Bernard Pras, eds., *Research Traditions in Marketing* (Boston: Kluwer, 1993), 265–79.
54. See, e.g., Samuel, *Freud on Madison Avenue*, and Allen, *Romance of Commerce and Culture*. See also Daniel Horowitz, "The Émigré as Celebrant of American Consumer Culture," in Susan Strasser et al., eds., *Getting and Spending: European and American Consumer Societies in the Twentieth Century* (Cambridge: Cambridge University Press, 1998), 149–66, and most recently Joseph Malherek, "Qualitative Capitalism and Continental Critique: Émigré Social Scientists Encounter the American Consumer, 1933–45," *Ideas in History* 6 (2012): 65–92.
55. See, e.g., Mitchell Ash and Alfons Söllner, eds., *Forced Migration and Scientific Change: Émigré German-Speaking Scientists and Scholars after 1933* (Washington, DC: German Historical Institute, 1996), and more recently Corinna Unger, *Reise ohne Wiederkehr? Leben im Exil 1933 bis 1945* (Darmstadt: Primus, 2009).
56. See, e.g., Lewis Coser, *Refugee Scholars in America: Their Impact and Their Experiences* (New Haven: Yale University Press,1984), and Margit Seckelmann, "'Mit seltener Objektivität': Fritz Morstein Marx—Die mittleren Jahre (1934–1961)," *Die Öffentliche Verwaltung* 67 (2014): 1029–48.
57. See Ash and Söllner, *Forced Migration and Scientific Change*, and Sibylle Quack, *Zuflucht Amerika. Zur Sozialgeschichte der Emigration deutsch-jüdischer Frauen in die USA 1933–1945* (Bonn: Dietz, 1995).

58. For the visual arts see Marion Deshmukh, "The Visual Arts and Cultural Migration in the 1930s and 1940s: A Literature Review," *Central European History* 41 (2008): 569–604, and for design, Alison Clarke and Elana Shapira, eds., *Émigré Cultures in Design and Architecture* (London: Bloomsbury Academic Press, 2017). See also Claus-Dieter Krohn, *Refugees in Exile: Refugee Scholars and the New School for Social Research* (Amherst: University of Massachusetts Press, 1993); Christian Fleck, *Transatlantische Bereicherungen. Zur Erfindung der empirischen Sozialforschung* (Frankfurt am Main: Suhrkamp, 2007), and Thomas Wheatland, *The Frankfurt School in Exile* (Minneapolis: University of Minnesota Press, 2009) for the social sciences.
59. See Berghoff and Spiekermann, *Immigrant Entrepreneurship*, and Hartmut Berghoff and Andreas Fahrmeir, "Unternehmer und Migration. Einleitung," *Zeitschrift für Unternehmensgeschichte* 58 (2013): 141–48.
60. See Ursula Seeber et al., eds., "Kometen des Geldes". Ökonomie und Exil [special issue], *Exilforschung. Ein Internationales Jahrbuch* 33 (2015). On emigré entrepreneurs between the continents, see Martin Münzel, *Die jüdischen Mitglieder der deutschen Wirtschaftselite, 1927–1955. Verdrängung, Emigration, Rückkehr* (Paderborn: Schöningh, 2006).
61. See Rodgers, *Atlantic Crossings*.
62. See, e.g., Madeleine Herren, *Internationale Organisationen seit 1865. Eine Globalgeschichte der internationalen Ordnung* (Darmstadt: Wissenschaftliche Buchgesellschaft, 2009).
63. See Giuliana Gemelli and Roy MacLeod, *American Foundations in Europe: Grant-Giving Policies, Cultural Diplomacy, and Trans-Atlantic Relations, 1920–1980* (Brussels: Peter Lang, 2003), and Isabella Löhr, "Solidarity and the Academic Community: The Support Networks for Refugee Scholars in the 1930s," *Journal of Modern European History* 12 (2014): 231–46.
64. See Dittmar Dahlmann and Reinhold Reith, eds., *Elitenwanderung und Wissenstransfer im 19. und 20. Jahrhundert* (Essen: Klartext 2008).
65. On biography in transnational history, see Volker Depkat, "Biographieforschung im Kontext transnationaler und globaler Geschichtsschreibung," *BIOS* 28 (2015): 3–18, and Jan Logemann, "Transatlantische Karrieren und transnationale Leben. Zum Verhältnis von Migrantenbiographien und transnationaler Geschichte," *BIOS* 28 (2015): 80–101.
66. See Matthew Pratt Guterl, "The Futures of Transnational History. Comment to AHR Forum Transnational Lives in the Twentieth Century," *AHR* 118 (2013): 130–39. On elite migration, see Nancy Green, *The Other Americans in Paris: Businessmen, Countesses, Wayward Youth, 1880–1941* (Chicago: Chicago University Press, 2014).
67. For the concept of "cultural brokers," see Mark Häberlein and Alexander Keese, eds., *Sprachgrenzen—Sprachkontakte—kulturelle Vermittler. Kommunikation zwischen Europäern und Außereuropäern* (Stuttgart: Franz Steiner, 2010).
68. On cultural translation, see Simone Lässig, "Übersetzung in der Geschichte—Geschichte als Übersetzung? Überlegungen zu einem analytischen Konzept

und Forschungsgegenstand für die Geschichtswissenschaft," *Geschichte und Gesellschaft* 38 (2012): 189–216, and Doris Bachmann-Medick, "The Translational Turn," *Translation Studies* 2 (2009): 2–16.

69. See Anselm Doering-Manteuffel, *Wie westlich sind die Deutschen? Amerikanisierung und Westernisierung im 20. Jahrhundert* (Göttingen: Vandenhoeck & Ruprecht, 1999).

70. On the concept of transatlantic intermediaries and their role in postwar Germany, see especially Arndt Bauerkämper et al., eds., *Demokratiewunder. Transatlantische Mittler und die kulturelle Öffnung Westdeutschlands 1945–1970* (Göttingen: Vandenhoeck & Ruprecht, 2005). On the importance of elite networks, see also Volker Berghahn, *Industriegesellschaft und Kulturtransfer: Die deutsch-amerikanischen Beziehungen im 20. Jahrhundert* (Göttingen: Vandenhoeck & Ruprecht, 2010). More generally on the role of networks in transnational history, see Berthold Unfried et al., eds, *Transnationale Netzwerke im 20. Jahrhundert* (Vienna: Akademische Verlagsanstalt, 2008).

71. On the cultural and intellectual Cold War, see, e.g., Reinhold Wagnleitner, *Coca-Colonization and the Cold War: The Cultural Mission of the United States in Austria after the Second World War* (Chapel Hill: University of North Carolina Press, 1994); Volker Berghahn, *America and the Intellectual Cold Wars in Europe: Shepard Stone between Philanthropy, Academy, and Diplomacy* (Princeton: Princeton University Press, 2001); and Giles Scott-Smith and Hans Krabbendam, eds., *The Cultural Cold War in Western Europe, 1945–1960* (London: Frank Cass, 2003).

72. See Udi Greenberg, *The Weimar Century: German Émigrés and the Ideological Foundations of the Cold War* (Princeton: Princeton University Press, 2014), and Daniel Bessner, *Democracy in Exile: Hans Speier and the Rise of the Defense Intellectual* (Ithaca: Cornell University Press, 2018).

73. See Stuart Ewen, *Captains of Consciousness: Advertising and the Social Roots of the Consumer Culture* (New York: McGraw-Hill, 1976).

74. On social engineering, see, e.g., Kerstin Brückweh et al., eds., *Engineering Society: The Role of the Human and Social Sciences in Modern Societies, 1880–1980* (Basingstoke: Palgrave Macmillan, 2012), and Thomas Etzemüller, ed., *Die Ordnung der Moderne: Social Engineering im 20. Jahrhundert* (Bielefeld: transcript Verlag, 2009).

75. See Scott, *Seeing like a State*. See also Ulrich Herbert, "Europe in High Modernity: Reflections on a Theory of the 20th Century," *Journal of Modern European History* 5 (2007): 5–21, and Lutz Raphael, "Ordnungsmuster der 'Hochmoderne'? Die Theorie der Moderne und die Geschichte der europäischen Gesellschaften im 20. Jahrhundert," in Lutz Raphael and Ute Schneider, eds., *Dimensionen der Moderne* (Frankfurt am Main: Peter Lang, 2008), 73–92.

76. On the 1930s as the "engineering decade," see Gordon, *Rise and Fall*, 535–65.

77. On the machine age and its cultural fascination with "efficiency," "planning," and "social control," see John Jordan, *Machine-Age Ideology: Social*

Engineering and American Liberalism, 1911–1939 (Chapel Hill: University of North Carolina Press, 1994).
78. Christina Cogdell, *Eugenic Design: Streamlining America in the 1930s* (Philadelphia: University of Pennsylvania Press, 2004).
79. David Engerman and Corinna Unger, "Towards a Global History of Modernization," *Diplomatic History* 33 (2009): 375–85, and Kiran Klaus Patel, *The New Deal: A Global History* (Princeton: Princeton University Press, 2016).
80. See, e.g., Ruth Oldenziel and Karin Zachmann, eds., *Cold War Kitchen: Americanization, Technology, and European Users* (Cambridge: MIT Press, 2011), and Stefan Schwarzkopf, "Advertising, Emotions, and 'Hidden Persuaders': The Making of Cold-War Consumer Culture in Britain from the 1940s to the 1960s," in Annette Vowinckel, Marcus M. Payk, and Thomas Lindenberger, eds., *Cold War Cultures: Perspectives on Eastern and Western European Societies* (New York: Berghahn, 2012), 172–90.
81. See Mark Solovey and Hamilton Cravens, eds., *Cold War Social Science Knowledge Production, Liberal Democracy, and Human Nature* (New York: Palgrave Macmillan, 2012).

CHAPTER ONE

1. Lucian Bernhard, "Putting Beauty into Industry," *Advertising Arts* (a section of *Advertising and Selling*), Jan. 8, 1930, 34.
2. See Sheldon and Arens, *Consumer Engineering*.
3. Earnest Elmo Calkins, "The New Consumption Engineer and the Artist" (1930), cited in Meikle, *Twentieth Century Limited*, 70.
4. Bernhard, "Putting Beauty into Industry," 36.
5. On Bernhard, see Hubert Riedel and Ursula Zeller, eds., *Bernhard: Werbung und Design im Aufbruch des 20. Jahrhunderts* (Stuttgart: Institut f. Auslandsbeziehungen, 1999).
6. On the contemporary notion of cultural cross-fertilization through forced migration, see Paul Tillich, "Mind and Migration," *Social Research* 4 (1937): 295–305, here 295.
7. Sheldon and Arens, *Consumer Engineering*, 15. See esp. chap. 1, "Out of Production-minded Chaos," 15–29, for the following discussion.
8. Gary Cross and Robert Proctor, *Packaged Pleasures: How Technology and Marketing Revolutionized Desire* (Chicago: University of Chicago Press, 2014).
9. See especially Philip Scranton, *Endless Novelty: Specialty Production and American Industrialization, 1865–1925* (Princeton: Princeton University Press, 1997), and Blaszczyk, *Imagining Consumers*.
10. On the emergence of mass marketing, see especially Strasser, *Satisfaction Guaranteed*.
11. Vicki Howard, *From Main Street to Mall: The Rise and Fall of the American Department Store* (Philadelphia: University of Pennsylvania Press, 2015),

and William Leach, *Land of Desire: Merchants, Power, and the Rise of a New American Culture* (New York: Pantheon, 1993).
12. Tracey Deutsch, *Building a Housewife's Paradise: Gender, Politics, and American Grocery Stores in the Twentieth Century* (Chapel Hill: University of North Carolina Press, 2012); Marc Levinson, *The Great A&P and the Struggle for Small Business in America* (New York: Farrar, Straus and Giroux, 2013), and Susan Spellman, *Cornering the Market: Independent Grocers and Innovation in American Small Business* (Oxford: Oxford University Press, 2016).
13. See, e.g., Regina Blaszczyk, *American Consumer Society, 1865–2005: From Hearth to HDTV* (Wheeling, IL: Harlan Davidson, 2009), 116–36.
14. Marina Moskowitz, *Standard of Living: The Measure of the Middle Class in Modern America* (Baltimore: Johns Hopkins University Press, 2004).
15. Roland Marchand, *Advertising the American Dream: Making Way for Modernity, 1920–1940* (Berkeley: University of California Press, 1996), and Laird, *Advertising Progress*.
16. Rob Schorman, "Claude Hopkins, Earnest Calkins, Bissell Carpet Sweepers and the Birth of Modern Advertising," *Journal of the Gilded Age and Progressive Era* 7 (2008): 181–219.
17. See, e.g., Charles McGovern, *Sold American: Consumption and Citizenship, 1890–1945* (Chapel Hill: University of North Carolina Press, 2006), 23–60.
18. Walter Friedman, *Birth of a Salesman: The Transformation of Selling in America* (Cambridge: Harvard University Press, 2004).
19. Roland Marchand, *Creating the Corporate Soul: The Rise of Public Relations and Corporate Imagery in American Big Business* (Berkeley: University of California Press, 2001). See also Richard Tedlow, *New and Improved: The Story of Mass Marketing in America* (New York: Basic Books, 1990).
20. On the impact of the Great Depression, see Gary Cross, *An All-Consuming Century: Why Commercialism Won in Modern America* (New York: Columbia University Press, 2002), 67–82.
21. Sheldon and Arens, *Consumer Engineering*, 29.
22. Kazuo Usui, *The Development of Marketing Management: The Case of the USA, c. 1910–1940* (Farnham, UK: Ashgate, 2009), esp. 65–77.
23. Sheldon and Arens, *Consumer Engineering*, 19.
24. Ibid.
25. Ibid.
26. Ibid., 164 and 107.
27. Ibid., 173.
28. Meikle, *Twentieth Century Limited*, chap.1. See also Blaszczyk, *American Consumer Society*, 126.
29. Smith, *Making the Modern*, and Michael Augspurger, *An Economy of Abundant Beauty: Fortune Magazine and Depression America* (Ithaca: Cornell University Press, 2004).
30. Meikle, *Twentieth Century Limited*, 21–23.
31. Miriam Beard, "Business Cultivates the Arts," *New York Times*, Jan. 31, 1926.

32. See, e.g., Adelheid v. Saldern, "Transatlantische Konsumleitbilder und ihre Übersetzung," in Heinz Haupt and Claudius Torp, eds., *Die Konsumgesellschaft in Deutschland* (Frankfurt: Campus, 2009), 389–402.
33. Elizabeth Cary, "Six Countries Exhibit," *New York Times*, May 13, 1928.
34. Cf. "Glassmaker's Art to Be Shown," *New York Times*, Nov. 4 1929.
35. "Supports Home Design," *New York Times*, Sept. 10, 1935.
36. On the early American reception of Bauhaus art, see esp. Margret Kentgens-Craig, *The Bauhaus and America: First Contacts, 1919–1936* (Cambridge: MIT Press, 1999).
37. Edward Alden Jewell, "Bringing Art to the Factory and to the Machine," *New York Times*, July 1, 1928.
38. Earnest Elmo Calkins, "Art as a Means to an End," *Advertising Arts*, Jan. 8, 1930, 17–23.
39. Abbott Kimball, "Beauty and the Balance Sheet," *Advertising Arts*, May 1931, 28–32.
40. Industrial Art Council Inc., "How Can You Use Art as a Profit-Maker in Your Business?" *Advertising Arts*, Jan. 8, 1930.
41. Joseph Sinel, "What Is the Future of Industrial Design?" *Advertising Arts*, July 9, 1930, 17–18.
42. Egmont Arens, "Package Engineering," *Advertising Arts*, Jan. 1931, 13–20.
43. Nathan Horwitt, "The Coffers of Taste, Where Fortunes Lie," *Advertising Arts*, Apr. 2, 1930, 8.
44. See, e.g., Dr. M. F. Agha, "What Makes a Magazine Modern?" and Amos Stote, "Jean Carlu," *Advertising Arts*, Oct. 1, 1930.
45. Walter Dorwin Teague, "Will It Last?" *Advertising Arts*, Mar. 1931, 13–19.
46. One of the champions of this cause of American design independence was German American banker and patron of the arts Otto Kahn, who claimed that "in importing good taste from Europe we have paid enough to liquidate all our war debts." See "Style Leadership Planned for Nation," *New York Times*, Feb. 4, 1933.
47. Walter Storey, "Beauty in Machine Products," *New York Times*, Mar. 6, 1932.
48. See, e.g., Nolan, *Visions of Modernity: American Business and the Modernization of Germany* (Oxford: Oxford University Press, 1994), and Egbert Klautke, *Unbegrenzte Möglichkeiten: "Amerikanisierung" in Deutschland und Frankreich, 1900–1933* (Stuttgart: Steiner, 2003). For the American side of this debate, see Adelheid v. Saldern, *Amerikanismus: Kulturelle Abgrenzung von Europa und US-Nationalismus im frühen 20. Jahrhundert* (Stuttgart: Steiner, 2013).
49. See esp. de Grazia, *Irresistible Empire*. On advertisers, see Alexander Schug, "Wegbereiter der modernen Absatzwerbung in Deutschland: Advertising Agencies und die Amerikanisierung der deutschen Werbebranche in der Zwischenkriegszeit," in *WerkstattGeschichte* 34 (2003): 29–51.
50. See, e.g., Erika Rappaport, *Shopping for Pleasure: Women in the Making of London's West End* (Princeton: Princeton University Press, 2000).

51. See Janet Ward, *Weimar Surfaces: Urban Visual Culture in 1920s Germany* (Berkeley: University of California Press, 2001).
52. Heinz-Gerhard Haupt, *Konsum und Handel: Europa im 19. und 20. Jahrhundert* (Göttingen: Vandenhoeck, 2003). On the notion of a "consumer revolution," see Paul Lerner, *The Consuming Temple: Jews, Department Stores, and the Consumer Revolution in Germany, 1880–1940* (Ithaca: Cornell University Press, 2015), 10–15.
53. See esp. Christian Kleinschmidt and Florian Triebel, *Marketing: Historische Aspekte der Wettbewerbs- und Absatzpolitik* (Essen: Klartext, 2004), and Berghoff, *Marketinggeschichte*. See also Stefan Schwartzkopf, "What Was Advertising? The Invention, Rise, Demise and Disappearance of Advertising Concepts in Nineteenth- and Twentieth-Century Europe and America," *Business and Economic History On-Line* 7 (2009).
54. On the professionalization of the German advertising industry, see Hirt, *Verkannte Propheten*.
55. See Ward, *Weimar Surfaces*, chap. 1, and Alexander Schug, "Werbung und die Kultur des Kapitalismus," in Haupt and Torp, *Die Konsumgesellschaft in Deutschland*, 361.
56. See esp. Uwe Spiekermann, "'Der Konsument muss erobert werden!' Agrar- und Handelsmarketing in Deutschland während der 1920er und 1930er Jahre," in Berghoff, *Marketinggeschichte*, 123–47, who also emphasizes the often overlooked role of agricultural marketing in pioneering market research and marketing strategies.
57. See Alexandra Artley, ed., *The Golden Age of Shop Design: European Shop Interiors, 1880–1939* (London: Architectural Press, 1975), and Lerner, *Consuming Temple*, chap. 4. See also Geoffrey Crossick and Serge Jaumain, eds., *Cathedrals of Consumption: The European Department Store, 1850–1939* (Aldershot, UK: Ashgate, 1999). See also Antje Bornemann et al., eds., *Konsum und Gestalt: Leben und Werk von Salman Schocken und Erich Mendelsohn vor 1933 und im Exil* (Berlin: Hentrich & Hentrich, 2016).
58. The German Bund der Schaufensterdekorateure counted four thousand members by 1930. See Ward, *Weimar Surfaces*, chap. 4. On the history of window dressing, see also Uwe Spiekermann, "Display Windows and Window Displays in German Cities: Towards the History of a Commercial Breakthrough," in Clemens Wischermann and Elliott Shore, eds., *Advertising and the European City: Historical Perspectives* (Aldershot, UK: Ashgate, 2000), 139–71.
59. See, e.g., Elisabeth v. Stephani-Hahn, *Schaufensterkunst: Lehrsätze und Erläuterungen*, 3rd ed. (Berlin: Schottländer, 1926).
60. Uwe Spiekermann, *Basis der Konsumgesellschaft: Entstehung und Entwicklung des modernen Kleinhandels in Deutschland, 1850–1914* (Munich: C. H. Beck, 1999).
61. Peter Borscheid, "Agenten des Konsums: Werbung und Marketing," in Heinz-Gerhard Haupt and Claudius Torp, eds., *Die Konsumgesellschaft in Deutschland, 1890–1990* (Frankfurt: Campus, 2009), 84.

62. Roman Rossfeld, *Schweizer Schokolade: Industrielle Produktion und kulturelle Konstruktion eines nationalen Symbols, 1860–1920* (Baden: Hier + Jetzt, 2007), and Angelika Epple, *Das Unternehmen Stollwerck. Eine Mikrogeschichte der Globalisierung* (Frankfurt: Campus, 2010).

63. See Gasteiger, *Der Konsument*, 38, and Holm Friebe, "Branding German: Hans Domizlaff's Markentechnik and Its Ideological Impact," in Pamela Swett, Jonathan Wiesen, and Jonathan Zatlin, eds., *Selling Modernity: Advertising in Twentieth-Century Germany* (Durham: Duke University Press, 2008), 82.

64. On the development of design in interwar Germany, see, e.g., Gerd Selle, *Design-Geschichte in Deutschland: Produktkultur als Entwurf und Erfahrung* (Cologne: DuMont, 1987), 114–217, and on the period until World War I see John Heskett, *Design in Germany, 1870–1918* (London: Trefoil Books, 1986).

65. See Till Buddensieg and Ian Boyd, *Industriekultur: Peter Behrens and the AEG, 1907–1914* (Cambridge: MIT Press, 1984), and Frederic Schwartz, "Commodity Signs: Peter Behrens, the AEG, and the Trademark," *Journal of Design History* 9 (1996): 153–84.

66. On the Werkbund, see Frederick Schwartz, *The Werkbund: Design Theory and Mass Culture before the First World War* (New Haven: Yale University Press, 1996), as well as Joan Campbell, *The German Werkbund: The Politics of Applied Arts* (Princeton: Princeton University Press, 1978).

67. Minutes of the 1907 Werkbund Congress in Werkbundarchiv Berlin, folder "Jahresberichte/Protokolle."

68. Already in 1912/13 the Werkbund organized a first exhibition of "German applied arts" and consumer goods to tour seven cities in the United States.

69. See de Grazia, *Irresistible Empire*, 250–60, and Hubert Riedel and Ursula Zeller, eds., *Lucian Bernhard: Werbung und Design im Aufbruch des 20. Jahrhunderts* (Stuttgart: Institut f. Auslandsbeziehungen, 1999).

70. See Jeremy Aynsley, "Gebrauchsgraphik as an Early Graphic Design Journal, 1924–1938," *Journal of Design History* 5 (1992): 53–71, and Steven Heller, "Graphic Design Magazine: Das Plakat," *U&lc* 25, no. 4 (1999). See also Steven Heller, *Merz to Emigre and Beyond: Avant-Garde Magazine Design of the Twentieth Century* (London: Phaidon Press, 2014).

71. Heather Hess, "The Wiener Werkstätte and the Reform Impulse," in Blaszczyk, *Producing Fashion*, 111–29.

72. See Anne Sudrow, "Der Typus als Ideal der Formgebung: Zur Entstehung der professionellen Produktgestaltung von industriellen Konsumgütern (1914–1933)," *Technikgeschichte* 76 (2009): 191–210.

73. On the Bauhaus see esp. Frederic Schwartz, "Utopia for Sale: The Bauhaus and Weimar Germany's Consumer Culture," in Kathleen James-Chakraborty, ed., *Bauhaus Culture: From Weimar to the Cold War* (Minneapolis: University of Minnesota Press, 2006), 115–38.

74. See "Aufzeichnung (Hans Riedel)," BhA, Nachlass Johannes Jacobus v. Linden, folder 10.

75. Claudia Regnery, *Die Deutsche Werbeforschung, 1900–1945* (Münster: Monsenstein & Vannerdat, 2003).
76. On the Vienna research institute, see below, chapter 2, and Ronald Fullerton, "Tea and the Viennese: A Pioneering Episode in the Analysis of Consumer Behavior," in Chris Allen and Deborah John, eds., *Advances in Consumer Research* (Provo: Association for Consumer Research, 1994), 418–21. On the Nürnberg center, see Georg Bergler, *Die Entwicklung der Verbrauchsforschung in Deutschland und die GfK bis zum Jahre 1945* (Kallmünz: Lassleben, 1960).
77. See Regnery, *Die Deutsche Werbeforschung*, 267, and on the GfK more generally, Wilfried Feldenkirchen and Daniela Fuchs, *Die Stimme des Verbrauchers zum Klingen bringen: 75 Jahre Geschichte der GFK Gruppe* (Munich: Piper, 2009).
78. On the *Verkaufspraxis*, see Uwe Spiekermann, "German Style Consumer Engineering: Victor Vogt's Verkaufspraxis, 1925–1950," in Logemann, Cross, and Köhler, *Consumer Engineering*.
79. See, e.g., "Kauf-Motive," *Verkaufspraxis* 1, no. 9 (June 1926): 3–7, or "Die fünf Sinne und die Werbung," *Verkaufspraxis* 2, no. 5 (Feb. 1927), 265.
80. Tino Jacobs, *Rauch und Macht: Das Unternehmen Reemtsma, 1920 bis 1961* (Göttingen: Wallstein, 2008).
81. Schwartzkopf, "What Was Advertising?" 11–12.
82. On the notion of marketing as a social technology, see Berghoff, *Marketinggeschichte*.
83. Claudius Torp, *Konsum und Politik in der Weimarer Republik* (Göttingen: Vandhoek & Ruprecht, 2011).
84. On marketing and advertising under the National Socialist regime, see recently Jonathan Wiesen, *Creating the Nazi Marketplace: Commerce and Consumption in the Third Reich* (Cambridge: Cambridge University Press, 2011), and Pamela Swett, *Selling under the Swastika: Advertising and Commercial Culture in Nazi Germany* (Stanford: Stanford University Press, 2013).
85. Hartmut Berghoff, "'Times change and we change with them': The German Advertising Industry in the 'Third Reich': Between Professional Self-Interest and Political Repression," *Business History* 46 (2003): 128–47.
86. Cf. Schug, "Wegbereiter."
87. See, e.g., Wolfang König, *Volkswagen, Volksempfänger, Volksgemeinschaft: "Volksprodukte" im Dritten Reich, vom Scheitern einer nationalsozialistischen Konsumgesellschaft* (Paderborn: Schöningh, 2004), and Hartmut Berghoff, "Enticement and Deprivation: The Regulation of Consumption in Pre-war Nazi Germany," in Martin Daunton and Matthew Hilton, eds., *The Politics of Consumption* (Oxford: Berg, 2001), 165–84.
88. Wiesen, *Creating the Nazi Marketplace*, p. 19.
89. See, e.g., Belinda Davis, *Home Fires Burning: Food, Politics, and Everyday Life in World War I Berlin* (Chapel Hill: University of North Carolina Press, 2000).

90. Stefan Schwarzkopf, "Markets, Consumers, and the State: The Uses of Market Research in Government and the Public Sector in Britain, 1925–1955," in Hartmut Berghoff et al., eds., *The Rise of Marketing and Market Research* (New York: Palgrave, 2012), 171–92.
91. See, e.g., Oliver Küschelm, "Why to Shop Patriotically: Buy Domestic / Buy National Campaigns in Austria and Switzerland during the Interwar Period," in Yann Decorzant et al., eds., *Le Made in Switzerland: Mythes, Fonctions et Réalités* (Basel: Schwabe, 2012), 109–35.
92. See, e.g., Helmut Gruber, *Red Vienna: Experiment in Working-Class Culture, 1919–1934* (New York: Oxford University Press, 1991), and Eve Blau, *The Architecture of Red Vienna: 1919–1934* (Cambridge: MIT Press, 1998).
93. See Orsi Husz, "The Morality of Quality: Assimilating Material Mass Culture in Twentieth-Century Sweden," *Journal of Modern European History* 10 (2012): 152–81, and Helena Mattson and Sven-Olof Wallenstein, eds., *Swedish Modernism: Architecture, Consumption and the Welfare State* (London: Black Dog, 2010).
94. Sheldon and Arens, *Consumer Engineering*, 95–96.
95. Carolyn Goldstein, *Creating Consumers: Home Economists in Twentieth-Century America* (Chapel Hill: University of North Carolina Press, 2012).
96. Lisa Jacobson, *Raising Consumers: Children and the American Mass Market in the Early Twentieth Century* (New York: Columbia University Press, 2004).
97. See, e.g., Frederick J. Schlink and Stuart Chase, *Your Money's Worth* (New York: Macmillan, 1927). Cf. Charles McGovern, *Sold American*, 162–85.
98. See Lawrence Glickman, *Buying Power: A History of Consumer Activism in America* (Chicago: University of Chicago Press, 2009), 189–218.
99. Sheldon and Arens, *Consumer Engineering*, 28–29.
100. See, e.g., Ronald Tobey, *Technology as Freedom: The New Deal and the Electrical Modernization of the American Home* (Berkeley: University of California Press, 1997). On the 1930s emergence of a consumer-oriented social policy, see Rachel Moran, "Consuming Relief: Food Stamps and the New Welfare of the New Deal," *Journal of American History* 76 (2011): 1001–22.
101. See Cohen, *Consumers' Republic*. On the importance of consumption for midcentury liberalism, see Kathleen Donohue, *Freedom from Want: American Liberalism and the Idea of the Consumer* (Baltimore: Johns Hopkins University Press, 2006).
102. Sheldon and Arens, *Consumer Engineering*, 55.

THE RISE OF CONSUMER ENGINEERING

1. See Cohen, *Consumers' Republic*. This democratic promise, to be sure, was at best partially fulfilled, as Cohen shows with new and old inequalities in America's mass consumer society along the fault lines of race, class, and gender.

CHAPTER TWO

1. See Alexis Sommaripa (Business Advisory and Planning Council, Dept. of Commerce) to Dexter Keezer (Dir. Consumer Advisory Board, NIRA), Jan. 16, 1934, PLA, folder "Biographie 1933–46."On the state-business collaboration of the committee, see Robert Collins, "Positive Business Responses to the New Deal: The Roots of the Committee for Economic Development, 1933–1942," *Business History Review* 52 (1978): 369–91.
2. Sommaripa to Keezer, Jan. 16, 1934, PLA, folder "Biographie 1933–46."
3. Paul Lazarsfeld, "The Art of Asking WHY in Marketing Research: Three Principles Underlying the Formulation of Questionnaires," *National Marketing Review* 1 (1935): 26–38.
4. Cohen, *Consumers' Republic*, 298. See also Harold Kassarjian, "Scholarly Traditions and European Roots of American Consumer Research," in Laurent, Lilien, and Pras, *Research Traditions in Marketing*, 265–79.
5. Advertising Research Foundation, "The Founding Fathers of Advertising Research," reprinted from *Journal of Advertising Research*, June 1977, PLA, folder "About PFL 2." The other "founding fathers" portrayed included George Gallup, Archibald Crossley, A. C. Nielsen, and Henry Brenner.
6. For exceptions, see, e.g., Samuel, *Freud on Madison Avenue*. On Dichter, see also Stefan Schwarzkopf and Rainer Gries, eds., *Ernest Dichter and Motivation Research: New Perspectives on the Making of Post-War Consumer Culture* (New York: Palgrave Macmillan, 2010). Ronald Fullerton has written in detail about the marketing work at the Wirtschaftspsychologische Forschungsstelle: Fullerton, "Tea and the Viennese," and Fullerton, "The Devil's Lure (?): Motivation Research,1934–1954," in Leighann C. Neilson, ed., *The Future of Marketing's Past: Proceedings of the 12th Conference on Historical Analysis and Research in Marketing* (Long Beach CA: Association for Historical Research in Marketing, 2005), 134–43.
7. Hans Zeisel, "Die Wiener Schule der Motivforschung," Keynote Address for the Congress of the World Association for Public Opinion Research (WAPOR) and the European Society for Opinion and Market Research (ESOMAR), Aug. 21, 1967, PLA, folder "Lazarsfeld Vienna Marktforschungsstelle" (my translation).
8. See especially Christian Fleck, *A Transatlantic History of the Social Sciences: Robber Barons, the Third Reich and the Invention of Empirical Social Research* (London: Bloomsbury Academic, 2011).
9. See Peggy Kreshel, "Advertising Research in the Pre-Depression Years: A Cultural History," *Journal of Current Issues and Research in Advertising* 15 (1993): 59–75; Douglas Ward, "Capitalism, Early Market Research, and the Creation of the American Consumer," *Journal of Historical Research in Marketing* 1 (2009): 200–223. On "scientific marketing," see esp. Usui, *Development of Marketing Management*, chap. 3.

10. Ralph Hower, *The History of an Advertising Agency: N. W. Ayer & Son at Work, 1869–1939* (Cambridge: Harvard University Press, 1939), 254–78.
11. Sam Meek, "New Client Presentation," 1928, JWT Archives, cited in Kreshel, "Advertising Research," 66.
12. On the popularity of applied psychology in advertising during the 1920s, see also Ludy Benjamin, "Science for Sale: Psychology's Earliest Adventures in American Advertising," in Jerome Williams, ed., *Diversity in Advertising* (New York: Psychology Press, 2004), 21–39.
13. See, e.g., Kreshel, "Advertising Research," and Ward, "Capitalism." See also Lawrence Lockley, "Notes on the History of Marketing Research," *Journal of Marketing* 14 (1950): 733–36. On the Nielsen Company panels, see Alfred Root and Alfred Welch, "The Continuing Consumer Study: A Basic Method for the Engineering of Advertising," *Journal of Marketing* 7 (1942): 3–21.
14. D. G. Brian Jones and Mark Tadajewski, "Market Research Corporation of America, 1934–1951," *CHARM Proceedings* (2011): 232–35.
15. Cattell had been a student of German psychologist Wilhelm Wundt at Leipzig. On consumer research before and during the Depression, see Igo, *Averaged American*, as well as Daniel Robinson, *The Measure of Democracy: Polling, Market Research and Public Life, 1930–1945* (Toronto: University of Toronto Press, 1999), 15–18 and 39–63.
16. Henry Link and Irving Lorge, "The Psychological Sales Barometer," *Harvard Business Review* 14 (1935): 193–204.
17. Usui, *Development of Marketing Management*, chap. 4.
18. Herbert Hess, "The New Consumption Era," *American Marketing Journal* 1 (1935): 16–25.
19. See John Wright and Parks Dimsdale, *Pioneers in Marketing* (Atlanta: Georgia State University, 1974).
20. Harold Hardy, "Collegiate Marketing Education since 1930," *Journal of Marketing* 19 (1955): 325–30.
21. See Wilford White, "The Teaching of Marketing," *National Marketing Review* 1 (1935): 1–8, and Hugh Agnew, "The History of the American Marketing Association," *Journal of Marketing* 5 (1941): 374–79.
22. David Revzan, *A Comprehensive Classified Marketing Bibliography* (Berkeley: University of California Press, 1951), 76–83.
23. See, e.g., Edmund McGarry, "The Importance of Scientific Method in Advertising," *Journal of Marketing* 1 (1936): 82–86, and Frank Coutant, "Scientific Marketing Makes Progress," *Journal of Marketing* 1, no. 3 (1937): 226–30.
24. Lyndon Brown, *Market Research and Analysis* (New York: Ronald Press, 1935), v and 4.
25. D. M. Phelps, *Marketing Research: Its Function, Scope and Method* (Ann Arbor: University of Michigan Bureau for Business Research, 1937).
26. See, e.g., Malcolm Taylor, "Progress in Marketing Research," *Journal of*

Marketing 1 (1936): 56–64, which compiled ongoing research projects by university and research institution and became a recurring feature in the journal.
27. Ferdinand Wheeler, ed., *The Technique of Marketing Research* (New York: McGraw-Hill, 1937).
28. See Oliver Kühschelm, "(Mis)Understanding Consumption: Expertise and Consumer Policies in Vienna, 1918–1938," in Clarke and Shapira, *Émigré Cultures in Design and Architecture*, 45–60.
29. For intellectual histories of early-twentieth-century Vienna, see especially Carl Schorske, *Fin-de-Siècle Vienna: Politics and Culture* (New York: Knopf, 1979), and more recently Deborah Coen, *Vienna in the Age of Uncertainty: Science, Liberalism, and Private Life* (Chicago: University of Chicago Press, 2007); David Luft, *Eros and Inwardness in Vienna: Weininger, Musil, Doderer* (Chicago: University of Chicago Press, 2003), and Chandak Sengoopta, *Otto Weininger: Sex, Science, and Self in Imperial Vienna* (Chicago: University of Chicago Press, 2000).
30. See, e.g., Hess, "Wiener Werkstätte and the Reform Impulse."
31. On Vienna society, see Emil Brix and Allan Janik, eds., *Kreatives Milieu: Wien um 1900: Ergebnisse eines Forschungsgesprächs der Arbeitsgemeinschaft Wien um 1900* (Munich: Oldenbourg, 1993). Both Lazarsfeld and Dichter came out of Vienna's vibrant Jewish community, which played a significant role in the cultural life of the city; cf. Marsha Rozenblit, *Die Juden Wiens, 1867–1914: Assimilation und Identität* (Vienna: Boehlau, 1989); Steven Beller, *Vienna and the Jews, 1867–1938: A Cultural History* (Cambridge: Cambridge University Press, 1989); and Steven Beller, "Who Made Vienna 1900 a Capital of Modern Culture?" in Brix and Janik, *Kreatives Milieu*, 175–80.
32. On Vienna and the development of psychology, see Sheldon Gardner and Gwendolyn Stevens, *Red Vienna and the Golden Age of Psychology, 1918–1938* (New York: Praeger, 1992).
33. See Franz Eder, "'Man lebte damals von der Hand in den Mund': Zur Konsumgeschichte Wiens von 1920 bis 1945," in Roland Domenig and Sepp Linhart, eds., *Wien und Tokyo, 1930–1945: Alltag, Kultur, Konsum* (Vienna, 2007), 13–32. On Red Vienna, see, e.g., Helmut Gruber, *Red Vienna: Experiment in Working-Class Culture, 1919–1934* (New York: Oxford University Press, 1991), and Eve Blau, *The Architecture of Red Vienna: 1919–1934* (Cambridge: MIT Press, 1998).
34. See Hans Zeisel, "Die Österreichische Wirtschaftspsychologische Forschungsstelle, 1925–1938," in L. Rosenmayr and S. Höllinger, eds., *Soziologie Forschung in Österreich* (Vienna: Boehlau, 1969), 43–46.
35. Joan Gordon, interview with Paul Lazarsfeld, Nov. 29, 1961, New York, pp. 39–43, PLA, folder "Columbia Oral History." On his and Marie Jahoda's youth in "Red Vienna," see also Christian Fleck, *Rund um "Marienthal": Von den Anfängen der Soziologie in Österreich bis zu ihrer Vertreibung* (Vienna: Verlag für Gesellschaftskritik, 1990), 119–34.

36. See, e.g., Paul Lazarsfeld, "Die sozialistische Erziehung und das Gemeinschaftsleben der Jugend," *Die sozialistische Erziehung* 3 (1923): 151–154, and Lazarsfeld, "Die Berufspläne der Wiener Maturanden des Jahres 1928," *Mitteilungen aus Statistik und Verwaltung der Stadt Wien* (1928).
37. Marie Jahoda, "Aus den Anfängen der Sozialwissenschaftlichen Forschung in Österreich," in Norbert Leiser, ed., *Das geistige Leben Wiens in der Zwischenkriegszeit* (Vienna: Österreichischer Bundesverlag 1980), 216–22.
38. Charlotte Bühler, "Die Wiener Psychologische Schule in der Emigration," *Psychologische Rundschau* 16 (1965): 187–96. The Bühlers eventually immigrated to the United States in 1940.
39. On Austrian emigrés in the social sciences, see Christian Fleck, "Rückkehr unerwünscht. Der Weg der österreichischen Sozialforschung ins Exil," in Friedrich Stadler, ed., *Vertriebene Vernunft I: Emigration und Exil österreichischer Wissenschaft, 1930–1940* (Vienna: Jugend und Volk, 1987), 182–212.
40. Announcement of constitution (Gründungserklärung), "Verein Österreichische Wirtschaftspsychologische Forschungsstelle," PLA, folder "Lazarsfeld Vienna Marktforschungsstelle."
41. See "Hundred Arrests in Anti-Socialist Drive: 'Secret Centre' Raided in Vienna," *Daily Herald* (London), Dec. 14, 1936.
42. See, e.g., Report "Stadtrandsiedlung Leopoldau," 1934, PLA, folder "Wirtschaftspsychologische Forschungsstelle I." On the RAVAG study and the institute's radio research, see Cornelia Epping-Jäger, "Kontaktaktion: Die frühe Wiener Ausdrucksforschung und die Entdeckung des Rundfunkpublikums," in Irmela Schneider and Isabell Otto, eds., *Formationen der Mediennutzung II* (Bielefeld: transcript, 2007), 55–71.
43. Paul Lazarsfeld, Marie Jahoda, and Hans Zeisel, *Die Arbeitslosen von Marienthal: Ein soziographischer Versuch über die Wirkungen langdauernder Arbeitslosigkeit* (Vienna, 1933). See also Fleck, *Rund um "Marienthal."*
44. See, e.g., Jahoda, "Aus den Anfängen."
45. See Alois Wacker, "Marie Jahoda und die Oesterreichische Wirtschaftspsychologische Forschungsstelle—zur Idee einer nicht-reduktionistischen Sozialpsychologie," *Psychologie und Geschichte* 8 (1998): 112–49, and Hans Zeisel, "The Vienna Years," in Robert Merton et al., eds., *Qualitative and Quantitative Social Research: Papers in Honor of Paul Lazarsfeld* (New York: Free Press, 1979), 10–15.
46. Paul Lazarsfeld, "Neue Wege der Marktforschung," Mitteilungen der industrie- und Handelskammer Berlin, Oct. 25, 1932, PLA, folder "Early Vienna."
47. Dr. Wagner, Presentation on Work of the Forschungsstelle, at Colloquium Prof. Bühler, n.d., PLA, folder "Vienna Marktforschungsstelle."
48. See, e.g., "Der Gegenstand der Wirtschaftspsychologie. Theoretische Überlegungen für ihre Praktiker: Wirtschaftspsychologie als Zweig der Sozialpsychologie," "Verkaufen als Reklame—Ergebnis einer psychologischen Untersuchung," Herta Herzog, speech before the Institut für Handelsforschung, Berlin 1933, PLA, folder "Wirtschaftspsychologische Forschungsstelle I."
49. See, e.g., Herta Herzog, "Stimme und Persönlichkeit," *Zeitschrift für Psy-*

chologie 130 (1933): 300–369; Paul Lazarsfeld, "Neue Wege der Marktforschung," Mitteilungen der Industrie- und Handelskammer Berlin, Oct. 25, 1932; Olga Kukulka, "Theorie der Begründung" (diss., Vienna, 1932); Lotte Ledermacher, "Zur Psychologie der Schuhmode," Referat, 1932, PLA, folder "Wirtschaftspsychologische Forschungsstelle I."

50. Herta Herzog, speech before the [Institut für Handelsforschung], Berlin, 1933, PLA, folder "Wirtschaftspsychologische Forschungsstelle I."
51. See, e.g., Th. Mautner Markhof KG to Forschungsstelle, Vienna, Aug. 8, 1933, and "Bestellschein Verkaufs- und Konsumbarometer," PLA, folder "Vienna Marktforschungsstelle."
52. Hans Zeisel, "Market Research in Austria," *Human Factor* 8 (1934): 29–32.
53. "Untersuchung über die Absatzchancen eines fertigen Kaffees auf dem Wiener Markt," report, 1933, PLA, folder "Lazarsfeld Vienna Marktforschungsstelle."
54. "Doctor in America," *Tide*, Nov. 1934.
55. "'Ich habe die Welt nicht verändert.' Gespräch mit Marie Jahoda," in Jürgen Manthey, ed., *Die Zerstörung einer Zukunft. Gespräche mit emigrierten Sozialwissenschaftlern* (Hamburg: Rowohlt, 1979), 103–44.
56. Igo, *Averaged American*.
57. Application letter to Rockefeller Foundation, 1932, draft, PLA, folder "Biography I."
58. Lazarsfeld to the President of the Forschungsstelle, Jan. 1934, PLA, folder "Lazarsfeld Vienna Marktforschungsstelle."
59. Leo Gold to Lazarsfeld, Zürich, Jan. 5, 1934, PLA, folder "Lazarsfeld Vienna Marktforschungsstelle."
60. Lazarsfeld to the President of the Forschungsstelle, Jan. 1934, PLA, folder "Vienna Marktforschung."
61. Paul Lazarsfeld, "Psychological Aspects of Market Research," *Harvard Business Review* (Oct. 1934): 54–71; Lazarsfeld, "The Factor of Age in Consumption," *Market Research* 3, no. 3 (Sept. 1935): 13–16; and Lazarsfeld, "Art of Asking WHY in Marketing Research." His later publications include "The Use of Detailed Interviews in Market Research," *Journal of Marketing* 2 (1937): 3–8; "The Outlook for Testing Effectiveness in Advertising," *Management Review* (Jan. 1936): 3–12; and "Who Are the Marketing Leaders?" in James McNeal, ed., *Dimensions of Consumer Behavior*, 2nd ed. (New York: Appleton-Century-Crofts, 1969), 161–68.
62. "Doctor in America," *Tide—The Newsmagazine for Advertising and Marketing*, Nov. 1934, 59–61.
63. Joan Gordon, interview with Paul Lazarsfeld, Nov. 29, 1961, New York, p. 26, PLA, folder "Columbia Oral History."
64. Paul Lazarsfeld, "Die NRA und der Konsument," *Der österreichische Volkswirt* 26, nos. 22, 24 (Feb. 1934). See also Lazarsfeld to Forschungsstelle, Feb. 1934, PLA, folder "Lazarsfeld Vienna Marktforschungsstelle." On the development of advertising psychology in German-speaking Europe, see Claudia

Regnery, *Die deutsche Werbeforschung, 1900–1945* (Münster: Monsenstein & Vannerdat, 2003).
65. John van Sickle (Rockefeller Foundation European Office) to Robert Lynd, Apr. 27, 1934, PLA, folder "Biography 1." The Rockefeller administrator endorsed Lazarsfeld's renewal but felt compelled to observe that, contrary to Lynd's description, "on the one point of his appearance . . . my recollection is that he definitely bears the marks of his race.".
66. "Study of the Psychological Aspects of Stock Market Activities, Cooperation between Psychological Corporation and PFL (Rockefeller Fellow)," c. 1934, PLA, folder "Biography I."
67. V. Pelz (Market Research Council) to Lazarsfeld, Feb. 28, 1934, PLA, folder "Biography I."
68. Osgood Lovekin (JWT) to Lazarsfeld, July 30, 1934, and memo, Paul Achilles on contacts with Benton & Bowles and General Foods, Dec. 28, 1934, PLA, folder "Biography 1933–46."
69. John Jenkins (General Electric) to Lazarsfeld, Sept. 10, 1934, about meeting in Schenectady, and K. G. Stuart (Eastman Kodak Co.) to Lazarsfeld, Dec. 24, 1934, PLA, folder "Biography 1933–46."
70. Lazarsfeld to David Craig, Jan. 18, 1935, PLA, folder "Biography 1933–46."
71. Telegram, Stacy May (Rockefeller) to Lazarsfeld, June 19, 1934, PLA, folder "Correspondence 1933–65." See also Lazarsfeld to Forschungsstelle, Feb. 1934, PLA, folder "Lazarsfeld Vienna Marktforschungsstelle."
72. Paul Achilles (Managing Director Psychological Corporation) to "Chalkley, Gallup etc." (draft), c. 1935, PLA, folder "Biography I."
73. See memo, "Market studies in Foreign Markets Covered by American Foreign Trade," Lazarsfeld to Mr. Gives at Bureau of Foreign and Domestic Commerce, Jan. 24, 1934, PLA, folder "Biography 1933–46."
74. See Paul Lazarsfeld, "Study of Consumer Buying," 1934, and Edward Chamberlin, "On the Circular of Mr. Lazarsfeld on the VALUE of Field Investigations of Consumer Habits (Consumer Advisory Board)," Jan. 25, 1934, PLA, folder "Biography 1933–46." Not everyone at the agency was convinced of Lazarsfeld's emphasis on psychological factors, however, and the project ran into funding problems. See memo, Caroline Ware to Chamberlin, Feb. 16, 1934, PLA, folder "Biography 1933–46."
75. See Paul Lazarsfeld, "Die NRA und der Konsument," *Der österreichische Volkswirt* 26, nos. 22, 24, Feb. 1934.
76. On Lazarsfeld as a pioneer of research institutes, see, e.g., Paul Neurath, "Paul Lazarsfeld und die Institutionalisierung der empirischen Sozialforschung: Ausfuhr und Wiedereinfuhr einer Wiener Institution," in Ilja Srubar, ed., *Exil, Wissenschaft, Identität: Die Emigration deutscher Sozialwissenschaftler* (Frankfurt: Suhrkamp, 1988), 67–105, and Charles Glock, "Organizational Innovation for Social Science Research and Training," in Robert Merton et al., eds., *Qualitative and Quantitative Social Research: Papers in Honor of Paul Lazarsfeld* (New York: Free Press, 1979), 23–36.

77. Rose Goldsen (Kohn), "Speech on 20th Anniversary," Apr. 27, 1957, and Paul Lazarsfeld, "Report on the University of Newark Research Center," 1936, PLA, folder "About Bureau 1."
78. Ibid.
79. Rowena Ripin and Paul Lazarsfeld, "The Tactile-Kinaesthetic Perception of Fabrics with Emphasis on Their Relative Pleasantness," *Journal of Applied Psychology* 21 (1937): 198–224.
80. Much has been written about the Office of Radio Research. A recent account from the history of communication studies is Jefferson Pooley, "An Accident of Memory: Edward Shils, Paul Lazarsfeld and the History of American Mass Communication Research" (PhD diss., Columbia University, 2006), esp. 230–237. On the radio project, see also Fleck, *Transatlantische Bereicherungen*, chap. 5, and Wheatland, *Frankfurt School in Exile*.
81. Lazarsfeld knew Stanton, too, and General Electric, among others, had expressed interest in his Austrian experiences with radio research. See GE Market Research Division to Lazarsfeld, Jan. 11, 1935, PLA, folder "Biography 1933–46."
82. On the "radio wars," see Susan Smulyan, *Selling Radio: The Commercialization of American Broadcasting, 1920–1934* (Washington, DC: Smithsonian Institution, 1994).
83. George Gallup to John Marshall, May 19, 1937, Rockefeller Archive Center, Rockefeller Foundation records, projects, RG 1.1, series 200.R, folder 271, folder 3234.
84. See Hadley Cantril to John Marshall, May 11, 1937, and John Marshall, memorandum, May 1937, Rockefeller Archive Center, Rockefeller Foundation records, projects, RG 1.1, series 200.R, folder 271, folder 3234.
85. Paul Lazarsfeld, "Men and Ideas: From Vienna to Columbia," *Columbia Forum* (Summer 1969): 31–36.
86. On Herta Herzog's career, see Elisabeth M. Perse, "Herta Herzog," in Nancy Signorelli, ed., *Women in Communication: A Biographical Sourcebook* (Westport, CT: Praeger, 1996), 202–11.
87. See Gordon Allport to Lazarsfeld, Dec. 22, 1933, PLA, folder "Biography 1933–46"; see also Hadley Cantril, *The Invasion from Mars: A Study in the Psychology of Panic* (Princeton: Princeton University Press, 1940), published with assistance from Hazel Gaudet and Herta Herzog. On the *War of the Worlds* broadcast, see Brad Schwartz, *Broadcast Hysteria: Orson Welles's War of the Worlds and the Art of Fake News* (New York: Hill & Wang, 2015).
88. See letters and memos in HkA, folders I 16.142–235 and VI 1, 296, 316, 345 and 351.
89. See Theodor Adorno, "Scientific Experiences of a European Scholar in America," and Lazarsfeld, "An Episode in the History of Social Research: A Memoir," in Donald Fleming and Bernard Bailyn, eds., *The Intellectual Migration: Europe and America, 1930–1960* (Cambridge: Harvard University Press, 1969), 270–337 and 338–370.

90. Ibid.
91. Rena Bartos, "Qualitative Research: What It Is and Where It Came From," *Journal of Advertising Research* 26 (1986): 3–6.
92. See Bureau of Applied Social Research, "Bibliography: From Its Founding to the Present," PLA, folder "BASR Studies."
93. Memo, "About Columbia University's Bureau of Applied Social Research," 1947, PLA, folder "About Bureau 1."
94. See Rose Goldsen (Kohn), "Speech on 20th Anniversary," Apr. 27, 1957, PLA, folder "About Bureau 1."
95. On the reports, see "B-Files" (BASR) list of reports at PLA. Hans Zeisel also published the report, "Progress in Radio Research," *Public Opinion Quarterly* 8 (1944): 432–34.
96. Cf. Wacker, "Marie Jahoda."
97. "About Columbia University's Bureau of Applied Social Research," 1947, PLA, folder "About Bureau 1."
98. See memo, Lazarsfeld to "Dear George," Mar. 2, 1950, PLA, folder "BASR."
99. See numerous reports between 1956 and 1958 by BASR researchers Bernard Levenson, Donna Smith, Lee Wiggins, et al. See also "A Case History in the Communication of Research Findings—From University Laboratory to Company Board Room," panel discussion at the American Association for Public Opinion Research (AAPOR) with Robert Ford (Bell Lab), Emmet Judge (Ford), Charles Glock (Columbia), David Wallace (Ford), Jackson Toby (Rutgers), May 9, 1957, PLA, folder "Edsel."
100. David Wallace, "Background and Objectives of the Edsel Panel Study," April 1961, PLA, folder "Edsel."
101. Charles Glock to David Wallace, Jan. 14, 1957, PLA, folder "Edsel."
102. On early concepts of the brand image, see Burleigh Gardner and Sidney Levy, "The Product and the Brand," *Harvard Business Review* 33 (1955): 33–39.
103. On the eventual problems with the project, the difficult collaboration with David Wallace, and concerns over too much corporate influence, see, e.g., the confidential memorandum by Lee Wiggins to David Sills and Clara Shapiro, June 20, 1958, PLA, folder "Edsel." Ernest Dichter, too, offered a motivational study on the Edsel that found the car to be ill-designed, lacking a masculine appeal or a clear personality. See "Edsel Pilot Study," Dec. 1957, DaW, #3185.
104. Ann Pasanella, interview with Paul Lazarsfeld, Apr. 19, 1975, transcript tape 4, p. 131, PLA, folder "Columbia Oral History."
105. See PLA, folder "AT&T (1969/70)," as well as "Status Report #2," July 8, 1971, Bates # 056708–15, and "Minutes of the Second Meeting of the Motivation Conference Committee," n.d., Bates # 056697–700, https://www.industrydocumentslibrary.ucsf.edu/. In 1972, Lazarsfeld participated in a Philip Morris–sponsored conference on motivations and smoking behavior. His contribution was eventually published as Paul Lazarsfeld, "The Social

Sciences and the Smoking Problem," in W. Dunn, ed., *Smoking Behavior: Motives and Incentives* (Washington: Winston & Sons, 1973), 283–86. See also Christian Fleck, "Paul F. Lazarsfeld im Dienste von Philip Morris," in Franz Kolland, ed., *Alter und Gesellschaft im Umbruch* (Vienna: Edition echoraum, 2013), 185–97.

106. Paul Lazarsfeld to John Howard, Oct. 27, 1971, PLA, folder "Bio 4."
107. "The Social Role of Advertising," draft, Apr. 5, 1971, PLA, folder "Bio 4."
108. On Dichter, see, e.g., Horowitz, "Émigré as Celebrant of American Consumer Culture"; Franz Kreuzer, *A Tiger in the Tank: Ernest Dichter: An Austrian Advertising Guru* (Riverside, CA: Ariadne Press, 2007); Rainer Gries, *Die Geburt des Werbeexperten aus dem Geist der Psychologie: Der 'Motivforscher' Ernest W. Dichter als Experte der Moderne* (Frankfurt: Campus, 2004); Samuel, *Freud on Madison Avenue*; and Schwarzkopf and Gries, *Ernest Dichter and Motivation Research*.
109. Allison Rowland and Peter Simonson, "The Founding Mothers of Communication Research: Towards a History of Gendered Assemblage," *Critical Studies in Media Communication* 31 (2014): 3–26.
110. On Herzog's postwar work in advertising, see Perse, "Herta Herzog," and Malcolm Gladwell, "True Colors: Hair Dye and the Hidden History of Postwar America," *New Yorker*, Mar. 22, 1999, 70–81.
111. See Martina Thiele, "What Do We Really Know about Herta Herzog?" report on a symposium in Vienna, H-Soz-Kult, Jan. 20, 2012.
112. See Gladwell, "True Colors."
113. Hans Zeisel, "A Simple Way of Making Charts for Reports and Presentations," *Journal of Marketing* 7 (1943): 264–65; Zeisel, "Progress in Radio Research," *Public Opinion Quarterly* 8 (1944): 432–34; and Zeisel et al., "The Advertising Value of Different Magazines," *Journal of Marketing* 13 (1948): 56–61.
114. See "Hans Ziesel [sic], Forensic Sociologist," *Journal of Advertising Research* (June 1977): 23–25.
115. Hans Zeisel, *Say It with Figures*, 5th rev. ed. (New York: Harper & Row, 1968).
116. Ibid., 11.
117. See Dichter interview in Advertising Research Foundation, "The Founding Fathers of Advertising Research," reprinted from *Journal of Advertising Research*, June 1977, PLA, folder "About PFL 2." See also Lazarsfeld to Zeisel, Sept. 18, 1967, PLA, folder "Biography 1961–67." The letter gives comments and corrections for the 1967 speech cited above: Lazarsfeld noted that Dichter would be in the audience and wanted Zeisel to stress his transfers and translations in market research during the 1930s, adding, "There you could tactfully also include another reference to 'Motivation Research.'"
118. There were, to be sure, American precursors, but Lazarsfeld was among the early champions of research centers in the social sciences. See, e.g.,

Lazarsfeld, "Die Entstehung der ersten Forschungsinstitute in den USA: Eigenentwicklungen und Transfer aus Europa?" Symposium Formen der Institutionalisierung der Wissenschaft: Das Forschungsinstitut," Cologne, Dec. 4–5, 1975, PLA, folder "Correspondence 69–75."
119. Paul Lazarsfeld, "Reflections on Business," *American Journal of Sociology* 65 (1959): 1.
120. Zeisel, "Die Wiener Schule der Motivforschung."

CHAPTER THREE

1. Edward Bernays, "Molding Public Opinion," *Annals of the American Academy of Political and Social Science* 179 (1935): 82–87. On Bernays, see Larry Tye, *The Father of Spin: Edward L. Bernays and the Birth of PR* (New York: Crown, 1998).
2. Bernays, "Molding Public Opinion," 86.
3. Igo, *Averaged American*.
4. Walter Lippman, *The Phantom Public* (New York: Macmillan, 1927).
5. See Zunz, *Why the American Century?* 48–65.
6. On changing conceptions of "the consumer" at midcentury, see, e.g., Gasteiger, *Der Konsument*.
7. Fleck, *Transatlantic History of the Social Sciences*.
8. On the psychologists in exile, see Jean Matter Mandler and George Mandler, "The Diaspora of Experimental Psychology: The Gestaltists and Others," in Donald Fleming and Bernard Bailyn, eds., *The Intellectual Migration: Europe and America, 1930–1960* (Cambridge: Harvard University Press, 1969), 371–419.
9. Igo, *Averaged American*, 140.
10. Hadley Cantril to John Marshall, May 11, 1937, Rockefeller Archive Center, Rockefeller Foundation records, Projects, RG 1.1, Series 200.R, box 271, folder 3234.
11. Ibid.
12. See Igo, *Averaged American*, 124. On the development of opinion polling, see also Robinson, *Measure of Democracy*, and Jean Converse, *Survey Research in the United States: Roots and Emergence, 1890–1960* (Berkeley: University of California Press, 1987).
13. Henry Link, *The New Psychology of Selling and Advertising* (New York: Macmillan, 1932). On the Listerine campaign, see Marchand, *Advertising the American Dream*, 18–20.
14. See Pooley, "Accident of Memory," 214.
15. Paul Lazarsfeld, "The Psychological Aspect of Market Research," *Harvard Business Review* 34 (1934): 54–71.
16. Paul Lazarsfeld, "The Use of Detailed Interviews in Market Research," *Journal of Marketing* 2 (1937): 3–8. See also Lazarsfeld and Arthur Kornhauser, *The Techniques of Market Research from the Standpoint of a Psychologist* (New York: American Management Association, 1935).

17. Robert Merton and Patricia Kendall, "The Focused Interview," *American Journal of Sociology* 51 (1946): 541–57.
18. See, e.g., Paul Lazarsfeld, memorandum on Panel Study, May 12, 1938, and Paul Lazarsfeld, "A Project to Study the Use of Panels," May 16, 1947, PLA, file "Rote Mappe: misc. scientific."
19. See Christopher Simpson, *Science of Coercion: Communication Research and Psychological Warfare, 1945–1960* (Oxford: Oxford University Press, 1994).
20. Summary of Mr. Lazarsfeld's Speech to the American Marketing Association, Dec. 27, 1952, PLA, file "Rote Mappe: misc. scientific."
21. On the wartime emergence of mass consumer politics, see Cohen, *Consumers' Republic*; Brinkley, *End of Reform*; and Collins, *More*.
22. On price controls, see Jacobs, *Pocketbook Politics*.
23. James Sparrow, *Warfare State: World War II Americans and the Age of Big Government* (Oxford: Oxford University Press, 2011).
24. On the comparative aspect of the home front experience, see Hartmut Berghoff, Jan Logemann, and Felix Römer, eds., *The Consumer on the Home Front* (Oxford: Oxford University Press, 2017).
25. See Timothy Glander, *Origins of Mass Communications Research during the American Cold War* (Mahwah, NJ: Lawrence Earlbaum, 2000), 54.
26. See, for example, Cynthia Henthorne, *From Submarines to Suburbs: Selling a Better America, 1939–1959* (Athens: Ohio University Press, 2006); John Bush Jones, *All-Out for Victory! Magazine Advertising and the World War II Homefront* (Hanover, NH: Brandeis University Press, 2009); Dawn Spring, *Advertising in the Age of Persuasion: Building Brand America, 1941–1961* (New York: Palgrave, 2011); and Inger Stole, *Advertising at War: Business, Consumers, and Government in the 1940s* (Urbana: University of Illinois Press, 2012).
27. On the wartime expansion in survey research, see Seymour Sudman and Norman Bradburn, "Organizational Growth of Public Opinion Research in the United States," *Public Opinion Quarterly* 51 (1987): 67–78, and especially Converse, *Survey Research*, chap. 5. See also Glander, *Origins of Mass Communications Research*, pp. 47–55.
28. Converse, *Survey Research*, 162–65.
29. Hans Skott, "Attitude Research in the Department of Agriculture," *Public Opinion Quarterly* 7 (1943): 280–92, and Henry Link, "An Experiment in Depth Interviewing," *Public Opinion Quarterly* 7 (1943): 267–79.
30. Paul Lazarsfeld, "The Controversy over Detailed Interviews—An Offer for Negotiation," *Public Opinion Quarterly* 8 (1944): 38–60. On the "open question" controversy, see also Converse, *Survey Research*, 195.
31. On the research activities of the Bureau during the 1940s, see Paul Lazarsfeld and Frank Stanton, eds., *Radio Research, 1942–1943* (New York: Hawthorne Books, 1944), and Paul Lazarsfeld and Frank Stanton, eds., *Communications Research, 1948–1949* (New York: Harper & Bros., 1949).
32. Paul Lazarsfeld and Robert Merton, "Studies in Radio and Film Propaganda," *Transactions of the New York Academy of Sciences* 6 (1943): 58–74.

33. Robert Merton, *Mass Persuasion: The Social Psychology of a War Bond Drive* (New York: Harper, 1946).
34. See, e.g., "The Negro and the War," National Archives and Records Administration (NARA), Washington, DC, RG 44 entry 164, box 1802. On the BASR studies conducted during the war, see Judith Barton, ed., *Guide to the Bureau of Applied Social Research* (New York: Clearwater, 1984).
35. Paul Lazarsfeld and Jehuda Katz, *Personal Influence: The Part Played by People in the Flow of Mass Communication* (Glencoe, IL: Free Press, 1955). See also Gertrude Robinson, "The Katz-Lowenthal Encounter: An Episode in the Creation of Personal Influence," *Annals of the American Academy of Political and Social Science*, 608 (2006): 76–97.
36. Paul Lazarsfeld, Bernard Berelson, and Hazel Gaudet, *The People's Choice*, 2nd ed. (New York: Columbia University Press, 1948).
37. Lazarsfeld and Katz, *Personal Influence*, 16.
38. Ibid., xv.
39. See, e.g., Todd Gitlin, "Media Sociology: The Dominant Paradigm," *Theory and Society* 6 (1978): 205–53.
40. See esp. Jefferson Pooley, "Fifteen Pages That Shook the Field: Personal Influence, Edward Shils, and the Remembered History of Mass Communication Research," *Annals of the American Academy* 609 (2006): 1–27, and Peter Simonson, "The Rise and Fall of the Limited Effects Model," in John Nerone, ed., *Media History and the Foundations of Media Studies* (Malden, MA: Blackwell-Wiley, 2012), 632–56.
41. Cf. Thymian Bussemer, "Gesucht und gefunden: Das Stimulus-Response-Modell in der Wirkungsforschung," *Publizistik* 48 (2003): 176–89. The prominence of the S-R approach in interwar media studies had been called into question by Hans-Bernd Brosius and Frank Esser, "Mythen in der Wirkungsforschung: Auf der Suche nach dem Stimulus Response Modell," *Publizistik* 43 (1998): 341–61. I am grateful to Benno Nietzel for pointing me toward this debate.
42. Harold Lasswell, "The Theory of Political Propaganda," *American Political Science Review* 21 (1927): 627–31, cited in Brosius and Esser, "Mythen in der Wirkungsforschung," 346.
43. Lazarsfeld and Katz, *Personal Influence*, 34–39.
44. See, e.g., Paul Lazarsfeld to Jane Hauser, n.d., PLA, "Blue Folder 38."
45. George Miller (Harvard) to Lazarsfeld, Apr. 18, 1961, PLA, "Blue Folder 38."
46. See "Diskussionsbemerkungen zur wirtschaftspsychologischen Arbeitsgemeinschaft" and "Beziehung Mensch—Ware," n.d., PLA, folder "WiFo I."
47. On the committee and its research, see Carl Guthe and Margaret Mead, "The Problem of Changing Food Habits: Report of the Committee on Food Habits," *Bulletin of the National Research Council* 108 (Washington: National Research Council, 1943), and Brian Wansik, "Changing Eating Habits on the Home Front: Lost Lessons from World War II Research," *Journal of Public Policy and Marketing* 21 (2002): 90–99.

48. "Contributions from the Field of Market Research (June 27 1941)," *Bulletin of the National Research Council* 108 (Oct. 1943): 141–48.
49. Kurt Lewin, "Forces behind Food Habits and Methods of Change," *Bulletin of the National Research Council* 108 (1943): 35–65.
50. See, e.g., Wolfgang Köhler, *The Task of Gestalt Psychology* (Princeton: Princeton University Press, 1969), and Kurt Koffka, *Principles of Gestalt Psychology* (New York: Harcourt Brace, 1935).
51. See, e.g., Norman Heller, "An Application of Psychological Learning Theory in Advertising," *Journal of Marketing* 20 (1956): 248–54.
52. Mandler and Mandler, "Diaspora of Experimental Psychology," p. 399.
53. I would like to thank Nora Binder for helpful discussions and her insights into Lewin's work and career in American exile. On Lewin, see Alfred Marrow, *The Practical Theorist: The Life and Work of Kurt Lewin* (New York: Basic Books, 1969). See also Dorwin Cartwright, foreword to Kurt Lewin, *Field Theory in Social Science: Selected Theoretical Papers* (New York: Harper & Bros., 1951), vii–xx; Morton Deutsch, "Kurt Lewin: The Tough-Minded and Tendered-Hearted Scientist," *Journal of Social Issues* 48 (1992): 31–43; and Steven Chaffee and Everett Rogers, eds., *The Beginnings of Communication Study in America: A Personal Memoir by Wilbur Schramm* (Thousand Oaks, CA: Sage, 1997), chap. 4.
54. Mandler and Mandler, "Diaspora of Experimental Psychology," 399–405.
55. E. Stivers and S. Wheelan, eds., *The Lewin Legacy: Field Theory in Current Practice* (New York: Springer-Verlag, 1986): 12–20; David Bargal et al., "The Heritage of Kurt Lewin: Theory, Research and Practice," *Journal of Social Issues* 48 (1992): 3–13.
56. See Kurt Lewin, *Dynamic Theory of Personality* (New York: McGraw-Hill, 1935), and Lewin, *Principles of Topological Psychology* (New York: McGraw-Hill, 1936).
57. Kurt Lewin, "Field Theory and Experiment in Social Psychology," *American Journal of Sociology* 44 (1939): 868–96.
58. See Lewin, "Forces behind Food Habits," for the discussion below.
59. Ibid., 59–60.
60. Kurt Lewin, *Resolving Social Conflicts: Selected Papers on Group Dynamics* (New York: Harper Bros., 1948), vii.
61. Kurt Lewin, "The Research Institute for Group Dynamics at Massachusetts Institute of Technology," *Sociometry* 8 (1945): 126–36.
62. Kurt Lewin, "Group Decision and Social Change," in Guy Swanson et al., eds., *Readings in Social Psychology* (New York: Henry Holt, 1952), 459–73.
63. Ibid., 466–467.
64. On this link, see, e.g., David Statt, *Using Psychology in Management Training: The Psychological Foundations of Management Training* (Routledge: London, 2000), 162–63. Anthony Pratkanis and Elliot Aronson, *Age of Propaganda: The Everyday Use and Abuse of Persuasion* (Holt: New York, 2001), 167–68, discusses Lewin's approach as a form of "self-persuasion."

65. Robert Pratt, "Marketing Applications of Behavioral Economics," in Burkhardt Strumpel and Ernst Zahn, eds., *Human Behavior in Economic Affairs* (Amsterdam: Elsevier, 1972), 189–212, esp. 196. See also Kurt Lewin et al., "Level of Aspiration," in Joseph Hunt, ed., *Personality and the Behavioral Disorders* (New York: Ronald Press, 1944), 333–78.
66. On the growing importance of the "attitude" concept in midcentury social research, see Marie Jahoda and Neil Warren, eds., *Attitudes* (Baltimore: Penguin Books, 1966).
67. See, e.g., George Katona, *Psychologie der Relationserfassung und des Vergleichens* (Leipzig: Barth, 1924).
68. On Katona, see especially Horowitz, "Émigré as Celebrant of American Consumer Culture," and Daniel Horowitz, *Anxieties of Affluence: Critiques of American Consumer Culture, 1939–1979* (Amherst: University of Massachusetts Press, 2004), 48–78. Further information on Katona's life and career can be found in Richard Curtin, "Curtin on Katona," in Henry Spiegel and Warren Samuels, eds., *Contemporary Economists in Perspective* (Greenwich, CT, 1984), 495–522; Burkhard Strumpel et al., eds., *Human Behavior in Economic Affairs: Essays in Honor of George Katona* (San Francisco, 1972), and Freie Universität Berlin, *Ehrenpromotion von Prof. Dr. George Katona, Ann Arbor, Michigan, am 15. Juni 1981: Gedenkschrift* (Berlin: Freie Universität, 1982).
69. George Katona, *Organizing and Memorizing: Studies in the Psychology of Learning and Teaching* (New York: Hafner, 1967 [orig. 1940]) with a foreword by Max Wertheimer.
70. George Katona, *War without Inflation: The Psychological Approach to Problems of War Economy* (New York: Columbia University Press, 1942).
71. See Paul Lazarsfeld, memorandum, May 25, 1942, BeL Katona, folder 11.
72. George Katona, *Price Control and Business: Field Studies among Producers and Distributors of Consumer Goods in the Chicago Area, 1942–44* (Bloomington, IN: Principia Press, 1945).
73. George Katona, "The Role of Frame of Reference in War and Post-War Economy," *American Journal of Sociology* 49 (1944): 342.
74. Ibid., 346.
75. Rensis Likert, "A Technique for the Measurement of Attitudes" (PhD diss., Columbia University, New York, 1932).
76. Examples of his commercial work during the 1930s are collected at BeL Likert, box 21.
77. See Joan Gordon, interview with Paul Lazarsfeld, Nov. 29, 1961, 130–37, PLA, folder "Columbia Oral History."
78. "Summary Statement of the Work of the Division of Program Surveys for the Calendar Year 1942," BeL Likert, box 1, folder 3.
79. BAE, Program Survey Division, "An Exploratory Study of Attitudes toward Buying and Shortages of Consumer Goods," Mar. 5, 1943, BeL Likert, box 2, folder 3; BAE, Program Survey Division, "What Housewives Eat for Breakfast," Oct. 16, 1944, BeL Likert, box 3, folder 15.

80. On the "Likert surveys" of bond drives, see also Sheldon Garon, *Beyond Our Means: Why America Spends While the World Saves* (Princeton: Princeton University Press, 2013) 204–10.
81. Angus Campbell and George Katona, "A National Survey of Wartime Savings," *Public Opinion Quarterly* 10 (1946): 373–81.
82. George Katona and Rensis Likert, "Relationship between Consumer Expenditures and Savings: The Contribution of Savings Research," *Review of Economics and Statistics* 28 (1946): 197–99.
83. On the history of the Institute for Social Research, see Anne Frantilla, *Social Science in the Public Interest: A Fiftieth Year History of the Institute for Social Research* (Ann Arbor: University of Michigan, 1998).
84. Kurt Lewin to Likert, Nov. 30, 1943, BeL Likert, box 24 (contains "Appendix: Project of a Research Institute on Group Dynamics").
85. "Consumer Confidence: A Key to the Economy" was the initial title idea for Katona's monograph: *The Powerful Consumer: Psychological Studies of the American Economy* (New York: McGraw-Hill, 1960).
86. George Katona, "Validity of Consumer Expectations," speech before the American Statistical Association, Dec. 30, 1949, BeL Katona, folder 14.
87. See Robert Pratt, "Marketing Applications of Behavioral Economics," in Burkhardt Strumpel and Ernst Zahn, eds., *Human Behavior in Economic Affairs* (Amsterdam: Elsevier, 1972), 189–212.
88. See Hamid Hosseini, "George Katona: A Founding Father of Old Behavioral Economics," *Journal of Socio-Economics* 40 (2011): 977–84.
89. See, e.g., George Katona, "Contribution of Psychological Data to Economic Analysis," *Journal of the American Statistical Association* 42 (1947): 449–59.
90. George Katona, *The Psychological Analysis of Economic Behavior* (New York: McGraw-Hill, 1951).
91. Harold Kassarjian, "Scholarly Traditions and the Roots of American Consumer Research," in Gilles Laurent et al., eds., *Research Traditions in Marketing* (Boston: Kluwer, 1994), 265–79.
92. Ibid., 271–72. On the impact of Katona and Gestalt psychology on marketing, see also Bruce Pietrykowski, *The Political Economy of Consumer Behavior: Contesting Consumption* (London: Routledge, 2009), 54–78.
93. Leon Festinger, *A Theory of Cognitive Dissonance* (Stanford: Stanford University Press, 1962 [orig. 1957]), esp. 78–79. Regarding its impact on the marketing field, see, e.g., Sadaomi Oshikawa, "Can Cognitive Dissonance Theory Explain Consumer Behavior?" *Journal of Marketing* 33 (1969): 44–49.
94. See Abraham Maslow, *Motivation and Personality* (New York: Harper 1954), esp. ix and 109–11.
95. See, e.g., Paul Lazarsfeld and Morris Rosenberg, eds., *The Language of Social Research* (Glencoe, IL: Free Press, 1955), and Leon Festinger and Daniel Katz, eds., *Research Methods in the Behavioral Sciences* (New York: Holt, 1953).
96. See, e.g., Albert Blankenship, "Needed: A Broader Concept of Marketing Research," *Journal of Marketing* 13 (1949): 305–10; Joseph Clawson, "Lewin's

Vector Psychology and the Analysis of Motives in Marketing," in Reavis Cox and Wroe Alderson, eds., *Theory in Marketing* (Chicago: Richard Irvin, 1949), 41–63; William Brown, "The Determination of Factors Influencing Brand Choice," *Journal of Marketing* 14 (1950): 699–706; Wroe Alderson, "Psychology for Marketing and Economics," *Journal of Marketing* 17 (1952): 119–35; Norman Heller, "An Application of Psychological Learning Theory to Advertising," *Journal of Marketing* 20 (1956): 248–54; Reavis Cox et al., eds., *Theory in Marketing*, 2nd series (Homewood, IL: Irwin, 1964), and Joseph Newman, ed., *On Knowing the Consumer* (New York: Wiley & Sons, 1966).
97. Steuart Henderson Britt, ed., *Consumer Behavior and the Behavioral Sciences* (New York: Wiley & Sons, 1966).
98. Edgar Crane, *Marketing Communication: A Behavioral Approach to Men, Messages, and Media* (New York: Wiley & Sons, 1965), 14–15.
99. John Howard and Jagdish Seth, *Theory of Buyer Behavior* (New York: Wiley & Sons, 1969).
100. See Kurt Lewin, "The Special Case of Germany," *Public Opinion Quarterly* 7 (1943): 555–66. Katona wrote a report for the U.S. High Commissioner for Germany (HICOG) Office of Political Affairs in which he surveyed the state of opinion and market research in postwar Germany; Katona, "Report," Oct. 9, 1950, BeL Katona, folder 17.
101. See Simpson, *Science of Coercion*, 55.
102. See, e.g., "Five Factors in Morale," in Goodwin Watson, *Civilian Morale: 2nd Yearbook of the SPSSI* (Boston et al., 1942), 30–47, a summary APA roundtable discussion that included Lewin, Lazarsfeld, and Likert.
103. Igo, *Averaged American*, 109.
104. See Horowitz, "Émigré as Celebrant of American Consumer Culture." See also George Katona and Eva Mueller, *Consumer Attitudes and Demand, 1950–1952* (Ann Arbor: Institute for Social Research, 1953), and George Katona and Eva Mueller, *Consumer Expectations, 1953–1956* (Ann Arbor: Institute for Social Research, 1956). On communications research during the Cold War more generally, see Benno Nietzel, "Propaganda, Psychological Warfare and Communication Research in the USA and the Soviet Union during the Cold War," *History of the Human Sciences* 29 (2016): 59–76.
105. Katona, *Powerful Consumer*, 173.
106. George Katona, "Consumer Behavior in Our Changing Environment," speech, Oct. 1961, BeL Katona, folder 20.
107. Ernest Dichter, *The Psychology of Everyday Living* (New York: Barnes & Noble, 1947), 221.
108. Edward Bernays, *Public Relations* (Norman: University of Oklahoma Press, 1952).
109. Kurt Lewin, "The Research Center for Group Dynamics at Massachusetts Institute for Technology," *Sociometry* 8 (1945): 126.

110. George Katona, "Advertising in a Mass Consumption Society," speech before the American Association of Advertising Agencies, Nov. 1963, BeL Katona, folder 21.
111. Ibid.

CHAPTER FOUR

1. Vance Packard, *The Hidden Persuaders* (New York: McKay, 1957), 1.
2. Ibid., 4.
3. Cheskin, too, stressed the importance of analyzing consumers as irrational actors with the tools of modern psychology. Louis Cheskin and L. B. Ward, "Indirect Approach to Market Reactions," *Harvard Business Review* 26 (1948): 572–80.
4. For a definition of entrepreneurship that emphasizes the synthesis and flow of information into and within organizations, see, e.g., Mark Casson, "Der Unternehmer. Versuch einer historisch-theoretischen Deutung," *Geschichte und Gesellschaft* 27 (2001): 524–44.
5. Andrew Godley and Mark Cassons, "'Doctor, Doctor . . .': Entrepreneurial Diagnosis and Market Making," *Journal of Institutional Economics* 11 (2015): 601–21.
6. Alfred Politz, "A.P. Book (Draft #1)," Dec. 22, 1971, PP St. Johns, box 1, folder 3.
7. Ibid., 5–6.
8. Ibid., 6.
9. Gartman, *Auto Opium*.
10. Roland Marchand, "Customer Research as Public Relations: General Motors in the 1930s," in Strasser, *Getting and Spending*, 85–109.
11. Gerben Bakker, "Building Knowledge about the Consumer: The Emergence of Market Research in the Motion Picture Industry," *Business History* 45 (2003): 101–27.
12. On "fashion intermediaries" and forecasting in various "fashion goods" industries, see Blaszczyk, *Imagining Consumers*.
13. Howard Hovde, "Recent Trends in the Development of Market Research," *American Marketing Journal* 3 (1936): 3–19.
14. Willard E. Freeland, "Scientific Management in Marketing," *Journal of Marketing* 4 (1940): 31–38.
15. Elma Moulton, *Marketing Research Activities of Manufacturers* (Washington, DC: US Department of Commerce, 1939). The study was conducted by the Bureau of Foreign and Domestic Commerce with an AMA advisory committee that included Frank Coutant, Alexis Sommaripa, Ferdinand Wheeler, Paul Converse, and others.
16. William W. Heusner, Charles M. Dooley, Gordon A. Hughes, and Percival White, "Marketing Research in American Industry: I," *Journal of Marketing* 11 (1947): 338–54; and Heusner et al., "Marketing Research in American

Industry: II," *Journal of Marketing* 12 (1947): 25–37. The study was conducted in cooperation with the National Association of Manufacturers.
17. Richard Crisp, *Marketing Research* (New York: McGraw-Hill, 1957), 9 and 26–30.
18. Heusner et al., "Marketing Research in American Industry: I," 348–52. See also National Industrial Conference Board (NICB), *Marketing, Business and Commercial Research in Industry* (New York: National Industrial Conference Board, 1955).
19. Crisp, *Marketing Research*, 31.
20. Gordon Hughes, "The Relationship of Advertiser and Agency in Research," *Journal of Marketing* 14 (1950): 579–80.
21. NICB, *Marketing, Business and Commercial Research*, 30–31.
22. Crisp, *Marketing Research*, 32.
23. Cf. ibid., 699–757.
24. Roger Barton, *Advertising Agency Operations and Management* (New York: McGraw-Hill, 1955), 69–71.
25. Steuart Henderson Britt, "Research and Merchandising in a Modern Advertising Agency," *Journal of Marketing* 13 (1949): 506–10.
26. Crisp, *Marketing Research*, 753–54.
27. Britt, "Research and Merchandising."
28. Crisp, *Marketing Research*, 758–70.
29. For a survey of the field, see Richard Crisp, *Company Practices in Marketing Research* (New York: American Management Association Research Report #22, 1953).
30. Donald Longman, "The Status of Marketing Research," *Journal of Marketing* 16 (1951): 197–201.
31. Heusner et al., "Marketing Research in American Industry: II," 30.
32. Paul Lazarsfeld, introduction to Fifth Edition, in Hans Zeisel, *Say It with Figures* (New York: Harper & Row, 1968), xiv.
33. On Politz, see Hugh S. Hardy, ed., *The Politz Papers: Science and Truth in Marketing Research* (Chicago: American Marketing Association, 1990). His papers are located at St. John's University in New York and at the American Heritage Center in Laramie, WY.
34. Alfred Politz, "Highlights of My Biography," Apr. 23, 1982, PP St. Johns, box 1, folder 1.
35. Alfred Politz, "Advertising Psychology and Technique," in *Weltmacht-Reklame* (August 1929), reprinted in Hardy, *Politz Papers*, 309–13.
36. Resumé, Mar. 7, 1940, PP St. Johns, box 1, folder 1.
37. See ibid. and letter, Politz to Edwin Levinson, Mar. 30, 1943, PP St. John's, box 1, folder 1.
38. Politz, "Highlights of My Biography," Apr. 23, 1982, PP St. Johns, box 1, folder 1.
39. On his client list, see "Some Highlights of Achievements," Dec. 1, 1960, PP St. John's, box 2, folder 2.

40. See "Alfred Politz Research Inc.," n.d. (c. 1964), and Jane Elligett, "Alfred Politz Research Inc.," Aug. 16, 1982, PP St. Johns, box 2, folder 1.
41. In 1977 he also made it into the Hall of Fame of Transit Advertising. See Alfred Politz, "Biography," n.d., PP St. Johns, box 1, folder 10; "Biography for Hall of Fame," June 15, 1974, PP St. Johns, box 2, folder 4; "Transportation Advertising Hall of Fame Awardee 1977," PP St. Johns, box 2, folder 5.
42. Advertising Research Foundation, "The Founding Fathers of Advertising Research," reprinted from *Journal of Advertising Research* 17, no. 3 (June 1977), PLA, folder "About PFL 2."
43. See "Politz Plan," *Tide*, June 15, 1945, reprinted in Hardy, *Politz Papers*, 249–51. On the development of sampling, see also Crisp, *Marketing Research*, 93–126. Politz also introduced "duration sampling" as a means to measure exposure rates of outdoor advertising (e.g., on buses). See, e.g., Robert J. Williams, "Recollections of Alfred Politz, Aug.–Sept. 1982," PP St. Johns, box 1, folder 2.
44. Alfred Politz and Edward Deming, "On the Necessity to Present Consumer Preferences as Predictions," *Journal of Marketing* 18 (1953): 1–5.
45. Alfred Politz, "A.P. Book (Draft #1)" and "Product Changes and Consumer Reactions," in Lincoln Clark, ed., *Consumer Behavior: Research on Consumer Reactions* (New York: Harper & Bros., 1958), 61–68.
46. Hardy, *Politz Papers*, 22.
47. Alfred Politz, "Does Today's Advertising Understand the Consumer?" speech, Boston Conference for Distribution, Oct. 19, 1959, PP St. Johns, box 3, folder 23.
48. Alfred Politz, "Market and Opinion Surveys—Their Role in Industry and Society," draft of speech, Oct. 1950, PP St. Johns, box 3, folder 8.
49. Alfred Politz, "Effective Market and Opinion Research," speech before Boston Conference on Distribution, Oct. 21, 1952, PP St. Johns, box 3, folder 14.
50. Politz fought for independent research institutes against the Advertising Research Foundation which favored corporate and ad agency research departments; see Hardy, *Politz Papers*, chap. 1.
51. "A.P. Book (Draft #1)," Dec. 22, 1971, PP St. Johns, box 1, folder 3, p. 7. See also Crisp, *Marketing Research*, 772–74.
52. Frank Coutant, "Scientific Marketing Makes Progress," *Journal of Marketing* 1, no. 3 (1937): 226–30.
53. Paul Converse, "The Development of the Science of Marketing: An Exploratory Survey," *Journal of Marketing* 10 (1945): 14–23, or Robert Bartels, "Can Marketing Be a Science?" *Journal of Marketing* 15 (1952): 319–28.
54. See, e.g., Albert Blankenship, "Psychological Difficulties in Measuring Consumer Preference," *Journal of Marketing* 6 (1942): 66–75, who especially notes the influence of Paul Lazarsfeld on the study of consumer motivations. See also Ferdinand Wheeler, "New Methods and Results in Market Research," *American Marketing Journal* 2 (1935): 35–39.

55. Ralph Cassady, "Statistical Sampling Techniques and Marketing Research," *Journal of Marketing* 9 (1945): 317–41.
56. Wroe Alderson and Reavis Cox, "Towards a Theory of Marketing," *Journal of Marketing* 13 (1948): 137–52. See also, e.g., Albert Blankenship, *Consumer and Opinion Research: The Questionnaire Technique* (New York: Harper, 1943); Albert Blankenship and Myron Heidingsfield, *Market and Marketing Analysis* (New York: Henry Holt, 1947); Darrell Lucas and Steuart H. Britt, *Advertising Psychology and Research* (New York: McGraw-Hill, 1950); and Crisp, *Marketing Research*. See also David Revzan, *A Comprehensive Classified Marketing Bibliography* (Berkeley: University of California Press, 1951).
57. Kenneth Hutchinson, "Marketing as a Science: An Appraisal," *Journal of Marketing* 16 (1952): 286–93.
58. Lyndon Brown, "Towards a Profession of Marketing," *Journal of Marketing* 13 (1948): 27–31.
59. Herbert Hess, "The New Consumption Era," *American Marketing Journal* 1 (1935): 16–25.
60. Edmund McGarry, "The Importance of Scientific Method in Advertising," *Journal of Marketing* 1 (1936): 82–86.
61. One contributor even proposed a revolving door model: H. R. Nissey, "Exchange Professorship Idea Applied to Commercial Research," *Journal of Marketing* 5 (1940): 43–44.
62. Laurence Jaffe and Richard Lessler, "Finding a Job in Business Research," *Journal of Marketing* 15 (1950): 33–40. See also Heusner et al., "Marketing Research in American Industry: II," 35–36.
63. William Marsteller, "Putting the Marketing Research Department on the Executive Level," *Journal of Marketing* 16 (1951): 56–60.
64. Henry Bund, "Does Marketing Research Know Its Own Market?" *Journal of Marketing* 15 (1950): 198–205.
65. Alfred Politz and Edward Deming, "On the Necessity to Present Consumer Preferences as Predictions," *Journal of Marketing* 18 (1953): 5.
66. Alfred Politz, "President of Alfred Politz Research Inc.," PP St. Johns, box 2, folder 1.
67. Perrin Stryker, "Motivation Research," *Fortune* 53 (June 1956): 230.
68. See, e.g., Barbara Stern, "The Importance of Being Ernest: Commemorating Dichter's Contribution to Advertising Research," *Journal of Advertising Research* (2004): 165–69, and for a critical assessment, Ronald Fullerton, "'Mr. Mass-motivations Himself': Explaining Dr. Ernest Dichter," *Journal of Consumer Behavior* 6 (2007): 269–382.
69. On Dichter, see, e.g., Stefan Schwarzkopf and Rainer Gries, eds., *Ernest Dichter and Motivation Research: New Perspectives on the Making of Post-War Consumer Culture* (New York, 2010). Extensive collections of Dichter's personal and business papers can be found at Hagley Museum and Library (http://www.hagley.org/2009/02/ernest-dichter-papers-at-the-hagley-library) and at the

University of Vienna (http://bibliothek.univie.ac.at/sammlungen/ernest_dichterarchiv.html).
70. See Research Dept. J. Sterling Getchell, "The Psychology of Car Buying," January 1940, HgA Dichter, box 1. See also various reports at DaW, nos. 1, 11, 13, and 14.
71. See Dichter's diary from 1942 to 1944, HgA Dichter, box 161.
72. "Summary of Meeting held by CBS Committee on Daytime Program Research," May 18, 1943, DaW, no. 22.
73. See, e.g., summaries of meetings of Committee on Daytime Program Research, Dec. 1942 to Feb. 1944, HgA Dichter, box 1.
74. See Lydia Strong, "They Are Selling Your Unconscious," *Saturday Review*, Nov. 13, 1954, and Office Memorandum SAC New York to director FBI, Apr. 23, 1958, case file 100–1234581, Institute for Motivational Research.
75. See "Psyche & Sales: Research Helps Industry Probe Buyer Motivations," *Wall Street Journal*, Sept. 13, 1954.
76. Stryker, "Motivation Research," provides these figures for Politz's and Dichter's firms in 1956. On his business practice, see also Ernest Dichter, *Motivation Research* (Tel Aviv: Israel Management Center, 1961), 61–62.
77. Ernest Dichter, "Creative Research Memorandum on the Psychology of Hot Dogs," n.d., DaW, no. 4122.
78. Ernest Dichter (CBS), "Depth Interviewing," talk at Market Research Council, Oct. 15, 1943, HgA Dichter, box 138.
79. Ernest Dichter, "Twelve Questions and Answers on Psychological Research," *Ad Quiz* 34 (Oct. 1946): 1–10, DaW, no. 4202.
80. Ernest Dichter, "Psychodramatic Research Project on Commodities as Intersocial Media," *Sociometry* 7 (1944): 432.
81. Ernest Dichter, "Do You Really Know Your Customers?" in American Management Association, ed., *Coordinating Markets and Sales Efforts* (New York, 1947), 14–26.
82. Ernest Dichter, "Psychology in Market Research," *Harvard Business Review* (Summer 1947): 437.
83. Ernest Dichter, "These Are the Real Reasons Why People Buy Goods," *Advertising and Selling* (July 1948), HgA Dichter, box 131.
84. Ernest Dichter, "Why Accounts Leave Home," *Advertising Agency* (November 1954), HgA Dichter, box 131, and "A Psychological View of Advertising Effectiveness," *Journal of Marketing* (July 1949): 61.
85. See, e.g., "You Either Offer Security or Fail," *Business Week*, June 23, 1951, and "Ernest Dichter of Croton: A Doctor for Ailing Products," *Printer's Ink*, June 26, 1959.
86. See *Motivations* 1, no. 1 (Mar. 1956): 7–10, HgA Dichter, box 127.
87. "What You Should Know about Perception," *Motivations* 1, no. 3 (1956): 5–7.
88. Institute for Mass Motivations, "Memo from the Desk of Dr. Ernest Dichter," Nov. 20, 1955, HgA Dichter, box 127.

89. "Psyche & Sales: Research Helps Industry Probe Buyer Motivations," *Wall Street Journal*, Sept. 13, 1954. See also Ronald Fullerton, "The Birth of Consumer Behavior: Motivation Research in the 1950s," *CHARM Proceedings* (2011): 61–68.
90. See Crisp, *Marketing Research*, 662–95.
91. George Smith, *An Introductory Bibliography of Motivation Research* (New York: Advertising Research Foundation, 1953).
92. See Stryker, "Motivation Research."
93. Pierre Martineau, *Motivation in Advertising: Motives That Make People Buy* (New York: McGraw-Hill, 1957).
94. Burleigh Gardner and Sidney Levy, "The Product and the Brand," *Harvard Business Review* (Mar.–Apr. 1955): 33–39.
95. Sidney Levy, "Roots of Marketing and Consumer Research at the University of Chicago," *Consumption, Markets and Culture* 6 (2003): 99–110.
96. Ronald Fullerton, "The Birth of Consumer Behavior: Motivation Research in the 1940s and 1950s," *Journal of Historical Research in Marketing* 5 (2013): 212–22.
97. See Crisp, *Marketing Research*, 181–82.
98. Alfred Politz, "Product Changes and Consumer Reactions," in Clark, *Consumer Behavior*, 61–68. See also Politz, "What Is Consumer Motivation Research? An Appraisal," 27th Boston Conference on Distribution, Oct. 1955, PP St. Johns, box 3, folder 14.
99. Ernest Dichter, *The Strategy of Desire* (Garden City: Double Day, 1960), 19.
100. Bartos, "Qualitative Research," 6.
101. Ibid., 343.
102. "You Either Offer Security or Fail," *Business Week*, June 23, 1951, and Ernest Dichter, "Kreativität in der Forschung," typescript, n.d., DaW, no. 5004.
103. Ernest Dichter, "What this ad business needs is . . . CREATIVE RESEARCH," *Printer's Ink*, June 4, 1954.
104. Dichter, *Psychology of Everyday Living*, esp. 211–20.
105. In German, Politz used the term "schöpferische Forschung": Alfred Politz, "Einige Erläuterungen und Bemerkungen über Werbeforschung," in Claude Hopkins, *Wissenschaftlich Werben: Wirtschaftlich Werben* (Stuttgart: Forkel Verlag, 1954), 168–69.
106. See Frank, *Conquest of Cool: Business Culture, Counterculture, and the Rise of Hip Consumerism* (Chicago: University of Chicago Press, 2006), and T. J. Jackson Lears, *Fables of Abundance: A Cultural History of Advertising in America* (New York: Basic Books, 1994).
107. Alfred Politz, "Does Today's Advertising Understand the Consumer?" 31st Annual Boston Conference on Distribution, Oct. 19, 1959, PP St. Johns, box 3, folder 23.
108. Alfred Politz, "A.P. Book (Draft #1)," pp. 15–16, PP St. Johns, box 1, folder 3.
109. Alfred Politz, "The Dilemma of Creative Advertising," *Journal of Marketing* 25 (1960): 1–6.

110. Alfred Politz, "Effective Market and Opinion Research," address before Boston Conference on Distribution, Oct. 21, 1952, PP St. Johns, box 3, folder 9.
111. Ibid., p. 2.
112. The growing importance of psychological needs was most succinctly expressed in Maslow's "pyramid of needs." See Abraham Maslow, "A Theory of Human Motivation," *Psychological Review* 50 (1943): 370–96.
113. Stefan Schwarzkopf, "Mobilizing the Depths of the Market: Motivation Research and the Making of the Disembedded Consumer," *Marketing Theory* 15 (2015): 39–57.
114. See Horowitz, "Émigré as Celebrant of American Consumer Culture," and on the notion of growth liberalism, see Robert Collins, *More: The Politics of Economic Growth in Postwar America* (Oxford: Oxford University Press, 2000).
115. See, e.g., Mark Tadajewski, "Promoting the Consumer Society: Ernest Dichter, the Cold War and FBI," *Journal of Historical Research in Marketing* 5 (2013): 192–211.
116. Ernest Dichter, *Psychology of Everyday Living* (New York: Barnes and Noble, 1947), esp. 233–36.
117. Dichter, *Motivation Research*, 69.
118. Dichter, *Strategy of Desire*.
119. See Kenneth Lipartito, "Subliminal Seduction: The Politics of Consumer Research in Post World War II America," in Berghoff et al., *Rise of Marketing and Market Research*, 215–35.
120. Raymond Bauer, "Limits of Persuasion: The Hidden Persuaders Are Made of Straw," *Harvard Business Review* 36 (1958): 105–90.
121. On the broader notion of "scientization" in the twentieth century, see Lutz Raphael, "Die Verwissenschaftlichung des Sozialen als methodische und konzeptionelle Herausforderung für eine Sozialgeschichte des 20. Jahrhunderts," *Geschichte und Gesellschaft* 22 (1996): 165–93.

"TASTEMAKERS" OR "WASTEMAKERS"?

1. Vance Packard, *The Waste Makers* (New York: McKay, 1960).
2. Cohen, *Consumers' Republic*, 294. On postwar obsolescence, see also Slade, *Made to Break*, esp. 151–85.

CHAPTER FIVE

1. Loewy introduced the first West German congress on postwar design and advertising organized by the "Verein Industriekultur e.V." Raymond Loewy, Essen speech, Sept. 21, LoW, box 3, folder 2. On the event itself, see Christopher Oestereich, *"Gute Form" im Wiederaufbau: Zur Geschichte der Produktgestaltung in Westdeutschland nach1945* (Berlin: Lukas Verlag, 2000), 222–23.

NOTES TO PAGES 134–137

2. Loewy, Essen speech, pp. 2–3.
3. On the design profession, see especially Meikle, *Twentieth Century Limited*, and Arthur Pulos, *The American Design Adventure: 1940–1975* (Cambridge: MIT Press 1988). On the broader context of design in the United States, see also Jeffrey Meikle, *Design in the USA* (Oxford: Oxford University Press, 2005).
4. See Allen, *Romance of Commerce and Culture*.
5. Harold Van Doren, *Industrial Design: A Practical Guide* (New York: McGraw-Hill, 1940), xvii.
6. Ibid., 3.
7. On the notion of corporate appropriation, see Shannan Clark, "The Popular Front and the Corporate Appropriation of Modernism," in Elspeth Brown, Catherine Gudis, and Marina Moskowitz, eds., *Cultures of Commerce: Representations and American Business Culture, 1877–1960* (New York: Palgrave, 2006), 51–73.
8. Reckwitz, *Die Erfindung der Kreativität*.
9. On the "creative revolution" in the U.S. advertising industry, see Frank, *Conquest of Cool*, and Corinna Ludwig, *Amerikanische Herausforderungen: Deutsche Großunternehmen in den USA nach dem Zweiten Weltkrieg* (Frankfurt: Peter Lang, 2016), 184–201.
10. See Smith, *Making the Modern*, and Michele Bogart, *Advertising, Artists, and the Borders of Art* (Chicago: University of Chicago Press, 1995).
11. Michael Augspurger, *An Economy of Abundant Beauty: Fortune Magazine and Depression America* (Ithaca: Cornell University Press, 2004), 7. On *Fortune* and *Time*, see also Smith, *Making the Modern*, 159–98.
12. Sheldon Cheney and Martha Candler Cheney, *Art and the Machine: An Account of Industrial Design in 20th Century America* (New York: Whittlesey House, 1936), x.
13. See Scranton, *Endless Novelty*, and Blaszczyk, *Imagining Consumers*.
14. See Tedlow, *New and Improved*, and Gartman, *Auto Opium*. See also Meikle, *Twentieth Century Limited*, 14–18.
15. Christopher Innes, *Designing Modern America: Broadway to Main Street* (New Haven: Yale University Press, 2005). See also Norman Bel Geddes, *Horizons* (Boston: Little, Brown, 1932).
16. On the new designers, see especially Meikle, *Twentieth Century Limited*, 39–67. See also Jeffrey Meikle, "Geniestreiche, Werksentwürfe, Beraterverträge: Zur Geschichte der amerikanischen Industriedesigner," in *Raymond Loewy: Pionier des amerikanischen Industrie-Designs* (Munich: Prestel, 1990), 51–62.
17. Cheney and Cheney, *Art and the Machine*, chap. 4.
18. Cf. Henry Dreyfuss, *Ten Years of Industrial Design, 1929–1939* (New York: Pynson, 1939).
19. This is only a partial selection of names from a 1936 list of "pioneers" in the field compiled in Cheney and Cheney, *Art and the Machine*, chap. 4.
20. Cf. Roland Marchand, "The Designers Go to the Fair: Walter Dorwin Teague and the Professionalization of Corporate Industrial Exhibits, 1933–1940,"

Design Issues 8, no. 1 (1991): 4–17, and Marchand, "The Designers Go to the Fair II: Norman Bel Geddes, the General Motors 'Futurama,' and the Visit to the Factory Transformed," *Design Issues* 8, no. 2 (1992): 22–40.
21. Ely Jacques Kahn, *Design in Art and Industry* (New York: Scribner, 1935), 173.
22. See Van Doren, *Industrial Design*. On the Laboratory School of Industrial Design, see Shannan Clark, "When Modernism Was Still Radical: The Design Laboratory and the Cultural Politics of Depression-Era America," *American Studies* 50 (2009): 25–61, and on the Pittsburgh program, see Jim Lesko, "Industrial Design at Carnegie Institute of Technology, 1934–1967," *Journal of Design History* 10 (1997): 269–92. The Walter Peter Baermann Papers can be found at the North Carolina State University Library's archive.
23. Society of Industrial Designers, *Education Bulletin No. 1: Courses in Industrial Design* (New York, 1946).
24. Museum of Modern Art and Society of Industrial Designers, "Conference on Industrial Design, A New Profession," Nov. 11–14, 1946, IIT, acc. 1998.01, box 1a.
25. For the following, see ibid., pp. 1–17.
26. Ibid., pp. 4–5.
27. Ibid., pp. 18–22.
28. Ibid., p. 28.
29. Ibid., pp. 33 and 36.
30. Ibid., p. 38.
31. Ibid., pp. 49–53.
32. Ibid., pp. 54–55.
33. Ibid., pp. 56–58.
34. Kahn, *Design in Art and Industry*, p. 201.
35. See von Saldern, *Amerikanismus*.
36. See Meikle, "Geniestreiche."
37. Cogdell, *Eugenic Design*.
38. See, e.g., Smith, *Making the Modern*, 353–84. He refers to a "diverse but coherent imagery of a modern present," 6.
39. On the history of design education in the United States, see Pulos, *American Design Adventure*, 164–95.
40. Alexander Dorner, "The Background of the Bauhaus," typescript, c. 1940, BhA, folder 125 "Walter Gropius."
41. On Muller-Munk, see most recently the exhibition catalogue by Rachel Delphia and Jewel Stern, *Silver to Steel: The Modern Designs of Peter Muller-Munk* (Munich: Carnegie Museum of Art and Prestel, 2015). See also Dorothy Grafly, "Peter Muller-Munk: Industrial Designer," in *Design* 47 (May 1946): 9, and Jan Logemann, "Peter Muller-Munk (1904–1967)," *Transatlantic Perspectives* http://www.transatlanticperspectives.org/entry.php?rec=7.
42. See Walter Peter Baermann Papers, "Bibliographical Note," North Carolina State University Archives: https://www.lib.ncsu.edu/findingaids/mc00244/. On Baermann, see also Pulos, *American Design Adventure*, 167–69.

43. On Saarinen, see "Eliel Saarinen," in *International Dictionary of Architects and Architecture* (Detroit: St. James Press, 1993). Reproduced in *Biography Resource Center* (Farmington Hills, MI: Gale, 2010). On Cranbrook, see Robert Clark, Robert Judson, and Andrea Belloli, *Design in America: The Cranbrook Vision, 1925–1950* (New York: Abrams, 1983).
44. See Otto Kuhler, *My Iron Journey: An Autobiography of a Life with Steam and Steel* (Denver: Intermountain Chapter, National Railway Historical Society, 1967).
45. Schlumbohm's papers are partially included in the Marc Harrison Papers at Hagley Museum and Library, http://findingaids.hagley.org/xtf/view?docId=ead/2193.xml.
46. On Lescaze, see "William Lescaze," in *International Dictionary of Architects and Architecture*, reproduced in *Biography Resource Center*. On Kiesler, see "Kiesler, Frederick John," in Michael Berenbaum and Fred Skolnik, eds., *Encyclopaedia Judaica*, 2nd ed. (Detroit: Macmillan Reference, 2007).
47. On Neutra, see, e.g., "Richard Neutra," in *International Dictionary of Architects and Architecture*, reproduced in *Biography Resource Center*, and Wendy Kaplan, ed., *California Design, 1930–1965* (Cambridge: MIT Press, 2011).
48. On the careers of prominent midcentury interior designers, including many emigrés, see Cherie Fehrman and Kenneth Fehrman, *Postwar Interior Design: 1945–1960* (New York: Van Nostrand Reinhard, 1987).
49. On Alberses' career, see American Institute of Architects, press release, Jan. 30, 1961, AAA Knoll, box 4, folder 10, on the occasion of her winning the institute's 1961 "gold medal" for design. See also Virginia Gardner Troy, *The Modernist Textile: Europe and America, 1890–1940* (Burlington, VT: Lund Humphries, 2006), esp. 132–37.
50. See MoMA press release, Apr. 12, 1946, "New China in Modern Design." On Eva Zeisel, a niece of emigré economic theorist Karl Polanyi, see Martin Eidelberg, *Eva Zeisel: Designer for Industry* (Montreal: Le Château Dufresne, Musée des Arts Décoratifs, 1984), and the 2002 documentary film *Throwing Curves: Eva Zeisel* directed by Jill Johnstone.
51. Cf. Volker Fischer, "'Gute Form' made in USA: Design zwischen Europa und Amerika," in Bernd Polster, ed., *West Wind: Die Amerikanisierung Europas* (Cologne: DuMont, 1995), 66–78, and Walter Dorwin Teague, *Design This Day: The Technique of Order in the Machine Age* (New York: Harcourt, Brace, 1940).
52. On Nelson, see Stanley Abercrombie, *George Nelson: The Design of Modern Design* (Cambridge: MIT Press, 1995), and George Nelson, *Building a New Europe: Portraits of Modern Architects* (New Haven: Yale University Press, 2007), a reproduction of the 1930s *Pencil Points* essays.
53. See Fischer, "'Gute Form' made in USA."
54. On the reception of the Bauhaus in the United States, see Kentgens-Craig, *Bauhaus and America*. On the transatlantic connections of the architectural modernists organized in the CIAM movement, see Constanze Domhardt, *The heart of the city. Die Stadt in den transatlantischen Debatten der CIAM 1933–1951* (Zürich: Gta Verlag, 2012).

55. On the variety of European design influences especially during the 1920s, see also Meikle, *Twentieth Century Limited*, 21–29.
56. See Arens and Sheldon, *Consumer Engineering*, 173–74. On Weber, see Christopher Long, *Kem Weber: Designer and Architect* (New Haven: Yale University Press, 2014).
57. See Christopher Long, "The Rise of California Modern Design, 1930–1941," in Kaplan, *California Design*, 61–89, esp. 66.
58. Raymond Loewy, speech at Harvard University, Mar. 22, 1950, HgA Loewy, box 21.
59. On Loewy's life, see his autobiography, Raymond Loewy, *häßlichkeit verkauft sich schlecht: die erlebnisse des erfolgreichsten formgestalters unserer Zeit* (Düsseldorf: Econ, 1953), and Internationales Design Zentrum Berlin, *Raymond Loewy: Pionier des amerikanischen Industrie-Designs* (Munich: Prestel, 1990). On his arrival in the United States, see also Loewy, "Railway Club," Pittsburgh, Apr. 1943, HgA Loewy, box 21.
60. *Harper's Bazaar* especially showcased European designs for an American audience. Between 1934 and 1958 Russian emigré graphic designer Alexey Brodovitch served as the magazine's art director.
61. See Loewy, *häßlichkeit*, 62.
62. Cf. "Prizes Taken by Hupmobile Car Designed by Raymond Loewy," HgA Loewy, box 23.
63. See "Raymond Loewy—Industrial Designer" (c.1937/38 for British audience), HgA Loewy, box 22, and Loewy, Railway Club, Pittsburgh, Apr. 1943, HgA Loewy, box 21.
64. "Raymond Loewy—Industrial Designer," HgA Loewy, box 22.
65. Ibid.
66. Loewy, "Railway Club," Pittsburgh, Apr. 1943, HgA Loewy, box 21.
67. Loewy, speech, Harvard University, Mar. 22, 1950, HgA Loewy, box 21.
68. Ibid.
69. Loewy, address, Society of Industrial Designers, Bedford Springs, 1953, HgA Loewy, box 21.
70. Loewy, "A.M.C. Speech," May 17, 1944, HgA Loewy, box 21.
71. Loewy Associates, *Picture Post* press release (c. 1945), HgA Loewy, box 22.
72. Raymond Loewy, "Introduction to Symposium—Office Staff," c. 1953, HgA Loewy, box 21.
73. Loewy, "Notes on Automobile Design and Industry in the Postwar Period" in HgA Loewy box 22.
74. Loewy, "Speech before S.A.E. (Society of Automotive Engineers)," Detroit 1947, HgA Loewy, box 22.
75. Loewy, "Introduction to Symposium—Office Staff," c. 1953, HgA Loewy, box 21
76. Ibid
77. Loewy, "Luncheon Speech," Montreal, c. 1958, HgA Loewy, box 21.
78. Loewy, *häßlichkeit verkauft sich schlecht*, 66–70.

79. Loewy, "Luncheon Speech," Montreal, c. 1958, HgA Loewy, box 21.
80. On the importance of his frequent trips to Europe, see Loewy, *häßlichkeit verkauft sich schlecht*, 205.
81. Raymond Loewy, "Introduction to Symposium—Office Staff," c. 1953, HgA Loewy, box 21.
82. Loewy, "Why Americans Buy Foreign Cars," 1954, HgA Loewy, box 22.
83. See, e.g., Patricia Johnston, "Art and Commerce: The Challenge of Modernist Advertising Photography," in Brown, Gudis, and Moskowitz, *Cultures of Commerce*, 91–113.
84. For a comprehensive discussion, see Bogart, *Advertising, Artists, and the Borders of Art*, esp. 125–70.
85. On the following, see Leo Lionni, *Zwischen Welten und Zeiten: Autobiographie* (Munich: Middlehauve, 1998).
86. Ibid., 259–60 and 312.
87. Ibid., 333–44.
88. On the Olivetti exhibition, see MoMA press release, "Design Exhibit for the Olivetti Company in Italy to Display," Oct. 16, 1952, MoMA, N620.M9 A353 (= MoMA Press Release Collection).
89. See, e.g., Antje Krause-Wahl, "American Fashion and European Art: Alexander Liberman and the Politics of Taste in *Vogue* of the 1950s," *Journal of Design History* 28 (2015): 67–82.
90. On Burtin, see Margaret Re, "Will Burtin," in *Transatlantic Perspectives*, http://transatlanticperspectives.org/entry.php?rec=161. His papers are available at the Will Burtin Archive of the Rochester Institute for Technology (http://library.rit.edu/depts/archives/willburtin/findingaid.html).
91. See "Toward One World of Allner," reprinted from *Fortune* at Work Summer Issue 1968, BhA, folder "Allner 1." On Allner's work for *Fortune*, see Nachlass Walter Allner, BhA, folder 3
92. See M. Peter Piening Papers at Syracuse University Archives: http://archives.syr.edu/collections/fac_staff/sua_piening_mp.htm.
93. See Deshmukh, "Visual Arts and Cultural Migration."
94. See introduction to the catalogue *Modern Art in Advertising* (Chicago: Paul Theobald, 1946).
95. On Bayer, see Bauhaus Archiv, ed., *Herbert Bayer: Das künstlerische Werk, 1918–1938* (Berlin: Mann, 1982), and Gwen Finkel Chanzit, *Herbert Bayer and Modernist Design in America* (Ann Arbor: UMI Research Press, 1987). See also Allen, *Romance of Commerce and Culture*, and Gwen Finkel Chanzit and Daniel Libeskind, eds., *From Bauhaus to Aspen: Herbert Bayer and Modernist Design in America* (Boulder: Johnson Books, 2005). More recently see Patrick Rössler, *Herbert Bayer: die Berliner Jahre—Werbegrafik 1928–1938* (Berlin: Vergangenheitsverlag, 2013), and Patrick Rössler and Gwen Chanzit, *Der einsame Großstädter: Herbert Bayer, eine Kurzbiographie* (Berlin: Vergangenheitsverlag, 2014).
96. See Lorraine Ferguson and Douglas Scott, "The Evolution of American

NOTES TO PAGES 153–157

Typography," *Design Quarterly* (1990): 23–54, and Bayer to Hans Wingler, Mar. 4, 1976, BhA Bayer, folder 2.

97. "Mittel zur Gestaltung der optischen Erscheinung einer Werbesache," typescript, Dec. 15, 1927, BhA Bayer, folder 1. See also Herbert Bayer, "Acceptance Speech Kulturpreis," Cologne 1969, BhA Bayer, folder 1. On the importance of psychology in advertising art at the Bauhaus, see also, e.g., Xanti Schawinsky, "Form in Advertising," *Industrial Arts* 1, no. 3 (1936): 232–37.

98. Herbert Bayer, "Analyse. Werkbundausstellung Paris," typescript, 1930, BhA Bayer, folder 1.

99. On Dorland, see Stefan Hansen, ed., *Moments of Consistency: Die Geschichte der Werbeagentur Dorland* (Bielefeld: Transcript-Verlag, 2004).

100. On the careers of Bayer and other Bauhaus artists during the National Socialist regime, see Winfried Nerdinger, ed., *Bauhaus Moderne im Nationalsozialismus: Zwischen Anbiederung und Verfolgung* (Munich: Prestel, 1993), especially the article by Ute Brüning, "Bauhäusler zwischen Propaganda und Wirtschaftswerbung," 24–47.

101. See Alfred Barr to Bayer, Sept. 28, 1937, and John McAndrew to Bayer, n.d., BhA Bayer, folder 2. Bayer's activities after immigration are documented in a comprehensive FBI background check; see United States Civil Service Commission (USCSC), case no. 1.22.57.4431, FBI New York Office, report, Sept. 4, 1957.

102. See USCSC, case No. 1.22.57.4431, Report of Investigation, July 8, 1957.

103. See Bayer to Gropius, Nov. 28, 1941, and Gropius to Luce, Dec. 1, 1941, BhA Gropius, folder 31.

104. "Herbert Bayer's Design Class," *A-D* 7 (June–July 1941), pp. 18–3, and Herbert Bayer, "Teaching program of the Light Workshop," n.d., n.p., BhA Bayer, folder 1. See also Clark, "Popular Front," 63–64.

105. USCSC, case no. 1.22.57.4431, Report of Investigation, July 8, 1957.

106. "Herbert Bayer, 85, Designer and Artist of Bauhaus School," *New York Times*, Oct. 1, 1985.

107. On Paepcke, see Greg Ruth, "Walter Paul Paepcke," in Jeffrey Fear, ed., *Immigrant Entrepreneurship: German-American Business Biographies, 1720 to the Present*, vol. 4, German Historical Institute, http://immigrantentrepreneurship.org/entry.php?rec=67, last modified May 27, 2014. On the CCA, see Philip Meggs, "The Rise and Fall of Design at a Great Corporation," in Georgette Balance and Steven Heller, eds., *Graphic Design History* (New York: Allworth, 2001), 283–92, and Allen, *Romance of Commerce and Culture*.

108. Art Institute of Chicago, *Modern Art in Advertising: An Exhibition of Designs for the Container Corporation of America* (Chicago, 1945).

109. USCSC, case no. 1.22.57.4431, Report of Investigation, June 14, 1957.

110. Herbert Bayer, "Container Corporation of America: Design as an Expression of Industry," *Gebrauchsgraphik / International Advertising Art* 23, no. 9 (1952): 1–57.

111. See Herbert Bayer, "Gestaltung und Industrie," typescript, Aspen, 1965, BhA Bayer, folder 1, and Herbert Bayer, "kunst—gebrauchskunst," typescript, Berlin, June 26, 1936, BhA Bayer, folder 1.

NOTES TO PAGES 158–161

112. Leo Lionni chaired the 1954 conference in Aspen. See Lionni, *Zwischen Zeiten und Welten*, 371–81, Pulos, *American Design Adventure*, 211–15, and Allen, *Romance of Commerce and Culture*, 269–79.
113. Art Institute of Chicago, *Modern Art in Advertising: An Exhibition of Designs for the Container Corporation of America* (Chicago: Theobald, 1945).
114. "Democracy in Design Shown in Exhibition of Useful Objects," MoMA, press release, call no. N620.M9 A353 (3 N47 R44), MoMA. On the museum's history and its role in popularizing modern art in the United States, see Sybil Gordon Kantor, *Alfred H. Barr and the Intellectual Origins of the Museum of Modern Art* (Cambridge: MIT Press, 2002), and Kristina Wilson, *The Modern Eye: Stieglitz, MoMA, and the Art of the Exhibition, 1925–1934* (New Haven: Yale University Press, 2009).
115. See, e.g., "Winners in Industrial design competition," Feb. 1, 1941, and "Exhibition on Modern Textiles," Aug 23, 1945, press releases, MoMA.
116. "Democracy in Design Shown," press release, Nov. 26, 1940, MoMA, and Edgar Kaufmann Jr., *What Is Good Design?* (New York: Museum of Modern Art, 1950), 7–8.
117. "Museum of Modern Art Exhibits Well-Designed Household Objects," Dec. 7, 1939, "Democracy in Design Shown in Exhibition of Useful Objects," Nov. 26, 1940, press releases, MoMA.
118. "Museum of Modern Art to Give Annual Design Awards," press release, Nov. 10, 1945, MoMA.
119. Several others came from Latin America and Asia. See Art Institute of Chicago, *Modern Art in Advertising*.
120. See, e.g., Kantor, *Alfred H. Barr*, 310–12, and "Exiled Art Purchased by Museum of Modern Art," press release, Aug. 7, 1939, MoMA.
121. "New China in Modern Design," press release, Apr. 12, 1946, MoMA. In 1949 Albers received an individual show as well; see "Museum to Show Imaginative and Experimental Textiles by Anni Albers," press release, Sept. 9, 1949, MoMA.
122. "Museum of Modern Art to Open Exhibition 'Are Clothes Modern?'" 1944, and "New Furniture Designs and Techniques Have Initial Showing at Museum of Modern Art," March 11, 1947, press releases, MoMA.
123. "MoMA and Merchandise Mart Announce Continuing Series of Exhibition in Joint Program 'Good Design,'" press release, Nov. 9 1949, MoMA.
124. "'Good Design' Show at Merchandise Mart to be Installed by Charles Eames," Dec. 5, 1949, and "The World's Greatest Market Place: The Merchandise Mart," n.d., press releases, MoMA.
125. "Brief Background Outline of Good Design," press release, Jan. 1, 1953, MoMA.
126. "'Good Design' Show at Merchandise Mart to Be Installed by Charles Eames," press release, Dec. 5, 1949, MoMA.
127. See, e.g., "Good Design to Open 13th Exhibition," press release, Jan. 4, 1954, MoMA.
128. "News from Good Design," press release, May 24, 1950, MoMA.

129. "News from Good Design," press release, Jan. 1950, MoMA.
130. "Museum to Show Experimental Graphic Design," Feb. 9, 1954, and "Typographic Design," Feb. 14, 1954, press releases, MoMA.
131. "Design Executed for the Olivetti Company in Italy to Go on Display," press release, Oct. 16, 1952, MoMA.
132. See e.g. Meikle, *Design in the USA*, 148–51.
133. See, e.g., Cogdell, *Eugenic Design*, 11.
134. Fernand Léger, "Relationship between Modern Art and Industry," in Art Institute of Chicago, *Modern Art in Advertising* (Chicago, 1945).
135. Raymond Loewy, "Address, Society of Industrial Designers," Bedford Springs, 1953, HgA Loewy, box 21.

CHAPTER SIX

1. "Flexible," *New Yorker*, May 26, 1945, 18–19.
2. See "US Production Lines to Refurnish Europe," *Daily Mirror*, Apr. 7, 1945, and "Europeans Flee Hitler, Set Up Shop Here, Like It So Well They'll Stay," *Wall Street Journal*, May 16, 1944.
3. On Kramer's biography, see Ferdinand Kramer, "Arbeiten in Deutschland und den USA," speech at Bauhaus Archiv, July 7, 1974, typescript, WbA Kramer, box 3–5. See also Bauhaus Archiv, ed., *Ferdinand Kramer: Architektur und Design* (Berlin: Bauhaus Archiv, 1982); Claude Lichtenstein, ed., *Ferdinand Kramer: Der Charme des Systematischen* (Gießen: Anabas, 1991); and Lore Kramer, *Texte: Zur aktuellen Geschichte von Architektur und Design* (Walldorf: Jochen Rahe, 1993).
4. "Flexible," *New Yorker*, May 26, 1945, 19.
5. On the Werkbund, see Campbell, *German Werkbund*, and Schwartz, *Werkbund*. On the transatlantic history of CIAM, see Domhardt, *The heart of the city*. On the impact of the Bauhaus in America, see Kentgens-Craig, *Bauhaus and America*; Kathleen James, *Bauhaus Culture: From Weimar to the Cold War* (Minneapolis: University of Minnesota Press, 2006), and Jill Pearlman, *Inventing American Modernism: Joseph Hudnut, Walter Gropius, and the Bauhaus Legacy at Harvard* (Charlottesville: University of Virginia Press, 2007).
6. Van Doren, *Industrial Design*, 79.
7. On the notion of the "elitism" of Bauhaus modernism, see, e.g., Hine, *Populuxe*. Shannan Clark, "When Modernism Was Still Radical: The Design Laboratory and the Cultural Politics of Depression Era America," *American Studies* 50 (2009): 35–61, discusses the decline of a radical social vision within the design profession.
8. Ferdinand Kramer, "Aus der Praxis eines Architekten in Deutschland und Amerika," typescript of a speech at Bauhaus Archiv Berlin, July 7, 1974, WbA Kramer, box 5.
9. On Kramer's involvement with "New Frankfurt," see Jochem Jourdan, "Neue Sachlichkeit und weiter: Zur Biographie Ferdinand Kramers," in Bauhaus

Archiv, *Ferdinand Kramer*, 7–10. On transatlantic exchanges in social housing, see Rodgers, *Atlantic Crossings*, 367–408.
10. CIAM to Kramer, Aug. 20, 1929, WbA Kramer, box 2.
11. Kramer, "Aus der Praxis eines Architekten," WbA Kramer, box 5.
12. Ferdinand Kramer, "Täglich 18,000 Stühle," *Frankfurter Zeitung*, Apr. 1929 (the article originally appeared in the Werkbund journal *Die Form*).
13. See Ruggero Tropeano, "Frankfurter Gebrauchsgerät: F.K. und das 'Neue Frankfurt,'" in Lichtenstein, *Ferdinand Kramer*, 26–31.
14. Ferdinand Kramer, "Zusammenarbeit mit der Industrie," *Die Form* (1930), WbA Kramer.
15. See König, *Volkswagen, Volksempfänger, Volksgemeinschaft*.
16. In the early 1930s, May helped create Magnitogorsk and other industrial cities created entirely from scratch. See Steven Kotkin, *Magnetic Mountain: Stalinism as a Civilization* (Berkeley: University of California Press, 1995).
17. Not all members of these institutions, to be sure, left Germany. Scholars have noted the degree to which especially Bauhaus modernism was coopted by the regime as former members of the school collaborated with the Nazi state. See, e.g., Winfried Nerdinger, ed., *Bauhaus-Moderne im Nationalsozialismus* (Munich: Prestel, 1993).
18. See Lore Kramer, "Die amerikanischen Kramermöbel: Kontinuität und Perspektiven im Exil," in Lichtenstein, *Ferdinand Kramer*, 60–69.
19. Ibid.
20. "New Demountable Furniture Saves Shipping Space," *Architectural Forum*, July 1943, 100.
21. Advertisement, Jordan Marsh Company, c. 1944, WbA Kramer, box 15.
22. Herbert Bayer was among the investors in this enterprise. See Gropius to Bayer, Dec. 3, 1947, BhA Gropius, folder GS 19/1.
23. "A New Line of Folding Weatherproof Outdoor Furniture," *Architectural Forum* (May 1945), 80; and "Various Current Uses of Aluminum," *Progressive Architecture*, Sept. 1948.
24. On Kramer's department store designs, see esp. Claude Lichtenstein, "F.K.s Vorschläge für amerikanische Warenhäuser 1945–47," in Lichtenstein, *Ferdinand Kramer*, 74–81.
25. F. Kramer, "Warenhäuser in U.S.A.," typescript speech, c. 1955, WbA Kramer, box 14.
26. Ferdinand Kramer, "A System of Visual Store Planning for Aldens, Inc.," typescript, May 1, 1947, WbA Kramer, box 14.
27. See Ferdinand Kramer, "Visual Planning, Equipment Selection and Arrangement for Small Department Stores," *Progressive Architecture USA* 5 (1948), WbA Kramer, box 14, and F. Kramer, "The Vizual Principle," BhA Kramer, folder 6.
28. On the Rainbelle, see Fred Oppenheimer, "Rainbelle," in Lichtenstein, *Ferdinand Kramer*, 70–73.
29. "Object 'Rainbelle' Umbrellas, Department of Design in Industry, Institute of Contemporary Art, Boston," BhA Kramer, folder 6.

30. See F. Kramer, "Industrial Design," typescript of a speech, TH Darmstadt, July 1955, WbA Kramer, box 3.
31. Cheney and Cheney, *Art and the Machine*, 36, 38–39.
32. See Domhardt, *The heart of the city*. On the vibrant transatlantic exchanges in the field of architecture and urban planning around the middle of the twentieth century, see Philip Wagner, *Stadtplanung für die Welt? Internationales Expertenwissen 1900–1960* (Göttingen: Vandenhoek, 2016). On the central role of emigrés, see Andreas Joch, "'Must Our Cities Remain Ugly?'— America's Urban Crisis and the European City: Transatlantic 'Perspectives on Urban Development, 1945–1970," *Planning Perspectives* 29 (2014): 165–87.
33. Alexander Dorner, "The background of the Bauhaus," typescript, n.d., BhA Gropius, folder 125.
34. See Schwartz, "Utopia for Sale," 119. On the critique of Bauhaus elitism, see, e.g., Gert Selle, "Ferdinand Kramer oder die Realutopie des sozialfunktionalen Design," in Bauhaus Archiv, *Ferdinand Kramer*, 15–19.
35. See "Aufzeichnung (Hans Riedel)," BhA, Nachlass Johannes Jacobus v. Linden, folder 10.
36. The Thonet company produced some of Breuer's designs. See Schwartz, "Utopia for Sale," 126–27.
37. Peter Blake, "One Day in the Late 1920s," typescript, 1949, BhA Breuer, folder 4. Blake, born Peter Jost Blach in 1920, emigrated to the UK in 1933 and moved to the United States in 1940.
38. Marcel Breuer, "In den vier Jahren hier haben wir," typescript of a speech, n.d., Weimar, BhA Breuer, folder 2.
39. On Breuer's journey into exile, see various letters in BhA Breuer, folder 6.
40. On the global impact of design emigration, see Henning Engelke and Tobias Hochscherf, "Between Avant-Garde and Commercialism: Reconsidering Émigrés and Design," introduction to a special issue of *Journal of Design History* 28 (2015): 1–14. See also Nikolaus Pevsner, *An Inquiry into Industrial Art in England* (Cambridge: Cambridge University Press, 1937).
41. See Karl-Heinz Füssl, *Deutsch-amerikanischer Kulturaustausch im 20. Jahrhundert* (Frankfurt: Campus, 2004), 133–45.
42. On Albers and the Black Mountain College, see Frederick Horowitz and Brenda Danilowitz, *Josef Albers: To Open Eyes: The Bauhaus, Black Mountain College, and Yale* (London: Phaidon, 2009), and Mervin Lane, ed., *Black Mountain College: Sprouted Seeds* (Knoxville: University of Tennessee Press, 1990).
43. See, e.g., Anna Vallye, "Design and the Politics of Knowledge in America, 1937–1967," PhD diss., Columbia University, 2011.
44. W. Gropius to H. Bayer, Dec. 15, 1938, BhA Gropius, folder 19/1.
45. Bayer to Gropius, June 15, 1944, BhA Gropius, folder 19/1.
46. See Gropius to Bayer, Jan. 5, 1944, and Bayer to Gropius, May 17, 1944, BhA Gropius, folder 19/1.

47. Bayer to Gropius, Nov. 26, 1947, BhA Gropius, folder 19/1. On the General Panel Corporation, see Alicia Imperiale, "The Packaged House of Konrad Wachsmann and Walter Gropius," in *Offsite: Theory and Practice of Architectural Production* (Washington, DC: Association of Collegiate Schools of Architecture, 2013), 46–50, and Gilbert Herbert, *The Dream of the Factory-Made House: Walter Gropius and Konrad Wachsmann* (Cambridge: MIT Press, 1984).
48. Sibyl Moholy-Nagy, *moholy-nagy: an experiment in totality* (Cambridge: MIT Press, 1969 [1950]), 139.
49. On the "American Bauhaus," see Allen, *Romance of Commerce and Culture*, chap. 2; Victor Margolin, *The Struggle for Utopia: Rodchenko, Lissitzky, Moholy-Nagy* (Chicago: University of Chicago Press, 1997), 214–20; Paul Betts, "New Bauhaus and School of Design, Chicago," in Jeannine Fiedler and Peter Feierabend, eds., *Bauhaus* (Cologne: Könemann, 2000), 66–73; and Bauhaus Archiv, *50 Jahre new bauhaus: Bauhausnachfolge in Chicago* (Berlin: Argon, 1987). The most comprehensive study is Alain Findeli, "Du Bauhaus à Chicago: Les années d'enseignement de Laszlo Moholy Nagy," PhD diss., Université de Paris VIII, 1989.
50. The Sybil and Laszlo Moholy-Nagy papers are at the Smithsonian Archives of American Art. On the Chicago Association for Arts and Industries, see Lloyd Engelbrecht, "The Association of Arts and Industries: Background and Origins of the Bauhaus Movement in Chicago," PhD diss., University of Chicago, 1973. On Sybil, see Judith Paine, "Sybil Moholy-Nagy: A Complete Life," *Archives of American Art Journal* 15 (1975): 11–16.
51. See, e.g., Meikle, *Twentieth Century Limited*. Meikle gives a more nuanced picture in "Negotiating Modernity: Moholy-Nagy and American Commercial Design," BhA, Nachlass Moholy-Nagy, folder 17.
52. Pamphlet (1937), "Association of Arts and Industries," Institute of Design Collection, UIC, box 1, folder 1.
53. Ibid.
54. On Moholy-Nagy's early career, see Margolin, *Struggle for Utopia*, and Kristina Passuth, *Moholy-Nagy* (London: Thames and Hudson, 1985). See also Moholy-Nagy, "Deklaration der ungarischen Aktivistengruppe MA," 1922, BhA Moholy Nagy, folder 1.
55. On his ideas on design education, see also Laszlo Moholy-Nagy, *Vision in Motion* (Chicago: Paul Theobald, 1947).
56. Laszlo Moholy-Nagy to Alvar Aalto, Feb. 9, 1932, IIT, acc. 2007.15 Alain Findeli, box 4.
57. Sibyl to Moholy-Nagy, Aug. 4, 1938, Laszlo Moholy-Nagy Papers, Smithsonian Archives of American Arts.
58. S. Moholy-Nagy, *moholy-nagy*, 151.
59. See Gropius to Powell (Assoc. of Arts and Industries), Aug. 19, 1938, IIT, acc. 2007.15 Alain Findeli, box 5.

60. Law offices Kelly and Cohler to Shawinsky, Helion and Bayer, Jan. 7, 1939, IIT, , acc. 2007.15 Alain Findeli, box 5.
61. Spiegel Inc. had been founded by German American immigrant entrepreneur Joseph Spiegel in the 1860s. See Frederic Kopp, "Modie J. Spiegel," in Giles R. Hoyt, ed., *Immigrant Entrepreneurship: German-American Business Biographies, 1720 to the Present*, vol. 3, German Historical Institute, http://www.immi grantentrepreneurship.org/entry.php?rec=140, last modified Feb. 18, 2014.
62. S. Moholy-Nagy, *moholy-nagy*, 149 and 163.
63. Modie Spiegel to Robert Jay Wolff, May 18, 1976, IIT, acc. 2007.15 Alain Findeli, box 5.
64. Internal memorandum, Chicago Association of Commerce, Dec. 12, 1944, IIT, acc. 2007.15 Alain Findeli, box 5.
65. E. F. Brooks (Sears Roebuck) to Walter Paepcke, Jan. 10, 1946, IIT, acc. 2007.15 Alain Findeli, box 5.
66. Moholy-Nagy to Zay Smith (United Airlines), July 26, 1945; Moholy-Nagy to Paepcke, July 26, 1945; and internal memo, Zay Smith to W. Patterson, July 16 1945, IIT, acc. 2007.15 Alain Findeli, box 5.
67. Letter B. L. Robbins (General Outdoor Advertising Co.) to Walter Paepcke, Jan. 22, 1946, in IIT Archives, Acc. 2007.15 Alain Findeli, box 5.
68. Walter Paepcke to B. L. Robbins (General Outdoor Advertising Company), Jan. 25, 1946, IIT, acc. 2007.15 Alain Findeli, box 5.
69. Paepcke to Bayer, June 15, 1945, BhA Gropius, folder GS 19/1.
70. Henry Heald (IIT) to Walter Paepcke, June 22, 1942, UIC, Institute of Design Collection, box 2, folder 36.
71. S. Moholy-Nagy to Herbert Bayer, May 24, 1945, Laszlo Moholy-Nagy papers, Smithsonian Archives of American Arts.
72. See Bayer to Gropius, June 5, 1945, and Gropius to Bayer, June 8, 1945, BhA Gropius, folder GS 19/1.
73. S. Moholy-Nagy to György Kepes, Aug. 29, 1948, AAA Kepes.
74. Moholy-Nagy to Walter Paepcke, July 31, 1946, IIT, acc. 2007.15 Alain Findeli, box 5.
75. W. Gropius to S. Giedion, Dec. 16, 1946, IIT, acc. 2007.15 Alain Findeli, box 5.
76. N.N., memo, "Background Information [on Institute of Design]," May 1, 1955, IIT, Institute of Design Records, box 1a.
77. George Keck, "History of the Institute of Design," Aug. 18, 1955, IIT, Institute of Design Records, box I 13b.
78. On the CCA, see Meggs, "Rise and Fall of Design at a Great Corporation."
79. Moholy-Nagy to Paepcke, May 11, 1940, and Inez Cunningham Stark to Paepcke, May 8, 1940, UIC, Institute of Design Collection, box 6, folders 183–84.
80. Walter Paepcke to Mr. J. Finlay, May 23, 1940, UIC, Institute of Design Collection, box 2, folder 36.

81. On the friends of the institute, contacts to corporations, and detailed contributor lists, see UIC, Institute of Design Collection, box 1, folder 14, and box 2, folders 36–43.
82. See, e.g., agenda, meeting of the board of directors, Nov. 14, 1944, UIC, Institute of Design Collection, box 1, folder 5.
83. See "Report of Progress of School of Design Chicago under the Grant of the Carnegie Corporation of New York of $10,000," Feb. 1, 1944, UIC, Institute of Design Collection, box 1, folder 14.
84. "The New Bauhaus: A Program for Art Education," *New York Times*, May 29, 1938.
85. "School of Design in Chicago," p.1, c. 1940, IIT, Institute of Design Records, box 1a.
86. "The institute of design chicago," p. 17, IIT, Institute of Design Records, box 2, folder 10.
87. See, e.g., class schedules, IIT, Institute of Design Records, box 1a.
88. Lectures on "Prefabrication," IIT, Institute of Design Records, box 1, folder 12.
89. Walter Paepcke to B. L. Robbins (General Outdoor Advertising Company), Jan. 25, 1946, IIT, acc. 2007.15 Alain Findeli, box 5.
90. S. Moholy-Nagy, *moholy-nagy*, 149.
91. See, e.g., "Day and Evening Classes: School of Design," fall semester 1941, IIT, Institute of Design Records, box 1, folder 14.
92. See, e.g., "Institute of Design—New Registrations (1945)," UIC, Institute of Design Collection, box 1, folder 14.
93. John Moss (head display department Marshall Field) to W. Street (Marshall Field), Aug. 22, 1944. Similarly positive was a letter by design head Egbert Jacobson (CCA) to Moholy-Nagy, Sept. 7, 1944, UIC, Institute of Design Collection, box 1, folder 23.
94. For the subsequent discussion, see "Report on Public Relations Activities: Institute of Design," Feb. 6, 1945, UIC, Institute of Design Collection, box 1, folder 19. See also "President's Report," Jan. 9, 1945, UIC, Institute of Design Collection, box 1, folder 13.
95. "Report on Public Relations," p. 2.
96. "Moholy-Nagy Reaffirms the Function of the Arts," *Furniture Manufacturer*, Oct. 1943, IIT, Institute of Design Records, box 1a.
97. Moholy-Nagy, "New Trends in Design," *Industrial Design* no. 4 (1943), IIT, Institute of Design Records, box 1a. See also Moholy-Nagy, "Design Potentialities," *Timber of Canada* (Sept. 1943), IIT, Institute of Design Records, box 1a.
98. Moholy-Nagy, "Modern Design from Chicago," *Timber of Canada* (Feb. 1943), IIT, Institute of Design Records, box 1a.
99. "Exhibition of the Advance Guard of Advertising," Katherine Kuh Gallery, Chicago, October 1941, IIT, acc. 2007.15 Alain Findeli, box 4.
100. "School of Design on Threshold of Fourth Year," *Chicago Sun*, Jan. 3, 1942.
101. S. Moholy-Nagy, *moholy-nagy*, 210–11. See also N.N., "Parker 51—Arrow of the Writing Warrior and the Fountain of Peace," a brief company history

that credits Moholy-Nagy with creating the "culture" of design in which the pen was conceived, http://www.wwpenclub.com/spx/art_5_theteacher.php.
102. Oral history interview with György Kepes, Aug. 18, 1968, Kepes Papers, Archives of American Art, Smithsonian Institution.
103. Kepes to Sybil Moholy-Nagy, Sept. 8, 1948, Kepes Papers, Archives of American Art, Smithsonian Institution.
104. See oral history interview with György Kepes, Aug. 18, 1968, Kepes Papers, Archives of American Art, Smithsonian Institution, and György Kepes, "Paperboard Goes to War," Container Corporation of America, 1942, IIT, Institute of Design Records, box 4.
105. Oral history interview with György Kepes, Aug. 18, 1968, Kepes Papers, Archives of American Art, Smithsonian Institution.
106. Kenneth Hudson (Dean, School of Fine Arts, Washington University) to Kepes, Mar. 30, 1945 Kepes Papers, Archives of American Art, Smithsonian Institution.
107. Jay Wolff, György Kepes—Biographical Sketch (April 1940), Kepes Papers, Archives of American Art, Smithsonian Institution.
108. Henry Dreyfuss to Kepes, Aug, 18 1970, Kepes Papers, Archives of American Art, Smithsonian Institution.
109. Paul Wilson (Director, School of Design) to Kepes, Jan. 13 1950, Kepes Papers, Archives of American Art, Smithsonian Institution.
110. György Kepes, *Language of Vision* (Chicago: Paul Theobald, 1944), esp. 11–13.
111. Ibid., 6–7.
112. Vallye, "Design and the Politics of Knowledge in America."
113. László Moholy-Nagy, *Vision in Motion* (Chicago: Paul Theobald, 1947), and Josef Albers, *Interaction of Color* (New Haven: Yale University Press, 1963).
114. Minutes of the conference "Industrial Design, a New Profession," held by the Museum of Modern Art for the Society of Industrial Designers, Nov. 11–14, 1946, IIT, Institute of Design Records, box 1a.
115. Ibid., 59–60.
116. Keck, "History of the Institute of Design."
117. See "Former Students Engaged in Educational Work" and "Teachers and Professionals Who Have Studied at the Summer Sessions of the Institute of Design," n.d., UIC, Institute of Design Collection, box 5, folder 150.
118. Society of Industrial Designers, *Education Bulletin*, no. 1, Courses in Industrial Design (1946).
119. Walter Gropius, "Design and Industry," address on the occasion of the addition of the Institute of Design to the Illinois Institute of Technology, Apr. 17, 1950, IIT, Institute of Design Records, box 1a.
120. See United States Civil Service Commission, Investigation of Herbert Bayer, case 1.22.57.4431, July 10, 1957.
121. See Blaszczyk, *Producing Fashion*.
122. "Message in a Bottle," *Time*, Feb. 18, 1946, 63.

CHAPTER SEVEN

1. "Blueprint for the Post-war World," *St. Louis Star Tribune*, Jan. 22, 1942. See also Robert Yoder, "Are You Contemporary? Check Your Mind against Moholy Nagy's, a Modernist Who Is So Far Ahead That He Is Almost Out of Sight," *Saturday Evening Post*, July 3, 1943, 16–17, 89.
2. Mary Douglass and Baron Isherwood, *The World of Goods* (New York: Basic Books, 1979), 5.
3. On the postwar transformation of corporate design, see, e.g., Harwood, *Interface*.
4. See Hine, *Populuxe*, and Kara Ann Marling, *As Seen on TV: The Visual Culture of Everyday Life in the 1950s* (Cambridge: Harvard University Press, 1994).
5. Cf. Jan de Vries, *The Industrious Revolution: Consumer Behavior and the Household Economy, 1650 to the Present* (Cambridge: Cambridge University Press, 2008), 33.
6. Innes, *Designing Modern America*.
7. On the impact of department store windows on the consumer imagination, see Leach, *Land of Desire*.
8. On the cultural impact of the New York World's Fair on American consumer culture, see Warren Susman, *Culture as History: The Transformation of American Society in the Twentieth Century* (New York: Pantheon Books, 1984), 211–29.
9. Blaszczyk, *Producing Fashion*, 6.
10. Gerald Johnson, "Effective Marketing Begins on the Design Board," *Journal of Marketing* 13 (1948): 32.
11. Raymond Loewy, talk, Harvard University, Mar. 22 1950, HgA Loewy, box 21.
12. On the MAYA stage, see Raymond Loewy, *Never Leave Well Enough Alone: The Personal Record of an Industrial Designer* (New York: Simon and Schuster, 1951), chap. 20.
13. Speech, Railway Assoc., Pittsburgh, Apr. 22, 1943, HgA Loewy, box 21.
14. Loewy Assoc. Staff, "Industrial Design in American Industry," speech, 1953, LoW, box 1.6.
15. William Snaith to Robert Parmelee (PR director at Plymouth and Brockton St. Railway Co.), May 13, 1966, LoW, box 67.6.
16. Loewy Assoc. Staff, "Industrial Design in American Industry," LoW, box 1.6.
17. Marchand, "Customer Research as Public Relations."
18. See Gartman, *Auto Opium*, 108.
19. Museum of Modern Art and Society of Industrial Designers, "Conference on Industrial Design: A New Profession," Nov. 11–14, 1946, 41–58.
20. Ibid., 45.
21. On Nelson, who was heavily influenced by European design modernism, see, e.g., Stanley Abercrombie, *George Nelson: The Design of Modern Design* (Cambridge: MIT Press, 1995).
22. Ibid., 46.

23. Ibid., 48.
24. Van Doren, *Industrial Design*, esp. 47–67.
25. Raymond Loewy, Essen speech, Sept. 21, 1955, LoW, box 3.3.
26. Ibid.
27. Raymond Loewy Associates, article on Reardon Packaging for *Industrial Packaging*, Nov. 1956, LoW, box 164.15.
28. Raymond Loewy Associates Chicago, "Consumer Survey of Trademark Recognition Characteristics (Meat and Meat Products)" for Armour, June 1955, LoW, box 140.8.
29. Organizational chart of Loewy/Snaith Inc., c. 1960, LoW, box 1.4.
30. Market Concepts Inc., "Information 1965," LoW, box 67.13.
31. The tachistoscope, also used by Landor's studio, was a shutter aperture device that large advertising agencies used for consumer testing. See Lee Adler, "What Big Agency Men Think of Copy Testing Methods," *Journal of Marketing Research* 2 (1965): 339–45, and Edward Godnig, "The Tachistoscope: Its history and Uses," *Journal of Behavioral Optometry* 14 (2003): 39–42.
32. Joseph Lovelace to Albert Hastorf, Feb. 16, 1965, LoW, box 1.11. Lovelace was research director at Loewy/Snaith for much of the 1960s and well into the 1970s.
33. "Taking a Product to Market" (draft 3), Sept. 28, 1970, LoW, box 67.6.
34. Sheldon and Arens, *Consumer Engineering*, 94–105.
35. See Harwood, *Interface*, 93–99.
36. Raymond Loewy, speech, University of Cincinnati, Apr. 20, 1956, HgA Loewy, box 21.
37. Herbert Krugman, "Package Research, Familiarity and Taste," First Symposium on Package Design Research, New York, Mar. 21, 1961, LoW, box 164.15.
38. Howard Treu and Edward Green, "Consumer Pre-Action" for Industrial Design, article draft, Mar. 1966, LoW, box 155.1.
39. Raymond Loewy, SPI luncheon speech, Nov. 9, 1967, LoW, box 3.3.
40. Study of metropolitan logistics for Broadway Department Stores (1947), LoW, box 141.10.
41. "Raymond Loewy / William Snaith, INC.," Jan. 28, 1971, LoW, box 67.6.
42. Loewy Assoc. staff, "Industrial Design in American Industry," speech, 1953, LoW, box 1.6.
43. Ibid.
44. "From Lipstick to Locomotives," *Cosmopolitan* (1950), HgA Loewy Scrapbook, vol. 20.
45. Cf. Loewy, *Never Leave Well Enough Alone*, 195–97.
46. On the use of consumer research at Landor's studio, see also Joseph Malherek, "Packaging Personality: Walter Landor and Consumer Product Design in Postwar America," *Australasian Journal of American Studies* 31 (2012): 57–70.
47. On Walter Landor, see Bernard Gallagher, "Walter Landor," in R. Daniel Wadhwani, ed., *Immigrant Entrepreneurship: German-American Business*

Biographies, 1720 to the Present, vol. 5, German Historical Institute, http://immigrantentrepreneurship.org/entry.php?rec=69, last modified Nov. 8, 2012.
48. Cf. Bobbye Tigerman, "Fusing Old and New: Émigré Designers in California," in Wendy Kaplan, ed., *California Design, 1930–1965* (Cambridge: MIT Press, 2011), 91–115.
49. Walter Landor, "Personal History," n.d., LaP, box 17.1, folder "WL Autobiographical."
50. "The Ultimate Image Maker," *San Francisco Focus*, Aug. 1992, LaP, box 1.1, folder "Landor Publicity."
51. See chap. 3 in Walter Landor, "Autobiography (Draft)," LaP, box 17.1, folder "WL Reminiscences."
52. "Milner Gray, a profile," *Arts & Artist*, Oct. 1986, 9–14, LaP, box 17.1, folder "Milner Gray."
53. Walter Landor, "Personal History," n.d., LaP, box 17.1, folder "WL Autobiographical."
54. Cf. Telesis, records of first discussion, Sept. 20, 1939, and subsequent records in LaP, box 17.1, folder "Telesis."
55. "TELESIS and its technicians," *Agenda* 1, no. 3 (July 1941), LaP, box 17.1, folder "Telesis."
56. "Telesis: The Group and the Exhibit," c. 1941, LaP, box 17.1, folder "Telesis." Landor was presented as exemplary for this approach; see "TELESIS and its technicians."
57. On the development of the firm, see, e.g., "Chronology of Significant Dates in the History of Landor Associates," LaP, box 1.1, folder "Landor Chronologies," and "Decade highlights," LaP, box 1.1, folder "Landor Chronologies."
58. Walter Landor, "Personal History," n.d., LaP, box 17.1, folder "WL Autobiographical."
59. Walter Landor, "good design does not 'date,'" *Designs*, Sept. 1947, LaP, box 17.1, folder "Articles by W.L."
60. Transcript of interview with Lewis Lowe (1990s), LaP, box 1.1, folder "Lewis Lowe."
61. "Why East Comes West for Design," *Good Packaging Yearbook* 19.7 (1958): n.p.
62. Walter Landor, "Design as Strategic Tool in Marketing and Long-Range Planning," speech, Taipei, Mar. 3, 1981, LaP, box 11.1, folder "Conventions / Taiwan Seminar."
63. "coffee can styling aimed at consumer's emotions," *Coffee & Tea Industries and the Flavor Field*, August 1958, LaP, box 15.
64. Hugh Schwartz, "Motivation Studies Are Package Design Oriented," *Candy Industry Packaging*, n.d. (c. 1960s), LaP, box 17.1.
65. Chapter on packaging research by Hugh Schwartz in "The Package in Marketing," book manuscript, LaP, box 19.2, folder "Packaging Design Book Files."
66. Ibid.

67. Walter Landor, "good design does not 'date,'" *Designs*, Sept. 1947, LaP box 17.1, folder "Articles by W.L."
68. Diana Twede, "History of Packaging," in Jones and Tadajewski, *Routledge Companion to Marketing History*, 115–28.
69. Walter Landor, "Design Moves Merchandise If It Moves People," *Good Packaging Yearbook* 18 (1957): 83.
70. Walter Landor, speech before American Marketing Association, National Packaging Conference, New York, May 1968, LaP, box 19.1, folder "AMA, 1968," and Walter Landor, "More Sales through Better Packaging," n.d., LaP, box 19.1, folder "More Sales through Better Packaging."
71. Walter Landor, "Predictions . . . of Things to Come in Packaging," reprint from *Sales Management*, Nov. 10, 1959, LaP, box 17.1, folder "Articles by W.L."
72. Walter Landor, "Effective Consumer Product Presentation through Packaging," speech before American Management Assoc., Dec. 1966, LaP, box 19.1, folder "Management."
73. On the history of brand identities, see Ross Petty, "A History of Brand Identity Protection and Brand Marketing," in Jones and Tadajewski, *Routledge Companion to Marketing History*, 97–114.
74. On Domizlaff's "brand technology," see Tino Jacobs, "Zwischen Intuition und Experiment: Hans Domizlaff und der Aufstieg Reemtsmas," in Berghoff, *Marketingeschichte*, 148–76. On Behrens's work at AEG, see Buddensieg and Boyd, *Industriekultur*.
75. On the rise of corporate public relations, see Richard Tedlow, *Keeping the Corporate Image: Public Relations and Business, 1900–1950* (Greenwich, CT: JAI Press, 1979), and Marchand, *Creating the Corporate Soul*.
76. See, e.g., Egbert Jacobson, ed., *Seven Designers Look at Trademark Design* (Chicago: Paul Theobald, 1952) (with contributions by three emigrés, Herbert Bayer, Bernard Rudofsky, and Will Burton), which directly referenced the work of Kurt Koffka along with Kepes's and Moholy-Nagy's adaptations.
77. Cf. John Balmer, "Corporate Identity and the Advent of Corporate Marketing," *Journal of Marketing Management* 14 (1998): 963–96. Lippincott claims to have pioneered use of the term "corporate identity." The firm was established by Pratt designers Donald Dohner and Gordon Lippincott in 1943. Walter Margulies became partner after Dohner's death later that year.
78. On the history of modern corporate design as an aspect of corporate personality, see Wally Olins, *The Corporate Personality: An Inquiry into the Nature of Corporate Identity* (London: Design Council, 1978), 151–64. See also Steve Baker, "Re-reading 'The Corporate Personality,'" *Journal of Design History* 2 (1989): 275–92.
79. Loewy/Snaith, "The Faceless Institution," draft, c. 1970, LoW, box 67.6.
80. Cf. Loewy, "Image et Prestige," 1963, LoW, box. 1.9.
81. On corporate logo design at Loewy's firm, see Michael Schirner, "Logo: Loewy's Markenzeichen," in Angela Schönberger, ed., *Raymond Loewy: Pioneer of American Industrial Design* (Munich: Prestel, 1990), 183–86.

82. Walter Landor, "Introduction to Aspects of Visual Communication," LaP, box 11.2, folder "Presidents Seminar Japan Sept. 1980."
83. Exhibition "Design Focus: Landor Associates," University of San Francisco McLaren Center, 1992, LaP, box 1.1, folder "Brochures 2." On Pabco, see also Ray Strout, Overview of Landor Decades, 1991, LaP, box 1.1 folder "Landor History."
84. "News about Selling: Good Design in Sales," n.d. (c. late 1940s), LaP, box 17.1, folder "Articles about W.L."
85. "Industrial Designer Offers Tips to California Trade," *Women's Wear Daily*, Oct. 24, 1947, LaP, box 17.1, folder "Articles about W.L."
86. "Walter Landor Tells the Story behind Corporate Face-Lifting," *Ad Age*, Sept. 3, 1958, LaP, box 17.1. folder "Articles about W.L."
87. Walter Landor, "Personal History," n.d., LaP, box 17.1, folder "WL Autobiographical," here p. 15.

BRIDGING TRANSLATLANTIC DIVIDES

1. De Grazia, *Irresistible Empire*.

CHAPTER EIGHT

1. Charlotta Heythum, "Ten Years—Knoll International in Germany," *Deutsche Bauzeitschrift*, Oct. 1961 (translated by Else Stone), AAA Knoll, box 4, folder 4.
2. "Knoll Opens Own Retail Outlet," *New York Times*, Oct. 14, 1971, 50.
3. On the history of Knoll Associates and its affiliated companies, see Brian Lutz, *Knoll: A Modernist Universe* (New York: Rizzoli, 2010), and Steven Rouland and Linda Rouland, *Knoll Furniture: 1938–1960*, 2nd ed. (Atglen, PA: Schiffer, 2005). On Hans Knoll, see "Hans G. Knoll," in *Current Biography Yearbook* (1955), 334–36, and "Hans G. Knoll," in *Who Was Who in America*, vol. 3: *1951–1960* (Chicago: Marquis' Who's Who, 1960), 485.
4. On the business history of transatlantic design and fashions, see, e.g., Veronique Pouillard, "Design Piracy in the Fashion Industries of Paris and New York in the Interwar Years," *Business History Review* 85 (2011): 319–44. For furniture, see Per Hansen, "Networks, Narratives and New Markets: The Rise and Decline of Danish Modern Furniture Design, 1930–1970," *Business History Review* 80 (2006): 449–83.
5. Alexandra Lange, "This Year's Model: Representing Modernism to the Postwar American Corporation," *Journal of Design History* 19 (2006): 234.
6. The role of immigrant entrepreneurship in American and transatlantic business history has recently been explored at the German Historical Institute Washington's project "Immigrant Entrepreneurship: German-American Business Biographies, 1720 to the Present," www.immigrant-entrepreneurship.org. This chapter draws in part on an article written for this project: Jan

Logemann, "Hans Knoll," in R. Daniel Wadhwani, ed., *Immigrant Entrepreneurship: German-American Business Biographies, 1720 to the Present*, vol. 5, German Historical Institute, https://www.immigrantentrepreneurship.org/entries.php?pg=7&start=120, last modified Mar. 25, 2014.

7. On the history of the Wilhelm and Walter Knoll companies, see *Walter Knoll: Design Reloaded* (Herrenberg: Walter Knoll AG, 2006). See also Lutz, *Knoll: A Modernist Universe*, chap. 2. On the Werkbund movement, see Campbell, *German Werkbund*.
8. Rouland and Rouland, *Knoll Furniture*, 4.
9. See chapter 6. On Plan Ltd. and its ties to the Walter Knoll company, see Barbara Tilson, "Plan Furniture, 1932–1938: The German Connection," *Journal of Design History* 3 (1990): 145–55.
10. Toby Rodes, *Einmal Amerika und zurück: Erinnerungen eines amerikanischen Europäers* (Stuttgart: Huber, 2009), 169.
11. Sybil Moholy-Nagy to Laszlo Moholy-Nagy, Sept. 7, 1937, Sybil and László Moholy-Nagy papers, reel 945, Smithsonian Institution, Archives of American Art.
12. Margaret Warren, "Top-Flight Rating Comes in 10 Years to Home Furnishing Team," *Christian Science Monitor*, Jan. 25, 1950, 6.
13. "Bickford's Rents 505 5th Avenue Unit," *New York Times*, Mar. 21, 1940, 50.
14. On the transformation of the American furniture industry, see Christian Carron, *Grand Rapids Furniture: The Story of America's Furniture City* (Grand Rapids, MI: Public Museum of Grand Rapids, 1998), and Norma Lewis, *Grand Rapids: Furniture City* (Charleston, SC: Arcadia, 2008).
15. On the predominance of specialized batch production in the furniture and other industries, see Scranton, *Endless Novelty*.
16. On the history of the Herman Miller company, see Hugh de Pree, *Business as Unusual: The People at Herman Miller* (Zeeland, MI: Herman Miller, 1986).
17. Ibid., 14–16.
18. *Walter Knoll*, 60.
19. See above, chapter 6. On the reception and success of the Bauhaus in the United States, see Kentgens-Craig, *Bauhaus and America*; James, *Bauhaus Culture*; and Pearlman, *Inventing American Modernism*.
20. On Hans Knoll's cooperation with Jens Risom and the early years of the company, see Lutz, *Knoll: A Modernist Universe*, 16–22. Wood-crafted modern Danish furniture came to enjoy tremendous popularity in the postwar United States. Risom's work in some ways foreshadowed this trend, even though his designs were closer to the functionalism of the Bauhaus. See Hansen, "Networks, Narratives and New Markets."
21. Florence Knoll, "After Cranbrook," AAA Knoll, box 1, folder 1. On Hans Knoll's talents as a salesman, see Fehrman and Fehrman, *Postwar Interior Design*, 38.
22. On the gendered connotation of interior design and Knoll's role in the professional transformation from interior decorating to interior design, see

Philip Hofstra, "Florence Knoll, Design and the Modern American Office Workplace," PhD diss., University of Kansas, 2008, 133–50.
23. "Barbara Southwick Wed," *New York Times*, Jan. 19, 1939, 26.
24. For another example of a prominent female design entrepreneur, see Susan Ingalls Lewis, "Lilly Daché, Milliner Deluxe: The Self-Production and Self-Promotion of a Fashion Icon," paper delivered at the Business History Conference 2011 in St. Louis.
25. On Florence Knoll and her pioneering career as an interior designer see Hofstra, "Florence Knoll," and Bobbye Tigerman, "'I Am Not a Decorator': Florence Knoll, the Knoll Planning Unit and the Making of the Modern Office," *Journal of Design History* 20, no. 1 (2007): 61–74. Her papers are available at the Smithsonian Institution's Archives of American Art: "Florence Knoll Bassett Papers," *Archives of the American Art Journal* 39 (1999): 59–61.
26. See above, chapter 5. On Cranbrook and some of its most accomplished designers, including Florence Knoll, see Paul Goldberger, "The Cranbrook Vision: The Metropolitan Museum Commemorates an American Giant among Schools of Design," *New York Times Magazine*, Apr. 8, 1984, 49–55.
27. On Bertoia's work for Knoll, see Katharine Elson, "Architect Designs New Furniture," *Washington Post*, Mar. 24, 1957, F16.
28. On the various Knoll designers, see Rouland and Rouland, *Knoll Furniture*, 16–29.
29. Sarah Booth Conroy, "So Cranbrook Married Bauhaus and We've Lived More Happily Ever After," *Washington Post*, Feb. 6, 1972, H1.
30. See "Modern Design Doesn't Pay or Does It?" *Interiors*, Mar. 1946, and Florence Knoll, "Early Designers," AAA Knoll, box 3, folder 6.
31. Lange, "This Year's Model," 242.
32. Cf. "Noted Furniture Designer to Open New England Showroom in Boston," *Christian Science Monitor*, May 8, 1950, 2; "Knoll, Chicago: New Tune in the Same Key," *Architectural Forum* (1951), 3–5.
33. "Knoll Associates: Showroom," *Architectural Record*, Nov. 1948, 1–2; "Furniture Showrooms in New York," *Architectural Review* 110 (Dec. 1951): 383–86.
34. William Freeman, "News of the Advertising and Marketing Fields," *New York Times*, June 19, 1955, F11. On the cooperation with Matter, see also items in AAA Knoll, box 3, folder 7.
35. See items in AAA Knoll, box 3, folder 8.
36. "Knoll: Im Haut- und Knochenstil," *Der Spiegel*, Apr. 13, 1960, 64–75.
37. See Hofstra, "Florence Knoll," 47.
38. See Tigerman, "'I Am Not a Decorator.'"
39. Cf. "Knoll, Inc.," in *International Directory of Company Histories*, vol. 80 (Detroit: St. James Press, 2006).
40. American Institute of Architects, press release, Jan. 30, 1961, AAA Knoll, box 4, folder 10.
41. Joseph Giovanni, "Florence Knoll: Form, Not Fashion," *New York Times*, Apr. 7, 1983.

42. On the "Knoll look," see Craig Miller, curator of the Cranbrook Show at the Metropolitan Museum of Art, New York, 1984, AAA Knoll, box 3, folder 5, esp. pp. 53–57.
43. Connecticut General Life Insurance Company, 92nd Annual Report (Hartford, CT, 1956).
44. Tigerman, "'I Am Not a Decorator,'" 72.
45. See, e.g., Nelson A. Rockefeller to Hans and Florence Knoll, Dec. 14, 1946, AAA Knoll, box 4, folder 2; and "A Humane Campus for the Study of Man: Palo Alto, California," *Architectural Forum*, Jan. 1955.
46. "CBS Offices by the Same Designer," *Architectural Forum*, Jan. 1955, AAA Knoll, box 1, folder 2.
47. "Knoll, Textile Man, Dies in Cuban Crash," *New York Times*, Oct. 10, 1955; "Hans G. Knoll," *Washington Post*, Oct. 12 1955, 16.
48. Toby Rodes, personal communication with Corinna Ludwig, July 21, 2011.
49. "Art Metal Buys Three Companies," *New York Times*, June 10, 1959, 51; Rita Reif, "Pioneer in Modern Furniture Is Charting Expansion Course," *New York Times*, June 17, 1959, 29.
50. Today, the Knoll Group is still a leading manufacturer of office furniture with offices in the United States and abroad. Knoll was bought by General Felt Industries Inc. in the early 1980s, which took the company public for a brief period following 1983. The initial stock offering raised $56 million, but by 1986 it was privately held once again. The company was sold to Westinghouse in 1990 but continued to struggle to make profits. By 1995, the Knoll Group had 4,000 employees and $576 in annual sales. See Funding Universe, "Knoll Group Inc.—Company History," http://www.fundinguniverse.com/company-histories/Knoll-Group-Inc-company-History.html, accessed Feb. 23, 2012.
51. "Modern Living: New Chairs," *Life*, May 20, 1946, 47–48.
52. Hine, *Populuxe*.
53. "Drum Beaters for Modern," *Life*, Mar. 2, 1953, 72–76.
54. See Sylvia Katz and Jeremy Myerson, "The First Lady of the Modern Office," *World Architecture* (1989): 76–78.
55. See Museum of Modern Art, The Collection: https://www.moma.org/collection/, accessed February 8, 2019.
56. "Hans G. Knoll," *Current Biography Yearbook*, 225.
57. Hofstra, "Florence Knoll," 35.
58. See "River Club Interests Society: New Organization on East Side to Have Complete Athletic, Social and Boating Facilities," *New York Times*, May 4, 1930, 11.
59. See Charles Eames to Florence Knoll, June 11, 1957, AAA Knoll, box 4, folder 3, as one example of the close personal relationships the Knolls maintained with designers documented in the Florence Knoll-Bassett Papers.
60. See, for example, Allen, *Romance of Commerce and Culture*. See also Alexandra Lange, "Tower, Typewriter, and Trademark: Architects, Designers, and the Corporate Utopia, 1956–1964," PhD diss. New York University, 2005.

NOTES TO PAGES 237-244

61. See Jonathan Woodham, *Twentieth Century Design* (Oxford: Oxford University Press, 1997), 151–54.
62. See below, chapter 9. On the broader strategy to use consumer goods and consumer design as political tools in the Cold War, see Greg Castillo, *Cold War on the Home Front: The Soft Power of Midcentury Design* (Minneapolis: University of Minnesota Press, 2010), and Castillo, "Domesticating the Cold War: Household Consumption as Propaganda in Marshall Plan Germany," *Journal of Contemporary History* 40 (2005): 261–88.
63. Warren, "Top-Flight Rating."
64. See Rodes, *Einmal Amerika*, 166–171, and *Walter Knoll*, 63.
65. De Pree, *Business as Unusual*, 69.
66. See Rodes, *Einmal Amerika*, 173.
67. Ibid., 178–79.
68. "Knoll: Im Haut- und Knochen-Stil," *Der Spiegel*, Apr. 13, 1960.
69. Ibid.
70. On Knoll's development in Germany, see also Heythum, "Ten Years."
71. Paul Betts, *The Authority of Everyday Objects: A Cultural History of West German Industrial Design* (Berkeley: University of California Press, 2007).
72. Heythum, "Ten Years."
73. Ibid., 5.
74. Ibid.
75. Rodes, *Einmal Amerika*, 173.
76. On the concept of a cultural Cold War, see Berghahn, *America and the Intellectual Cold Wars in Europe*, and Giles Scott-Smith, "The Congress for Cultural Freedom: Constructing an Intellectual Atlantic Community," in Michael Mariano, ed., *Defining the Atlantic Community: Culture, Intellectuals, and Policies in the Mid-Twentieth Century* (London, 2010), 132–45.
77. "Florence Bassett's World on Display," *Miami Herald*, Mar. 12, 1972, 1L.
78. See Oestereich, *"Gute Form" im Wiederaufbau*, 181 and 185.

CHAPTER NINE

1. Raymond Loewy, Harvard University speech, Mar. 22, 1950, HgA Loewy, box 21.
2. Inger Stole, "'Selling' Europe on Free Enterprise: Advertising, Business and the US State Department in the late 1940s," *Journal of Historical Research in Marketing* 8 (2016): 44–64.
3. See Sheryl Kroen, "Negotiations with the American Way: The Consumer and the Social Contract in Post-war Europe," in John Brewer and Frank Trentmann, eds., *Consuming Cultures, Global Perspectives: Historical Trajectories, Transnational Exchanges* (Oxford: Berg, 2006), and especially Castillo, *Cold War on the Home Front*.
4. Volker Berghahn, "The Debate on 'Americanization' among Economic and Cultural Historians," *Cold War History* 10 (2010): 107–30.

5. See, e.g., Schröter, *Americanization of the European Economy*, and Dominique Barjot, *Catching Up with America: Productivity Missions and the Diffusion of American Economic and Technological Influence after the Second World War* (Paris: Presses de l'Université de Paris-Sorbonne, 2002).
6. See Richard Kuisel, *Seducing the French: The Dilemma of Americanization* (Berkeley: University of California Press, 2007), and Jean-Jacques Servan-Schreiber, *Le défi américain* (Paris: Denoel, 1967).
7. See, e.g., Kaspar Maase, *BRAVO Amerika: Erkundungen zur Jugendkultur der Bundesrepublik in den fünfziger Jahren* (Hamburg: Junius, 1992), and Poiger, *Jazz, Rock, and Rebels*.
8. See Wagnleitner, *Coca-Colonization and the Cold War*, and Michael Hochgeschwender, *Freiheit in der Offensive?: Der Kongreß für kulturelle Freiheit und die Deutschen* (Munich: Oldenbourg, 1998).
9. See Logemann, *Trams or Tailfins?*
10. See, e.g., Kleinschmidt, *Der produktive Blick*.
11. See Doering-Manteuffel, *Wie westlich sind die Deutschen?*
12. Arndt Bauerkämper, Konrad Jarausch, and Marcus Payk, eds., *Demokratiewunder: Transatlantische Mittler und die kulturelle Öffnung Westdeutschlands, 1945–1970* (Göttingen: Vandenhoek, 2005). On intellectual transfers, see also Axel Schildt, ed., *Von draußen: Ausländische intellektuelle Einflüsse in der Bundesrepublik bis 1990* (Göttingen: Wallstein, 2016).
13. On the elite networks supporting Westernization efforts, cf. Berghahn, *America and the Intellectual Cold Wars in Europe*, 2002).
14. See Julia Angster, "Wertewandel in den Gewerkschaften. Zur Rolle der gewerkschaftlichen Remigranten in der Bundesrepublik der 1950er Jahre," in Claus-Dieter Krohn and Patrick von Mühlen, eds., *Rückkehr und Aufbau nach 1945* (Marburg: Metropolis,1998), 111–38, and Scott Krause, "Neue Westpolitik: The Clandestine Campaign to Westernize the SPD in Cold War Berlin, 1948–1958," *Central European History* 48 (2015): 79–99, with regard to Berlin. See more broadly Alfons Söllner, "Ernst Fraenkel und die Verwestlichung der politischen Kultur in der Bundesrepublik Deutschland," *Leviathan* 30 (2002): 132–54, and Arnd Bauerkämper, "Americanisation as Globalisation? Remigrés to West Germany after 1945 and Conceptions of Democracy: The Cases of Hans Rothfels, Ernst Fraenkel and Hans Rosenberg," *Leo Baeck Institute Year Book* 49 (2005):153–70. On "Americanization" in economics scholarship, see Jan-Otmar Hesse, "'Ein Wunder der Wirtschaftstheorie': Die 'Amerikanisierung' der Volkswirtschaftslehre in der frühen Bundesrepublik," *Jahrbuch des Historischen Kollegs* (2007): 79–113, and for the public service see e.g. Magrit Seckelmann, "'Mit seltener Objektivität': Fritz Morstein Marx—Die mittleren Jahre (1934–1961)," *Die Öffentliche Verwaltung* 67 (2014): 1029–48.
15. On Italian emigration, see Renato Camurri, ed., *Mussolini's Gifts: Exiles from Fascist Italy*, special issue of *Journal of Modern Italian Studies* 15, no. 5 (2010).
16. On "remigration," see Claus-Dieter Krohn and Axel Schildt, *Zwischen den Stühlen? Remigranten und Remigration in der deutschen Medienöffentlichkeit der*

Nachkriegszeit (Hamburg: Christians, 2002). See also Marita Krauss, *Heimkehr in ein fremdes Land: Geschichte der Remigration nach 1945* (Munich: Beck, 2001).
17. Marita Krauss, "Exilerfahrung und Wissenstransfer. Gastprofessoren nach 1945," in Dahlmann and Reith, *Elitenwanderung und Wissenstransfer*, 35–54.
18. See FBI file on Paul Lazarsfeld (FOIA request), U.S. Government memorandum, May 17, 1968.
19. Paul Lazarsfeld, "Round Letter," July 20, 1959, PLA, folder "Wifo 3."
20. On the Vienna Institute of Advanced Study, see Christian Fleck, "Wie Neues nicht entsteht: Die Gründung des Instituts für Höhere Studien in Wien durch Ex-Österreicher und die Ford Foundation," *Österreichische Zeitschrift für Geschichtswissenschaften* 11 (2000): 129–78.
21. See, e.g., Axel Schildt, "Reise zurück aus der Zukunft. Beiträge von intellektuellen USA-Remigranten zur atlantischen Allianz in den 50er Jahren," in Wulf Köpke et al., eds., *Exil und Remigration* (Munich: Edition Text + Kritik, 1991) (= *Exilforschung. Ein internationales Jahrbuch* 9), 25–45.
22. See correspondence Adorno with Beate and Ferdinand Kramer, BjA Adorno, BR 0811 and BR 0812.
23. See Prof. Neufert (Darmstadt) to Kramer, July 13, 1955, Rektor TH Darmstadt to Kramer Dec. 7, 1955, and Kramer to Otl Aichinger (Ulm), Dec. 9, 1957, WbA Kramer, box 2. Between 1955 and 1958 Kramer lectured in Darmstadt, Aachen, Basel, Zurich, and Nuremberg among other places.
24. Moholy-Nagy to Adorno, Nov. 17, 1954, BjA Adorno, Br 1030.
25. On the OSS, see Alfons Söllner, *Zur Archäologie der Demokratie in Deutschland*, vol. 1 (Frankfurt am Main: Fischer, 1986), and Tim Müller, *Krieger und Gelehrte. Herbert Marcuse und die Denksysteme im Kalten Krieg* (Hamburg: Hamburger Edition, 2010).
26. Alvin Hansen and Richard Musgrave, *Fiscal Problems of Germany: Report Prepared during the Summer of 1951* (Berlin: ECA Team on Fiscal Problems of Germany, 1951).
27. Cf. reports from 1950 and 1953, BeL Katona, folder 17.
28. Margaret Re, "Will Burtin (1908–1972)," www.transatlanticperspectives.org, 2014.
29. See Adorno to Lazarsfeld, Apr. 3, 1954, BjA Adorno, Br 0867/11–12, and Adorno to Katona, Aug. 9, 1953, BjA Adorno, Br 0718.
30. See, e.g., George Katona, correspondence 1961–72, BeL Katona, folder 3.
31. Raymond Loewy, *Häßlichkeit verkauft sich schlecht: Die Erlebnisse des erfolgreichsten Formgestalters unserer Zeit* (Düsseldorf: Econ-Verlag, 1953). See also Ernest Dichter, *Handbuch der Kaufmotive: der Sellingappeal von Waren, Werkstoffen und Dienstleistungen* (Düsseldorf: Econ-Verlag, 1964), and George Katona, *Die Macht des Verbrauchers* (Düsseldorf: Econ-Verlag, 1962).
32. George Katona, "Amerikanische Erfahrungen mit der Untersuchung von Verhalten, Einstellungen und Meinungen," paper delivered at Frankfurt workshop "Die Erforschung der öffentlichen Meinung," Sept. 16, 1950, BeL Katona, folder 16.

33. See Gasteiger, *Der Konsument*, 93–95.
34. Schröter, *Americanization of the European Economy*, 97–123.
35. See Hirt, *Verkannte Propheten*. See also Wiesen, *Creating the Nazi Marketplace*, and Swett, *Selling under the Swastika*.
36. Katona, report to the HICOG Office of Political Affairs, Oct. 9 1950, BeL Katona, folder 17.
37. Cf. de Grazia, *Irresistible Empire*.
38. See, e.g., Ralf Jessen and Lydia Nembach-Langer, *Transformations of Retailing in Europe after 1945* (Farnham, UK: Ashgate, 2012), and Sibylle Brändli, *Der Supermarkt im Kopf: Konsumkultur und Wohlstand in der Schweiz nach 1945* (Vienna: Böhlau, 2000).
39. Lydia Langer, *Revolution im Einzelhandel: Die Einführung der Selbstbedienung in Lebensmittelgeschäften der Bundesrepublik Deutschland (1949–1973)* (Cologne: Böhlau, 2013).
40. See, e.g., Sabine Effosse, *Le crédit à la consommation en France, 1947–1965* (Paris: IGPDE 2014), Britta Stücker, "Konsum auf Kredit in der Bundesrepublik," *Economic History Yearbook* 48 (2007): 63–88, and Gunnar Trumbull, *Consumer Lending in France and America: Credit and Welfare* (Cambridge: Cambridge University Press, 2014).
41. Harm Schröter, "Die Amerikanisierung der Werbung in der Bundesrepublik Deutschland," *Jahrbuch für Wirtschaftsgeschichte* (1997): 93–115.
42. Clemens Zimmermann, "Marktanalysen und Werbeforschung der frühen Bundesrepublik: Deutsche Traditionen und US-Amerikanische Einflüsse, 1950–1965," in Manfred Berg and Phillip Gassert, eds., *Deutschland und die USA in der Internationalen Geschichte des 20. Jahrhunderts* (Stuttgart: Steiner, 2004), 473–91.
43. J. Walther Thompson, "List of American Advertising Agencies, Clients and Accounts in Western Germany," May 1960, HaC Lanigan), box 1.
44. Cf. Hirt, *Verkannte Propheten*.
45. Sean Nixon, *Hard Sell: Advertising, Affluence and Transatlantic Relations, c. 1951–69* (Manchester: Manchester University Press, 2013), 11.
46. See Kleinschmidt, *Der produktive Blick*; Susanne Hilger, *'Amerikanisierung' deutscher Unternehmen: Wettbewerbsstrategien und Unternehmenspolitik bei Henkel, Siemens und Daimler-Benz (1945/49–1975)* (Wiesbaden: Franz Steiner, 2004); and Ludwig, *Amerikanische Herausforderungen*.
47. On the organized transatlantic diffusion of economic and technological knowledge, see Barjot, *Catching Up with America*.
48. See, e.g., report, "FOA Team—TA Project 09–278 (Trade Journals)— Industrial Press Work," BuArch, B 140/316. On the Gesellschaft zur Förderung des deutsch-amerikanischen Handels, see Ludwig, *Amerikanische Herausforderungen*, 34–36.
49. Cf. the essays by Norbert Grube, Stefan Schwarzkopf, and Judith Coffin in Kerstin Brückweh, ed., *The Voice of the Citizen Consumer: A History of Market Research, Consumer Movements, and the Political Public Sphere* (Oxford:

Oxford University Press, 2011). On the British case, see also Stefan Schwarzkopf, "Markets, Consumers, and the State: The Uses of Market Research in Government and the Public Sector in Britain, 1925–1955," in Hartmut Berghoff, Philip Scranton, and Uwe Spiekermann, eds., *The Rise of Marketing and Market Research* (New York: Palgrave Macmillan, 2012), 171–91.

50. George Katona, report, "Brief Resume of My Trip to Germany," Oct. 9, 1950, BeL Katona, folder 17.
51. On the history of market and opinion research in Germany, see Christoph Weiler, *Das Unternehmen 'Empirische Sozialforschung': Strukturen, Praktiken und Leitbilder der Sozialforschung in der Bundesrepublik Deutschland* (Munich: Oldenbourg, 2004), 130–43.
52. Cf. Rolf Berth, *Marktforschung zwischen Zahl und Psyche: Eine Analyse der befragenden Marktbeobachtung in Deutschland* (Stuttgart: Gustav Fischer, 1959), 191–225.
53. JWT, "Commentary on Economic Research in Western Germany," May 1960, HaC Lanigan, box 1.
54. Hilger, *Amerikanisierung*, p. 188.
55. On the GfK, see, e.g., Gasteiger, *Der Konsument*, 41–64, as well as Wilfried Feldenkirchen and Daniela Fuchs, *Die Stimme des Verbrauchers zum Klingen bringen: 75 Jahre Geschichte der GFK Gruppe* (Munich: Piper, 2009).
56. Hanns Kropff, "Methoden und Techniken der amerikanischen Verbrauchs- und Meinungsforschung, angewandt auf Leserschaftsanalysen," 1954, GfK, GAA 03.2.1 EM2.
57. N.N., Inventur der amerikanischen Werbeforschung (i.A. Arbeitskreis Werbeforschung), GfK, S 1954 004.
58. See Katona, "Brief Resume of My Trip to Germany," and correspondence Noelle-Neumann with Adorno, BjA Adorno, Br. 1089.
59. See Chlodwig Kapferer, *Market Research Methods in Europe* (Paris: OEEC, 1956), and BuArch, folder "Marktforschung" B 140/493.
60. Gasteiger, *Der Konsument*, 19 and 32.
61. See, e.g., "Untersuchungen zur Motivation des Konsumentenverhaltens bei den Marken Wipp und Sunil," 1957, GfK, S 1957 023–1, a motivational brand research study for Henkel utilizing word association tests. Georg Bergler, "Marktforschung und Motivforschung," 727, cited in Gasteiger, *Der Konsument*, 75.
62. See, e.g., Berth, *Marktforschung zwischen Zahl und Psyche*.
63. See, e.g., George Smith, *Warum Kunden kaufen: Motivforschung in Werbung und Verkauf* (Munich: Moderne Industrie, 1955), and Richard Crisp, *Absatzforschung* (Essen: Giradet, 1959). See also Clemens Sauermann, "Der Psychotest in der Marktforschung," *Zeitschrift für Markt- und Meinungsforschung* 1 (1957/58): 19–24.
64. Hans Schad, "Überlegungen zum Methodenstreit," *Zeitschrift für Markt- und Meinungsforschung* 1 (1957/58): 217–22.
65. Gerhard Kleining, "Bedeutungsanalyse. Ein Verfahren der qualitativen

Absatzforschung," *Zeitschrift für Markt- und Meinungsforschung* 2 (1958/59): 343–56.
66. See Gasteiger, *Der Konsument*, 68–104.
67. Hanns Kropff, "Die Hypothese als Grundlage der wissenschaftlichen Forschung im Marketing," *Zeitschrift für Markt- und Meinungsforschung* 2 (1958/59): 309–20. See also Bernd Semrad, "Ein Jude im Wien des frühen zwanzigsten Jahrhunderts: Der junge Ernst Dichter," in Rainer Gries and Stefan Schwarzkopf, eds., *Ernest Dichter: Doyen der Verführer* (Vienna: Mucha, 2007), 33–40.
68. Hugh Hardy, *The Politz Papers: Science and Truth in Marketing Research* (Chicago: American Marketing Association, 1990), chap. 1.
69. Schad, "Überlegungen zum Methodenstreit," 221, my translation.
70. See, e.g., Notger Küng, "Marktforschung im Umbruch," *Zeitschrift für Markt- und Meinungsforschung* 2 (1958/59): 505–10. On the increasing importance of psychological research, see also the recollections of Klaus Haupt, "Die Entwicklung der psychologischen Marktforschung in Deutschland," *Transfer: Werbeforschung & Praxis* (Jan. 2009): 24–33.
71. On their role in the German automobile industry, see Ingo Köhler, *Auto-Identitäten: Marketing, Konsum und Produktbilder des Automobils nach dem Boom* (Göttingen: Wallstein, 2018).
72. "Dr. Dichter . . . repeated yet again his research on why men buy various types of cars," the JWT employee groaned. See Constance Ivie to Denis Lanigan, Mar. 29, 1960, HaC Lanigan, box 1.
73. Ibid.
74. Ernst Dichter, "Psychologie Commerciale," *Comité National de l'Organisation Française* (May 1937): 136–40, DaW, #4613.
75. See "Notiz," *Zeitschrift für Markt- und Meinungsforschung* 1 (1957/58): 49. See also speeches and press clippings from European events in Austria, France, Germany, Spain, the United Kingdom, and the Netherlands, HgA Dichter, boxes 138 and 149.
76. See, e.g., Ernest Dichter, "How to Market in the U.S.," DaW, #4978, and "Eine Motivstudie über die psychologische Wirksamkeit der Schauma Schampoo Werbung," 1958, DaW, #1219. A full list of companies for which Dichter proposed and/or conducted studies can be found at http://www.ernest-dichter.net/.
77. See "Ernest Dichter International in Europa," brochure, n.d., DaW, #4457.
78. Ernest Dichter, "Germany and the Germans in the Eyes of Americans: A Proposal for a Motivation Study," Sept. 1969, HgA Dichter, box 100.
79. See Hubert Bonin and Ferry de Goey, eds., *American Firms in Europe: Strategy, Identity, Perception and Performance, 1880–1980* (Geneva: Librairie Droz, 2009).
80. Ernest Dichter, "Needed: A Psychological Bridge over the Ocean," typescript, July 12, 1967, HgA Dichter, box 170.
81. Ibid.

82. Ernest Dichter, "What Are the Unsatisfied Needs of Europe Today?" article for *Business Abroad* (typescript), n.d., HgA Dichter, box 167.
83. Dichter, "Needed: A Psychological Bridge over the Ocean."
84. Ernest Dichter, "The World Customer," in Steuart Britt, ed., *Consumer Behavior and the Behavioral Sciences* (New York: Wiley & Sons, 1966), 67–68.
85. Ibid.
86. "Proposal for a Motivational Research Study of the German Market for United States Exports," Sept. 1968, HgA Dichter, box 97.
87. Ernest Dichter, "Sales to the Women of Europe Can Be Killed with Kindness," *Business Abroad*, Dec. 12, 1966.
88. Ernest Dichter, "A Guide to Business Diplomacy in Europe," *Motivations* 14 (1956). On Europe's "love-hate relationship" with the United States, see also "Establishing a Direct Contact between European Investors and American Companies," HgA Dichter, box 100, and "Changing America's Image Abroad," speech at Overseas Press Club, New York, Oct. 1957, DaW, #4682.
89. Ernest Dichter, "Proposal for a Motivational Research Study of the German Market," Sept. 1968, HgA Dichter, box 97.
90. Ernest Dichter, "Business Abroad Article—(Rough Copy)," Mar. 28, 1967, HgA Dichter, box 169.
91. Ibid., 5.
92. Ibid., 6.
93. Ibid., 6–7.
94. See, e.g., George Katona, *Die Macht des Verbrauchers* (Düsseldorf: Econ-Verlag, 1962).
95. Gasteiger, *Der Konsument*, 142–43.
96. George Katona, "The Impact of Installment Credit," Eurofinas Association of European Finance Houses, Vienna, June 25–26, 1962, BeL Katona, folder 56.
97. George Katona, Burkhard Strumpel, and Ernest Zahn, *Aspirations and Affluence: Comparative Studies in the United States and Western Europe* (New York: McGraw-Hill, 1971).
98. Raymond Loewy, "Essen Speech," Sept. 21, 1955, LoW, box 3.2. See also above, chapter 5.
99. Veronique Pouillard, "Keeping Designs and Brands Authentic: The Resurgence of the Post-war French Fashion Business under the Challenge of US Mass Production," *European Review of History* 20 (2013): 815–35. See also "Fashion: Counter revolution," *Time*, Sept. 15 1947, and "Fashion: The American Look," *Time*, May 2, 1955.
100. See Per Hansen, "Networks, Narratives and New Markets: The Rise and Decline of Danish Modern Furniture Design, 1930–1970," *Business History Review* 80 (2006): 449–83. See also Marilyn Hoffman, "Design in Scandinavia," *Christian Science Monitor*, Jul. 9, 1954, and Elisabetta Merlo and Francesca Polese, "Turning Fashion into Business: The Emergence of Milan as an International Fashion Hub," *Business History Review* 80 (2006): 415–47.

101. On the history of German design at midcentury, see Betts, *Authority of Everyday Objects*. See also Gert Selle, *Geschichte des Design in Deutschland* (Frankfurt: Campus, 1997), and Oestereich, *"Gute Form" im Wiederaufbau*.
102. On the reconstitution of the Werkbund, see Betts, *Authority of Everyday Objects*, 73–108, and Gerda Breuer, ed., *Das Gute Leben: Der Deutsche Werkbund nach 1945* (Tübingen: Wasmuth, 2007). On the 1949 exhibition, see also Castillo, "Domesticating the Cold War."
103. Carl Otto, "Formentwicklung von Industrieprodukten in Amerika" (1950), BuArch B 102/34493.
104. Herwin Schaefer cited in Castillo, *Cold War on the Home Front*, 34–35.
105. On the history of the Design Council, see in BuArch B 102/34493, folder "Rat für Formgebung," and Betts, *Authority of Everyday Objects*, 210. See also Rat für Formgebung, *Designkultur 1953–1993: Philosophie—Strategie—Prozess* (Frankfurt: Design Council, 1993).
106. Betts, *Authority of Everyday Objects*, 137.
107. Loewy, *Häßlichkeit verkauft sich schlecht*.
108. Walter Kersting to Ministry of Economics, Sept. 30, 1951, BuArch B 102/34493.
109. At times, they found American witnesses for such concerns. See, e.g., Gordon Lippincott, "Unternehmertum und Formgebung," *form* 5 (1959): 22–23.
110. See, e.g., Hans Klaus, "Die Gute Form. Die Bedeutung der industriellen Formgestaltung," *VDI Nachrichten*, May 5,1951, 3, BuArch B 102/34493.
111. See *Internationale Diskussion über Industrielle Formgebung im Darmstädter Gespräch 1952—"Mensch und Technik,"* vol. 3 (Darmstadt: Neue Darmstädter Verlagsanstalt, 1952), 29–33. The French delegation included Jacques Vienot of Loewy's Parisian subsidiary.
112. The ministry collaborated with the Committee on Fashion Goods of the industry association BDI; see BuArch B 102/34493 and B 102/21241.
113. Philip Rosenthal, "'Industrial Design'—Kunst oder Geschäft? Industrie und Kunst müssen Kompromiss finden," *Die Neue Zeitung*, Sept. 27/28, 1952, 10.
114. On the dissemination of organic and international style design, see Petra Eisele, "Zwischen 'Organic design' und Nierentisch: Die 'organische Form' im Wirtschaftswunderland," in Breuer, *Das gute Leben*, 157–65, and Volker Fischer, "'Gute Form' made in the USA—Design zwischen Europa und Amerika," in Bernd Polster, ed., *West Wind: Die Amerikanisierung Europas* (Cologne: DuMont, 1995), 66–78.
115. On design in German industry practice, see Oestereich, *"Gute Form" im Wiederaufbau*, 171–87.
116. Karl Otto, *Industrielle Formgebung in den USA* (Berlin: Berliner Industriebank, 1963).
117. Berthold Semper, "Formgebung im Spannungsfeld von Lehre und Praxis," *Graphik: Werbung und Formgebung* 9, no. 1 (1956): 38–45.
118. See "Industrial Design Education USA" and "Das Pratt-Institut in New York," *form* 6 (1959): 32–35.

119. On the Ulm Institute, see Betts, *Authority of Everyday Objects*, 139–77, and Selle, *Geschichte des Design*, 290–300. On the complicated political history of the school, see Rene Spitz, *Hfg Ulm: der Blick hinter den Vordergrund: Die politische Geschichte der Hochschule für Gestaltung, 1953–1968* (Stuttgart: Axel Menges, 2002).
120. See, e.g., a special issue on American design of *form* 5 (1959).
121. See "Josef Albers 70 Jahre," *Architektur und Wohnform* 67, no. 3 (1959).
122. See Eberhard Hölscher, "Walter Allner. Werbegraphik aus Amerika," *Gebrauchsgraphik* 25, no. 12 (1954): 2–5, and Franz Wills, "Saul Bass," *Graphik: Werbung + Formgebung* 9, no. 1 (1956): 31–34.
123. See, e.g., special issue on packaging of *Graphik* 10, no. 3 (1957), with several articles on U.S. innovations, and R. Leslie, "Frank Gianninoto," *Graphik: Werbung + Formgebung* 9, no. 7 (1956): 348–51.
124. Joseph Bourdez to Walter Landor, July 30, 1958, LaP, box 18.2, folder "Holland."
125. Landor to Mr. v.d. Decken, Sept. 23, 1962 LaP, box 18.2, folder "Hamburg."
126. See, e.g., reviews in *Der Markenartikel* 3 (1954) and *Die Absatzwirtschaft* 3 (1954): 114–16. On the wider European press coverage, see HgA Loewy, boxes 6 and 22 (for speeches in Britain).
127. On the French and British subsidiaries, see Patrick Farrell, "Dreimal von Neuem: 1934, 1947, 1969—Loewy's Büros in London," and Evert Endt, "Eine französische Affäre: Die 'Compagnie de l'Esthetique Industrielle' in Paris," in Internationales Design Zentrum Berlin, *Raymond Loewy: Pionier des Amerikanischen Industrie-Designs* (Munich: Prestel, 1990), 173–81 and 161–72.
128. Raymond Loewy, "Image et Prestige," speech, Jan. 1963, LoW, box 1.9.
129. See Albrecht Graf Goertz, *You have got to be lucky "Reflections: Design is International, it has no boundaries"* (Munich: Goertz, 2003).
130. Oestereich, *"Gute Form" im Wiederaufbau*, esp. 188–216. Germany, too, saw increased attention to the role of market research in design, cf. 193.
131. Ferdinand Kramer, "Industrielles Design," talk at TH Darmstadt, July 1955, WbA Kramer, box 3.
132. Peter Muller-Munk, "Die Beziehung der Formgebung (Industrial Design) zum modernen Marketing Konzept," speech, Vienna 1960, reprinted in Otto, *Industrielle Formgebung in den USA*, 94–95.
133. See, e.g., Ruth Oldenziel and Karin Zachmann, eds., *The Cold War Kitchen: Americanization, Technology and European Users* (Cambridge: MIT Press, 2009).
134. See Castillo, "Domesticating the Cold War," and Castillo, *Cold War on the Home Front*.
135. Castillo, "Domesticating the Cold War," 266.
136. See Paul Betts, "The Bauhaus as Cold War Legend: West German Modernism Revisited," *German Politics and Society* 14 (1996): 75–100, and Castillo, "Domesticating the Cold War," 272.
137. Gay McDonald, "Selling the American Dream: MoMA, Industrial Design and Post-War France," *Journal of Design History* 17 (2004): 397–412.

138. See William Foster, "Amerikanische Gebrauchsformen—Einführung," in Landesgewerbeamt Stuttgart, ed., *Industrie und Handwerk schaffen: Neues Hausgerät aus den USA* (Stuttgart: Landesgewerbeamt, 1951), 1–3.
139. Edgar Kaufmann, introduction, in Landesgewerbeamt, *Industrie und Handwerk*, 6–10.
140. See, e.g., National Small Industries Corporation, *Design Today in America and Europe* (New Delhi, 1958), a catalogue for an exhibition organized by MoMA and the Ford Foundation in cooperation with the German Design Council.
141. Cf. Castillo, *Cold War on the Home Front*, 36.
142. See Jack Masey and Conway Morgan, *Cold War Confrontations: US Exhibitions and Their Role in the Cultural Cold War* (Zurich: Lars Müller, 2008), 90–91.
143. The TES itself was also headed by an emigré, art historian Annemarie Henle Pope. On Burtin, see Margaret Re, "Will Burtin," and Masey and Morgan, *Cold War Confrontations*, 104–6.
144. Will Burtin, "Werbepackung in Amerika. Eine Ausstellung der Amerika-Häuser in Deutschland," exhibition catalogue, n.d. I am grateful to Margaret Re for sharing her copy of the catalogue with me.
145. Masey and Morgan, *Cold War Confrontations*, 14, and Re, "Will Burtin."
146. James Plaut, "Industrielle Formgestaltung in den Vereinigten Staaten," *Perspektiven* 9 (Fall 1954): 112–30.
147. Castillo, *Cold War on the Home Front*, 33.
148. On the elite focus of US Cold War policy, see, e.g., Jessica Gienow-Hecht, "American Cultural Policy in the Federal Republic of Germany, 1949–1968," in Detlef Junker, ed., *The United States and Germany in the Era of the Cold War, 1945–1990*, vol. 1 (Cambridge: Cambridge University Press, 2004), 401–8, and Berghahn, *America and the Intellectual Cold Wars in Europe*.
149. Richard Neutra, "Rundfunkgespräch. Europa-Amerika und die Welt des Architekten," Oct. 31, 1954, BhA, folder "Neutra 2."

CONSUMER ENGINEERING

1. See, e.g., Gruen, *Heart of Our Cities*. See also Victor Gruen, "Public Planning and Land Ownership," paper presented at "The Crisis of Controls—A Symposium," Victor Gruen Foundation for Environmental Planning, Beverly Hills, CA, Sept.18, 1973, and "The Downfall and Rebirth of City Cores on Both Sides of the Atlantic," lecture for the Gruen Foundation, 1972, Deutsche Nationalbibliothek, Exilarchiv, Spalek Files.
2. On intellectual critiques of consumption, see *Anxieties of Affluence*. On the transnational character of the postwar debate, see Daniel Horowitz, *Consuming Pleasures: Intellectuals and Popular Culture in the Cold War* (Philadelphia: University of Pennsylvania Press, 2012).

NOTES TO PAGES 267-271

3. On the collaboration of Adorno with Lazarsfeld's radio project, see above, chapter 2. See also Wheatland, *Frankfurt School in Exile*.
4. Leo Löwenthal and Norbert Guterman, *Prophets of Deceit* (New York: Harper & Brothers and American Jewish Committee, 1949).
5. See, e.g., Christoph Weischer, *Das Unternehmen "Empirische Sozialforschung": Strukturen, Praktiken und Leitbilder der Sozialforschung in der Bundesrepublik Deutschland* (Munich: R. Oldenbourg Verlag, 2004), esp. 219-22. See also Thomas Wheatland, "Franz L. Neumann: Negotiating Political Exile," in Logemann and Nolan, *More Atlantic Crossings?* 111-38.
6. See, e.g., Kaspar Maase, "A Taste of Honey: Adorno's Reading of American Mass Culture," in John Dean and Jean-Paul Gabilliet, eds., *European Readings of American Popular Culture* (Westport, CT: Greenwood Press, 1996), 201-11, and David Jenemann, *Adorno in America* (Minneapolis: University of Minnesota Press, 2007).
7. Theodor Adorno, "Culture Industry Reconsidered," *New German Critique* 6 (1975 [orig. 1963]): 12.
8. Adorno to Lazarsfeld, Jan. 24, 1938, BjA Adorno, Br 0867/1-7.
9. Horkheimer to Adorno, Nov. 24, 1944, in Gunzelin Schmid Noerr, ed., *Max Horkheimer, Gesammelte Schriften*, vol. 17: *Briefwechsel, 1941-1948* (Frankfurt am Main: Fischer, 1996), 606, cited in Nora Binder, "Von Nazis und Nieren. Demokratische Gruppendynamik bei Kurt Lewin," in Nora Binder and Bernd Kleeberg, eds., *Die Wahrheit zurichten. Über Sozio- und Psychotechniken* (Tübingen: Mohr Siebeck 2018).
10. See, e.g., Herbert Marcuse, *Eros and Civilisation: A Philosophical Inquiry into Freud* (Boston: Beacon Press, 1955), and Herbert Marcuse, *One-Dimensional Man: Studies in the Ideology of Advanced Industrial Society* (Boston: Beacon Press, 1964). Fromm's publications include *Escape from Freedom* (New York: Holt, 1941) and *The Sane Society* (New York: Holt, 1955).
11. Fromm's impact is discussed in Neil McLaughlin, "Critical Theory Meets America: Riesman, Fromm, and the Lonely Crowd," *American Sociologist* 32 (2001): 5-26.
12. See, e.g., Victor Papanek, *Design for the Real World: Human Ecology and Social Change* (New York: Pantheon, 1970). On Rudofsky, see Andrea Bocco Guarneri, *Bernard Rudofsky: A Humane Designer* (Vienna: Springer, 2003).
13. On the consumer movement as a transnational force, see Matthew Hilton, *Prosperity for All: Consumer Activism in an Era of Globalization* (Ithaca: Cornell University Press, 2010). For the pioneering movement in the United States, see Glickman, *Buying Power*.
14. See, e.g., Christopher Klemek, *The Transatlantic Collapse of Urban Renewal: Postwar Urbanism from New York to Berlin* (Chicago: University of Chicago Press, 2012).
15. See Stefan Schwarzkopf, "From Fordist to Creative Economies: The De-Americanization of European Advertising Cultures since the 1960s," *European Review of History* 20 (2013): 859-79.

16. For an early account of the shift to flexible specialization in production, see Michael Piore and Charles Sabel, *The Second Industrial Divide: Possibilities for Prosperity* (New York: Basic Books, 1984).
17. See Gasteiger, *Der Konsument*.
18. Frank, *Conquest of Cool*. For Germany's "creative revolution" in advertising, see Cornelia Koppetsch, *Das Ethos des Kreativen: Eine Studie zum Wandel von Arbeit und Identität am Beispiel der Werbeberufe* (Konstanz: UVK, 2006), 159–80.
19. Stephan Malinowski and Alexander Sedlmaier, "'1968' als Katalysator der Konsumgesellschaft. Performative Regelverstöße, kommerzielle Adaptionen und ihre gegenseitige Durchdringung," *Geschichte und Gesellschaft* 32 (2006): 238–67.
20. Cf. Stefan Schwarzkopf, "Mobilizing the Depths of the Market: Motivation Research and the Making of the Disembedded Consumer," *Marketing Theory* 15 (2015): 39–57.
21. See, e.g., "Furniture Makers Go All Out for Modern," *New York Times*, Apr. 17, 1971, 64.
22. Köhler, *Auto-Identitäten*, 197.
23. On the connection between the crisis and an increased marketing effort among automobile companies, see esp. Ingo Köhler, "Imagined Images, Surveyed Consumers: Market Research as a Means of Consumer Engineering, 1950s–1980s," in Logemann, Cross, and Köhler, *Consumer Engineering*.
24. See esp. Lutz Raphael and Anselm Döring-Manteuffel, *Nach dem Boom* (Göttingen: Vandenhoeck & Ruprecht, 2008). See also Marc Levinson, *An Extraordinary Time: The End of the Postwar Boom and the Return of the Ordinary Economy* (London: Random House, 2016).
25. See, e.g., Carola Westermeier, "Werbung und Wertewandel. Diskurse über Verbraucher und deren Verhalten," in Bernd Dietz and Jörg Neuheiser, eds., *Wertewandel in der Wirtschaft und Arbeitswelt* (Berlin: De Gruyter Oldenbourg, 2016), 239–60.
26. On German companies in the U.S. market, see Ludwig, *Amerikanische Herausforderungen*.
27. Peter Bart, "Advertising: 'Foreign Relations' Assessed," *New York Times*, Apr. 29, 1962, 156.
28. See Schwarzkopf, "From Fordist to Creative Economies."
29. See, e.g., "Modern Living: Those Designing Europeans," *Time*, Feb. 17, 1975, and Art Seidenbaum, "Unstyled Americans," *Los Angeles Times*, Feb. 21, 1975.
30. Besides Interpublic, the French-based Publicis Group (Leo Burnett, Saatchi & Saatchi, etc.), the British WPP (Young & Rubicam, JWT), Omnicon (Doyle, Dane & Bernbach, BBDO), and the Japanese Dentsu network have emerged as the largest global players since the 1980s. See also Mark Tungate, *Adland: A Global History of Advertising* (London: Kogan Page, 2007).
31. See Ludwig, *Amerikanische Herausforderungen*, chap. 7.

32. See, e.g., Peter Bart, "Advertising: Europe Is Wary of Uniformity," *New York Times*, Aug. 28, 1963, 58. For global marketing and the example of the beauty industry, see Geoffrey Jones, *Beauty Imagined: A History of the Global Beauty Industry* (Oxford: Oxford University Press, 2011).
33. Jan Nederveen Pieterse, *Globalization and Culture: Global Melange* (Oxford: Rowman and Littlefield, 2004).
34. On Esslinger and his transatlantic career, see Sean Nye, "Hartmut Esslinger," in R. Daniel Wadhwani, ed., *Immigrant Entrepreneurship: German-American Business Biographies, 1720 to the Present*, vol. 5, German Historical Institute, http://www.immigrantentrepreneurship.org/entry.php?rec=236, last modified April 29, 2015.

Index

Aalto, Alvar (architect), 159
Achilles, Paul (market researcher), 59
A. C. Nielsen (market research), 47, 76, 107, 252
Adler (automobiles), 175
Adler, Alfred (psychologist), 50–51
Adorno, Theodor W. (sociologist), 62, 64, 70, 79, 175, 220, 247–49, 272
advertising: advertising agencies, 4, 10–11, 21, 24, 26, 29–30, 47, 49, 55, 61–66, 76, 99–112, 117–21, 125, 132–36, 151–55, 177, 186, 199, 206, 210, 251, 255, 265, 274, 277–78; advertising psychology, 3, 109; subliminal advertising, 128
Advertising and Selling, 20, 27, 60
Advertising Arts, 27–28, 35
Advertising Council, 244
Advertising Research Foundation, 63, 69, 122
AEG (electric), 31, 213, 239, 261
Agnew, Hugh (market researcher), 49
Aichinger, Otl (designer), 348
Air France (transportation), 265
Albers, Anni (designer), 144, 159, 174, 231, 233, 267
Albers, Josef (designer), 165, 174, 190, 205, 248, 264
Albini, Franco (designer), 230
Alderson & Sessions (market research), 106
Alfred Politz Research Inc. (market research), 100, 102, 106, 108–11, 115, 200, 254, 273; Universal Marketing Research, 254
Allied Purchasing Corporation (retail), 168
Allner, Walter (designer), 152–53, 264
Allport, Gordon (psychologist), 61, 88
American Advertising Guild, 155
American Bauhaus, 157, 167, 176–88, 190–91, 193, 225, 263–64. *See also* Institute of Design, Chicago
American Broadcasting Company (ABC) (media), 64
American Century, 7, 260
American Institute of Architects, 233
American Institute of Decorators, 161
American Institute of Public Opinion (market research), 48, 56, 59–62, 76, 111–12
Americanization, 3, 6, 14, 17, 21, 220, 237, 240, 244–45, 250, 255, 260, 269
American Jewish Committee, 64
American Journal of Sociology, 71
American Marketing Association (AMA), 49, 78, 103, 111–12
American Marketing Journal, 49
American Meat Institute, 84
American way of life, 97–98, 131, 149, 244, 260
Amsterdam, 30, 151, 266
Apple Inc. (technology), 278

359

INDEX

Archipenko, Alexander (designer), 179, 188
Architectural Forum, 144, 235
Arens, Egmont (designer), 8–10, 22, 24–26, 33, 35–37, 95, 100, 128, 137–40, 145, 158, 161, 196, 204, 263
Armour & Co. (food), 200, 202, 268
Arnheim, Rudolf (psychologist), 64
Art Deco, 30, 142–43, 146
Art Directors Club, Chicago, 188
Art Institute of Chicago, 153, 157–58
Arts and Craft, 143, 173, 226, 260
Arts Club, Chicago, 236
Ascoli, Max (political scientist), 246
Associated Merchandising Company (retail), 203
Association of Arts and Industry, Chicago, 177, 179–80
AT&T (communications), 44, 66
Auerbach, Alfred (designer), 161
Augspurger, Michael (historian), 135
Austria, 2, 4, 21, 26, 32, 35, 43, 51–60, 73, 114, 117, 136, 194, 244, 247, 268

Baermann, Walter (designer), 137–41, 143
Bahlsen (food), 31
Bally (shoes), 54
B&O Railroad (transportation), 181, 230
Barnard College, 51
Barr, Alfred (curator), 184, 236
Bartos, Rena (advertiser), 63, 71
Basel, 230
Bass, Paul (designer), 264, 268
Bauer, Raymond, 128
Bauhaus, 4, 11, 27–28, 32, 132–34, 137, 140, 142–45, 152–54, 158–59, 164–67, 172–205, 207, 223–24, 260–67, 278
Bayer, Herbert (designer), 3, 11, 153–59, 161, 175–77, 180, 182, 186–89, 192, 213, 248, 268, 277
Beall, Lester (designer), 268
Beard, Miriam (historian), 26
behavioral economics, 42, 89–90
behaviorist stimulus-response model (S-R model), 46–47, 55, 74, 82, 85, 94–96, 121
Behrens, Peter (designer), 2, 31, 173, 213
Bendix, Reinhard (sociologist), 123
Benson & Mather (advertising), 254
Benton & Bowles (advertising), 58
Bel Geddes, Norman (designer), 136, 168, 171–72, 187, 194–95

Belgium, 143
Bell Laboratories, 137
Bellman, Hans (designer), 230
Bell Telephone (communication), 186
Bergdorf Goodman (retail), 106
Bergler, Georg (market researcher), 253
Berlin, 3, 6, 29–31, 51–54, 75, 84–85, 90, 102, 108–9, 141–45, 153, 177–78, 185, 188, 239, 266
Berliner Tageblatt, 109
Bernays, Edward (public relations), 35, 73–75, 82, 98
Bernbach, William (advertiser), 176, 277
Bernhard, Lucian (designer), 20–21, 28, 31
Bertoia, Harry (designer), 229–30, 236
Betts, Paul (historian), 239, 262
Bill, Max (designer), 263
Boston, 55, 158, 170, 230–31
Boudreau, James (designer), 141
Black, Misha (designer), 206–7
Black Mountain College, 165, 175, 232
Blake, Peter (architect), 174, 268
Blaszczyk, Regina (historian), 11, 191, 195
Bloomingdale's (retail), 168
BMW (automobiles), 265
brand: brand image, 10, 42, 65, 121, 123–26, 128, 132, 213–15, 254, 264, 275–76; brand systems design, 214
Braun (electronics), 238, 240, 255, 261, 265
Bredendieck, Hin (designer), 179, 188, 190
Bremen, 175
Breuer, Marcel (designer), 145, 158–59, 174–75, 182, 192, 221, 230, 241, 264
Bristol Myers (hygiene), 111, 170
Britain, 34, 163, 206, 225, 251, 262, 265
British Petroleum (energy), 201
Brodovitch, Alexey (designer), 28
Brooklyn College, 188
Brose, Hans (advertiser), 251
Brown, Lyndon (marketing scholar), 49
Brussels, 238
Budapest, 54, 90, 188, 232
Buderus (heating), 167
Bühler, Charlotte (psychologist), 50–51, 83, 247
Bühler, Karl (psychologist), 50–51, 83
Bund, Henry (marketing scholar), 114
Bureau of Applied Social Sciences (BASR), 63–70, 77, 80–82, 92, 96–97, 117, 246, 253, 274
Bureau of Business Research, 55

INDEX

Burgess, Ernest (sociologist), 123
Burke Marketing Research Inc. (market research), 106
Burtin, Will (designer), 152, 161, 248–49, 268
Business Abroad, 256
Business Week, 121

California, 2, 137, 143–45, 205, 207, 214
California Institute of Technology, 137, 143
Calkins, Earnest Elmo (advertiser), 20, 26–27, 33, 136
Calkins & Holden (advertising), 26, 136
Canada, 238
C&A (textiles), 255
C. A. Pillsbury (food), 258
Caplovitz, David (sociologist), 71
Carlu, Jean (designer), 28, 257, 259, 186, 189
Carnegie Endowment, 16, 184, 234
Carnegie Institute of Technology, 137, 142
Cassandre, A. M., 151, 153, 157, 159, 231
Casson, Mark (management scholar), 11
Castillo, Greg (historian), 269
Castleton China (housewares), 159, 267
Cattel, James (psychologist), 48
Catton Rich, Daniel (artist), 153
Chase, Stuart (consumer advocate), 36
Cheney, Sheldon and Martha (designers), 137, 172
Chermayeff, Serge (designer), 160, 182, 188, 225
Cheskin, Louis (psychologist), 100
Chicago, 6, 26, 50, 67, 84–85, 91, 100, 106, 123, 138, 140, 144, 150, 153, 157–61, 170, 177–91, 199, 225–26, 230–31, 236, 253, 263
Chicago Association of Commerce, 181
Chicago Sun-Times, 161, 187
Chicago Tribune, 106, 123
Chrysler Corporation (automobiles), 117
Coca-Cola (food), 24, 111, 257
Cohen, Lizabeth (historian), 1, 36, 41, 45, 131
Cogdell, Christina (historian), 18, 142
Coiner, Charles (art director), 151–52
Cold War, 7, 17, 19, 42, 82, 95, 191, 219, 243–44, 251, 273; cultural Cold War, 17, 19, 127, 237–40, 244, 248, 266–69; and social sciences, 19, 97, 246
Collier's, 151

Cologne, 32, 152, 261–62, 268
Color Research Institute of America, 100
Collura, Francesco (designer), 268
Columbia Broadcasting System (CBS) (media), 61–62, 64–65, 117–18, 152, 157, 234–35
Columbia University, 51, 53, 55–56, 63–66, 70–71, 80, 92, 122, 253, 274
Compagnie de l'Esthetique Industrielle (design), 150, 260, 265
Compton Advertising (advertising), 109–10, 117
Computer Services Corporation (technology), 273
Congres Internationaux d'Architecture Moderne (CIAM), 145, 164–67, 172–75, 185, 187, 192, 207
Connecticut General Life Insurance Company (finance), 233
consulting firm, 10, 46, 106–7, 128, 195, 273
Consumer Advisory Board (National Recovery Administration), 44, 59
consumer aspirations, 89, 94–95, 194, 259, 276; levels of aspiration, 89
consumer attitudes, 12, 57–58, 61, 65, 70–76, 78, 80, 89–97, 106, 113, 118, 122–23, 147, 198, 250, 253, 257
consumer confidence, 90, 94, 97, 259; Index of Consumer Sentiment, 94, 274
consumer engineering, 5–10, 12, 16, 18, 20–27, 33, 35, 37, 45–47, 65, 70, 72, 101, 103, 113, 121, 126–28, 132, 135, 140, 144, 198, 212, 219–20, 223, 245, 253–54, 260, 263, 269–70, 271–79; *Consumer Engineering* (Sheldon and Arens), 8–9, 27, 36, 95, 95, 100, 145, 158, 202
consumer motivations, 3, 5, 9, 12, 43, 45, 48, 53, 57, 67–68, 75–78, 80, 89, 95, 122–24, 255, 276
consumer psychology, 5–6, 9, 21–22, 32, 45–46, 70, 74–75, 89, 95–98, 112, 135, 155, 170, 186, 193, 209, 211, 213, 215, 265–66, 272, 276
consumer response design, 210, 214
consumers: African-American consumers, 23–24, 80; women as consumers, 2, 54, 86, 88, 121, 140, 256–57; youth market, 51, 60, 275
Container Corporation of America (CCA) (packaging), 152, 155–56, 183, 188, 268
Contempora Studio (design), 21

361

INDEX

Cornell University, 58, 85; School of Home Economics, 85
Corning Glass (housewares), 136, 148, 186, 231, 263
corporate identity, 151, 161, 201, 203, 205, 213–15, 223, 263
Cosmopolitan, 121
Council on Industrial Design, UK, 262
Coutant, Frank (market researcher), 112
Cowles Commission, 91
Craig, David (market researcher), 58
Cranbrook Academy, 143–45, 182, 190, 229–30, 232–33, 236
Crane, Edgar (marketing scholar), 96
creative research, 13, 115, 119, 124, 126, 132, 195
creative revolution (in advertising), 9–10, 13, 125, 135, 176, 275
Crisp, Richard (marketing scholar), 122
Cross, Gary (historian), 23
Crossley, Archibald (market researcher), 48, 61, 76
Cuba, 235, 238
cultural brokers/cultural translators/cultural translation, 15, 17, 19, 220, 243, 245–46, 254, 269, 278
cultural diplomacy, 237, 248–49, 262
cultural hybridity, 14, 278
Cunningham Stark, Inez (art patron), 183
Curtis Publishing Company (media), 48, 199

Daché, Lilly (designer), 344
Daily Mirror, 163
Daimler Benz (automobiles), 239, 252
Dartmouth College, 201
De Bijenkorf (retail), 30
Denmark, 28, 228, 230; Danish modern, 261
department store, 23, 26–27, 30, 58, 131, 139, 154, 158–59, 166, 168, 170, 176–77, 184, 191, 195, 203, 226, 247, 261, 264
depth interview, 80, 119, 120, 122, 124, 210
depth psychology, 51, 66, 100–101
Deskey, Donald (designer), 137, 172
Dessau, 32, 137, 173–74, 177, 183
De Stijl, 30, 173
Detroit, 1, 148, 229, 234
Deutsche Bauzeitung, 222, 224, 240
Deutsche Volkswirt, Der, 90
De Vries, Jan (historian), 194

Dichter, Ernest (market researcher), 3, 10, 12, 15, 43, 45–46, 52, 63–67, 70, 72, 96, 98, 101, 114–28, 195, 209–11, 245, 248–59, 273–75, 278
DIVO Society (market research), 252
Doblin, Jay (designer), 183
Doering-Manteuffel, Anselm (historian), 17
Dohner, Donald (designer), 137
Domizlaff, Hans (marketer), 33–34, 213; Markentechnik, 33
Dorland (advertising), 153, 155
Dorner, Alexander (designer), 137, 142, 173
Douglas, Mary (anthropologist), 194
Douglas Aircraft Co. (aircraft), 155
Dow Chemical Co. (chemicals), 234
Doyle Dane Bernbach (DDB), 277
Dreyfuss, Henry (designer), 136, 179, 189, 204, 207
Dr. Oetker (foods), 30
Drucker, Peter (marketing scholar), 4
DuPont (chemicals), 35, 44, 60, 111, 202
Düsseldorf, 240, 255

Eames, Charles (designer), 145, 160, 182, 227, 229, 240
Earl, Harley (designer), 160, 197
Eastman Kodak (photography), 58, 60, 136, 183, 268
Economic Cooperation Administration (ECA), 248, 266–67
Econ-Verlag (publishing), 249
Ehrlich Anderson, Thelma (social scientist), 67
Ehrmann, Marli (designer), 144, 267
Eiermann, Egon (architect), 238
Einaudi, Mario (political scientist), 246
Einstein, Albert (physicist), 108
elites, 7, 11, 15–17, 19, 26, 29, 74, 161, 173, 184, 191, 195, 219–20, 223–24, 235, 238–39, 241–45, 254, 260, 262, 267–69, 282; elite migration, 15–17
Emergency Committee for Displaced Scholars, 86
Emil Busch (optics), 109
Emnid Institut (market research), 252
entrepreneurship, 7–11, 13, 59, 132; immigrant entrepreneur, 4, 15, 145, 183, 224; knowledge entrepreneur, 99–101, 107–8, 114–15, 121, 126–28, 149, 151, 254
Essen, 133, 162, 260, 262
Esslinger, Hartmut (designer), 278
European Economic Community, 247

European Society for Opinion and Market Research (ESOMAR), 72, 253
exile, 4, 16–17, 65, 83, 90, 173–76, 188, 222, 246–48, 272

fashion, 9, 11, 26–27, 37, 106, 135, 139, 146, 162, 165, 171, 174, 192, 215, 223, 229, 260, 273, 277
fashion intermediaries, 111, 161, 191, 195
Federal Bureau of Investigation (FBI), 191
Federated Department Stores (retail), 168
Festinger, Leon (psychologist), 84, 96
Field, Marshall, III, (publisher), 161, 176–77
Filene, Edward (entrepreneur), 166
Fiske, Marjory (social scientist), 62, 64
focused interview, 64, 77
focus group, 63, 67–68, 98, 119, 126, 209
forced migration, 15–16
Ford, Henry (entrepreneur), 7, 22
Ford Foundation, 19, 96, 246–47, 268
Fordism, 5, 8, 13, 21, 23, 29, 136, 166, 274, 276
Ford Motor Company, 24, 65, 76, 152; Edsel, 65, 127
Form, Die, 28
Fortune, 114, 122, 135–36, 151–52, 154, 264, 268
Foster, William (administrator), 267
frame of reference, 86, 89, 259
France, 4, 21, 28, 133, 143, 150, 157, 176, 232, 237, 244, 255, 262, 269, 273
Frankfurt, 84, 85, 90, 164–67, 169, 171, 173, 240, 247, 249, 252, 255, 266, 272
Frankfurt Kitchen, 166
Frankfurt School. *See* Institute for Social Research, Frankfurt (Frankfurt School)
Freeland, Willard (market researcher), 103
Freud, Sigmund (psychologist), 46, 50–51, 66, 70, 74, 95, 100, 119, 122, 124, 254
Fromm, Erich (psychologist), 96, 272–73
Fry, Maxwell (architect), 174
Fullerton, Ronald (historian), 123
functionalism, 11, 18, 27, 142, 147, 161, 164, 172, 183, 232; Zweckform, 167
Furniture Manufacturer, 186

Gallup, George (opinion researcher), 48, 61, 76, 81
Gardner, Burleigh (market researcher), 100, 123
Gasteiger, Nepomuk (historian), 253
gatekeeper, 83, 86, 89
Gaudet, Hazel (social scientist), 62, 67
Gebrauchsgraphik, 28, 31, 157
General Electric (electric), 35, 58, 146, 148, 152, 198, 202, 263, 268
General Foods (food), 58, 76
General Mills (food), 105–7
General Motors (automobiles), 44, 103, 197, 224, 263; Futurama display, 2, 136, 168, 195; Opel, 259
General Outdoor Advertising Co. (advertising), 181
General Panel Corporation (construction), 168, 176
General Services Administration (GSA), 235, 240
Genoa, 151
Georgia Institute of Technology, 190
German Design Council (Rat für Formgebung), 239, 261–62, 278
Germany, 4, 17, 21–22, 28, 30, 32–33, 41, 68, 75, 80, 85, 88–90, 109, 133, 143, 149, 153, 155, 157, 163–67, 172, 175, 185, 213, 222, 224–28, 237–40, 244–78; Ministry for Economics, 262; Ministry for Interior, 239; Ministry for Marshall Plan, 261
Gerstel, Fred (designer), 168
Gesellschaft für Konsumforschung (GfK) (market research), 32, 252–53
Gesellschaft für Marktforschung (GfM) (market research), 252
Gestalt psychology, 74–75, 84–86, 90–91, 95–96, 121, 124–25, 189, 201, 213, 276
Gestetner (office machines), 146
Gianninoto, Frank (designer), 264
Giedion, Sigfried (architecture historian), 165–66, 173, 186, 189
Giese, Fritz (economist), 32
Gimbel Bros. (retail), 202
Gladwell, Malcolm (journalist), 69
globalization, 14, 17, 220, 277–78
Godley, Andrew (historian), 11
Goffman, Erving (sociologist), 123
Goldsen, Rose (social scientist), 67
Goldsmiths College, 206
Good Packaging, 209, 211–12
Gothenburg, 253
Graf Goertz, Albrecht (designer), 265
Grand Rapids, 226, 235

INDEX

graphic design, 3, 26, 30–31, 132–33, 135, 151–53, 156, 161, 177, 189, 206, 213, 231, 264, 275, 277
Gray, Milner (designer), 206–7
Great Depression, 6, 8, 11, 21–24, 33, 36–37, 41, 44, 52, 75, 103, 124, 134–35, 137, 226
Greyhound Bus Co. (transportation), 196
Gropius, Ise (writer), 159
Gropius, Walter (architect), 142, 144–45, 154–55, 158–59, 164–68, 173–79, 182, 185, 186, 190–92, 205, 207, 230, 239, 241, 263–64
Grossman, Greta (designer), 267
Gruen, Victor (architect), 1–4, 205, 207, 232, 271

Hamburg, 239, 252
Hamburg-Amerikanische Packetfahrt-Actien-Gesellschaft (HAPAG), 165
Hannover, 142
Haraszty, Eszter (designer), 232
Harnoncourt, Rene (curator), 267
Harper, Marion (advertiser), 67–69
Harper's Bazaar, 146
Harrison, Wallace (designer), 230
Harvard Business Review, 57, 71
Harvard University, 61, 83, 145, 147, 182, 189, 230, 243; Harvard Business School, 55; Harvard School of Design, 138, 145, 165, 175, 198, 263
Harwood, John (historian), 202
Hayek, Friedrich (economist), 247
Hearst Sunday Papers (media), 76
Heinz Corporation (food), 234
Heller, Robert (designer), 137
Henkel (hygiene), 251–52
Henle Pope, Annemarie (art historian), 355
Henrion, Henri (designer), 152
Herman Miller Co. (furniture), 226–27, 231, 236–37
Herzog-Massing, Herta (market researcher), 46, 52–53, 62–72, 77, 96, 100, 106, 118, 122–23, 247, 253, 272
Hess, Eckhard (psychologist), 67
Hess, Herbert (marketing scholar), 113
Hewlett Packard (electronics), 208
Heythum, Antonin (designer), 263
high modernity, 6, 18, 273, 275–76
Hill & Knowlton (public relations), 278
Hochschule für Gestaltung, Ulm, 263

Hodgson Homes (construction), 257
Hoechst (chemicals), 239
Hoffmann, Josef (designer), 26, 28, 31
Hohlwein, Ludwig (designer), 31
Hollywood, 5, 29
home economics, 36, 85, 87
Hood Bassett, Harry (banker), 235
Hoover (electric), 137
Horkheimer, Max (social scientist), 62, 167, 177, 247, 272
Horney, Karen (psychologist), 96
Horwitt, Nathan (advertiser), 28
Hotchkiss, George (marketing scholar), 49
Hilbersheimer, Ludwig (architect), 263
Hine, Thomas (historian), 235
Hirschmann, Ira (retail manager), 27
Hudnut, Joseph (architect), 138, 190
Hughes, Gordon (market researcher), 105
humaneering, 9, 25, 100, 196, 202
Hungary, 3, 144, 178, 181, 188, 232
Hutchinson, Kenneth (marketing scholar), 113

Illinois Institute of Technology (IIT), 182–83, 230, 274
India, 267
Industrial Art Council, 27
industrial design, 2–3, 9–11, 26, 28, 131, 133–38, 142, 145–47, 158, 162, 164–65, 171, 173, 175, 183, 190–91, 196, 198–99, 204, 208–9, 211, 215, 223, 226, 227, 247, 249, 260–69
Industrial Design, 187
Industrial Designers Institute, 160
Industrial Designers Society of America, 137
Industrial Design Partnership (design), 206–7, 209
Industrieform e.V., 260, 262
Innes, Christopher (historian), 136
Institute for Advanced Study, Vienna, 247
Institute for Contemporary Art, Boston, 170
Institute for Experimental Psychology, Berlin, 75, 84–85
Institute for Motivational Research, 66, 115, 118–21, 209–10, 252, 273
Institute for Social Research, Frankfurt (Frankfurt School), 4, 62, 64–65, 68, 167, 220, 247, 252, 272
Institute for Social Research, Michigan, 93, 96–97, 274; Survey Research Center (SRC), 93, 96–97

INDEX

Institute für Demoskopie, Allensbach, 251–52
Institut für Marktpsychologie, Mannheim, 252
Institute of Design, Chicago, 138, 140, 144, 160, 176–88, 192, 225, 263, 273
Interiors, 235
International Business Machines Corporation (IBM), 106, 148, 263
International Design Conference, Aspen, 157, 268
International Harvester (machines), 214
International Marketing Institute, 209
Interpublic Group (advertising), 69, 277
Ireland, 265
Italy, 4, 26, 80, 145, 150–51, 223, 230, 232, 237, 246, 248–49, 256–57, 261, 264, 267, 277

Jacobson, Egbert (designer), 157, 161, 183, 189
Jahoda, Marie (sociologist), 51, 55, 65
Jantzen Knitting Mills (textiles), 225
Jeanneret, Pierre (designer), 230
Jenkins, John (psychologist), 58
Jensen, Gustav (designer), 28
Jewell, Edward (art critic), 27
Jewish emigrés, 2, 15, 34, 46, 50, 64, 109, 115, 151, 167, 205
J. H. Cross (advertising), 76
Johnson, Gerald (designer), 195
Jordan Marsh Co. (furniture), 169
Journal of Advertising Research, 45, 69, 111
Journal of Applied Psychology, 61
Journal of Marketing, 49–50
J. Walter Thompson (JWT) (advertising), 47, 58, 63, 71, 74, 76, 106, 136, 152, 155, 251–52, 254, 272, 277

Kahn, Ely Jacques (architect), 137, 141, 167
Kassarjian, Harold (marketing scholar), 95
Katona, George (economist), 3, 43, 75, 89–98, 127–28, 245, 248–53, 259–60, 274, 276
Katz, Jehuda (sociologist), 81, 83
Kaufmann, Edgar, Jr. (curator), 158–59, 236, 267
Kaufmann's Department Store (retail), 58, 184
Kawosky, Theodore (psychologist), 201
Kay, Herbert (market researcher), 209, 211
Keck, George (designer), 183, 185, 190

Kepes, György (designer), 156–57, 159, 179, 182–90, 213, 264, 267
Kersting, Walter (designer), 262
Keynesianism, 5, 18, 94
Key-Oberg, Rolf (designer), 267
Khrushchev, Nikita (politician), 266
Kienzle (clocks), 109
Kiesler, Frederic (architect), 144, 172
Kimball, Abott (writer), 27
Kimberly-Clark (hygiene), 202
Klein Institute for Aptitude Studies, 118
Knoll, Florence (designer), 160, 223–41, 273
Knoll, Hans (entrepreneur), 15, 224–32, 235–41, 278
Knoll, Walter, 225, 237
Knoll Associates (furniture), 192, 220, 222–42, 263, 267, 269, 273; Knoll International, 222–23, 237–41, 261; Knoll Look, 223, 231–37; Knoll Textiles, 232; Planning Unit, 223, 232–37
knowledge industry, 10
knowledge transfers, 14–16, 45, 75, 95
Koffka, Kurt (psychologist), 84–85, 91, 96, 189
Köhler, Ingo (historian), 12, 276
Köhler, Wolfgang (psychologist), 84–85, 143
Kraft Foods (food), 177, 184
Kramer, Beate (designer), 167
Kramer, Ferdinand (designer), 15, 163–74, 187, 192, 247, 265–66, 272
Kropff, Hans (marketing scholar), 252–53
Krugman, Herbert (market researcher), 202
Krupp (steel), 239
Kuhler, Otto (designer), 143
Kunstgewerbeschule, Vienna, 137

Laboratory School of Industrial Design, 137
Lady's Home Journal, 23
Landor, Walter (designer), 10, 132, 145, 196, 205–16, 223, 266, 268, 278
Landor Associates (design), 205, 208–16, 241, 264–65, 278
Lasswell (communications scholar), 79, 82
Lazarsfeld, Paul (sociologist), 3, 12, 16, 42, 45–47, 50–72, 76–83, 91–92, 96–97, 100, 108–9, 155, 117–18, 122, 128, 144, 245–48, 252–55, 259, 272, 274, 276
Le Corbusier (architect), 28, 144, 172, 230, 231
Léger, Fernand (designer), 152–53, 162
Lehman Brothers (finance), 168

365

INDEX

Leichter, Käthe (social scientist), 50
Leipzig, 30
Leistikow, Hans (designer), 166
Lennen & Mitchell (advertising), 146
Lescaze, William (architect), 137, 142, 172
Leslie, Robert (designer), 153
Lever Bros. (hygiene), 76, 122
Levy, Sidney (sociologist), 123
Lewin, Kurt (psychologist), 43, 51, 75, 83–91, 93–98, 250, 253, 259, 272, 276
Lichtblau, Emil (designer), 268
Life, 64, 152, 154–55, 241
Likert, Rensis (sociologist), 58–59, 80, 84, 92–93
Lindner, Richard (designer), 159
Link, Henry (psychologist), 59, 76–77, 80
Lionni, Leo (designer), 151–52, 156, 161, 248
Lippincott, Gordon (designer), 146
Lippincott & Margulies (design), 213
Lissitzky, Eliezer (designer), 172
Listerine advertisements, 76
Loewy, Raymond (designer), 3, 10–11, 132–62, 171, 175, 183, 191, 195–205, 207–9, 213–16, 223, 241, 243, 245, 248–49, 260, 266–68, 277–78
Loewy Associates (design), 146–50, 160, 183, 196–205, 265, 268, 273
London, 29, 56, 59, 65, 109, 145–46, 174, 176, 182, 188, 206–7, 209, 228, 230, 247, 251, 265, 277–78
Lone Star Brewing Company (food), 214
Look, 170
Loos, Adolf (designer), 30
Lorenz, Konrad (zoologist), 51
Los Angeles, 137, 203, 271
Löwenthal, Leo (sociologist), 64, 272
Luce, Henry (publisher), 7, 135, 152, 154, 175
Lufthansa (airline), 255
Lynd, Helen (sociologist), 51
Lynd, Robert (sociologist), 51, 55, 58, 61
Lysinski, Edmund (psychologist), 32

MacAlister, Paul (architect), 160
machine age, 18, 26–27, 142, 144, 150, 166, 185, 194, 273
Machover, Karen (psychologist), 67
Macy's (retail), 26, 142
Madison Avenue, 42, 66, 69, 106, 111, 117, 136, 155, 226, 231, 251, 254, 277

MAN (automobiles), 239
Marcuse, Herbert (sociologist), 272
Market Analysts Inc. (market research), 117
marketing department, 9, 24, 42, 71, 95, 105, 108, 192, 274–75; in-house research, 62, 67, 76, 100, 103–5, 110, 118, 122, 197, 199, 200–201, 203, 209, 231
marketing management, 12, 25, 96, 108, 221, 250, 274–75
market research, 3, 5–6, 9, 11, 12–15, 24–25, 32, 44–70, 76–77, 84, 92, 98, 99–108, 111–15, 118–21, 124, 126–28, 139, 149, 157, 170, 172, 182, 193–221, 245, 247, 250–55, 266, 274–78
Market Research, 48, 57
Market Research Corporation of America, 48
Market Research Council, 58, 119
Marschak, Jacob (economist), 91
Marshall Field & Co. (retail), 177, 186, 192
Marshall Plan, 14, 244, 266
Martineau, Pierre (market researcher), 66, 106, 123, 253
Maslow, Abraham (psychologist), 96, 202
Massachusetts Institute for Technology (MIT), 88, 93, 188, 190; Center for Advanced Visual Studies, 188, 190; Center for Group Dynamics, 88, 93
mass market, 4, 9–10, 23–26, 32–33, 130–31, 139, 145, 151, 161–63, 167, 173, 191, 196, 203, 223, 228, 235, 239, 241, 260, 270, 272
Matter, Herbert (designer), 161, 187, 231
May, Ernst (architect), 166–67, 172
Mayo, Elton (psychologist), 83
McCann-Erickson (advertising), 62, 67–69, 84, 106–7, 122, 210, 251–52, 254, 272, 274, 277; Jack Tinker Partner, 68; Marplan, 67, 69. *See also* Interpublic Group (advertising)
McClure's, 151
McCormick, Harold (entrepreneur), 171
McFadden (publishing), 64, 82
McGarry, Edmund (marketing scholar), 113
McGill University, 190
McGraw-Hill (publishing), 64
Mead, Margaret (anthropologist), 84
Meikle, Jeffrey (historian), 11, 142
Merchandise Mart, Chicago, 159
merchandising, 21, 25, 48, 105–7, 118, 138, 147, 196, 199, 202, 208, 211–12, 265
Merton, Robert (sociologist), 64, 74, 77, 80

INDEX

Metropolitan Museum of Art, New York, 26, 142
Metropolitan Opera, New York, 136
Meyer, Hannes (architect), 174
Mies van der Rohe, Ludwig (architect), 142, 144, 160, 182, 223–24, 230, 241, 263
Milan, 26, 230, 238, 266
Milk Research Council of New York, 60
Miller, George (psychologist), 83
Mobil Oil (energy), 111
modernism (in art), 19, 26, 27–28, 31, 50, 134–35, 141–45, 150, 155, 158–59, 172–76, 192, 205, 207, 220, 223–28, 241, 258, 260–62, 267, 269, 275, 277–78
Modern Packaging, 211
Moholy-Nagy, László (designer), 140–41, 160, 165, 172–92, 193, 198, 225, 230, 264, 277
Moholy-Nagy, Sibyl (art historian), 176–77, 179, 182, 225, 248
Monoprix (retail), 265
Montgomery Ward (retail), 161
Morgenstern, Oskar (economist), 52, 247
Moscow, 266
motivation(al) research, 12, 46, 65–66, 69–70, 77, 106–7, 114–23, 126, 209–10, 213, 220, 252–57, 274; costs, 119
Muller-Munk, Peter (designer), 137, 142–43, 263, 266–67
Munich, 31, 143, 145, 165, 205, 225, 266
municipal socialism, 50, 166
Musée des Arts Décoratifs, Paris, 240
Museum of Modern Art, New York (MoMA), 138, 141, 145, 152–54, 158–61, 175, 184, 190, 198, 236, 261–62, 267–68; Good Design exhibit, 158–60, 191, 236, 267
Musgrave, Richard (economist), 248
Muthesius, Hermann (architect), 31

Nakashima, George (designer), 230
Närmil (food), 54
Nast, Condé (publisher), 28, 146, 152
National Academy of Design, 226
National Biscuit Co. (Nabisco) (food), 202
National Broadcasting Company (NBC) (media), 64, 76
National Marketing Review, 49
National Research Council, Committee on Food Habits, 83, 86
National Retail Dry Goods Association, 161

National Socialism (Nazism), 2, 15–16, 22, 34, 42, 51, 85, 90, 97, 109, 149, 153, 167, 174, 225, 230, 244, 249–50, 252, 261–62
National Youth Administration, 60
Natzler, Otto (designer), 268
Nelson, George (designer), 138, 144, 157, 160, 198, 227
Netherlands, 262, 264
Neurath, Paul (sociologist), 64
Neutra, Richard (architect), 137, 144, 205, 269–71
Newark University, 60–63
New Deal, 6, 18, 33, 36, 42, 44–45, 47, 55, 59–60, 72, 75, 190, 192
New School for Social Research, 85, 90, 96; University in Exile, 90
New York, 2, 6, 17, 21, 26, 28, 50, 55, 58–60, 62–63, 67, 70, 72, 85, 90, 92, 100, 106, 109, 117–18, 136–38, 142–46, 150, 153–60, 165, 167–70, 175, 182, 186, 205, 207, 223–38, 247, 249, 255, 261, 272
New Yorker, 163–64, 168
New York Times, 26, 223, 233
New York University, 55, 58
Nitsche, Erik (designer), 161
Nixon, Richard (politician), 266
Noelle-Neumann, Elisabeth (social scientist), 251–52
Noguchi, Isamu (designer), 230
Noyes, Eliot (designer), 145–58
Nuremberg, 32, 252
N. W. Ayers (advertising), 151

obsolescence, 8–9, 13, 19, 21, 27, 89, 131, 135, 139–41, 149, 172, 215, 256, 258, 265, 269, 271
Odol (hygiene), 30
Office of Radio Research (OFF), 60–62, 67, 71
Ogilvy, David (advertiser), 48
Ogilvy & Mathers (advertising), 278
opinion leader, 67, 74, 81–82, 96–98, 276
Oscar Mayer Co. (food), 183, 200
Otto, Carl (designer), 261

packaging, 10, 23, 28, 30, 48, 111, 119, 132, 135, 145, 156, 187, 193, 199–201, 205–15, 241, 257, 264–65, 268; packaging design, 10, 111, 145, 187, 264–65, 268
Packaging Machine Manufacturers Institute, 212

367

INDEX

Packard, Vance (author), 43, 99–101, 114, 126–27, 131, 254, 258; *The Hidden Persuaders*, 99–100; *The Waste Makers*, 131
Paepcke, Walter (entrepreneur), 157, 161, 181–84, 268
Pahlmann, William (designer), 161, 189
panel survey, 61, 64–69, 77–78, 81, 98, 106–7, 109, 252
Papanek, Victor (designer), 273
Paris, 6, 26, 29, 52, 115, 117, 141, 146, 150, 161, 176, 231, 237, 240, 247, 255, 260, 265–67
Paris Exposition (1925), 26, 161
Parker Pen Co. (pens), 109, 181, 187
Parlin, Charles (marketing scholar), 49
Parzinger, Tommi (designer), 267
Patten, Ray (designer), 198
Paul, Bruno (designer), 26, 31, 142–43, 145
Pavlov, Ivan (psychologist), 47, 95
Pelikan (pens), 30
Pencil Points, 144
Pennsylvania Railroad (transportation), 146, 196
Perspektiven, 268
persuasion, 10, 42, 47, 78–89, 97, 100, 126, 128; group persuasion, 87, 89
Pfaff (sewing machines), 238, 261
Philadelphia, 28, 106, 152
Philip Morris (tobacco), 66, 208
Pietzsch, Martin (architect), 177
Pittsburgh, 58–59, 137, 142–43, 158, 184
Plakat, Das, 28, 31
Planck, Max (physicist), 108
Plan Ltd. (furniture), 225, 228
Platt, Joseph (designer), 204
Plaut, James (art historian), 268
Politz, Alfred (market researcher), 10, 15, 43, 45, 96, 101–2, 106–18, 121, 123–27, 139, 195, 252, 253, 259, 273
Politz, Martha (market researcher), 110
Ponti, Gio(vanni) (designer), 238
populuxe designs, 11, 194, 235, 262
Prague, 54
Pratt Institute, 137–38, 141, 183, 230, 248
Prestini, James (designer), 140
price controls, 79, 91
Printer's Ink, 24, 121
Proctor, Robert (historian), 23
Proctor & Gamble (hygiene), 66, 105, 110, 152
product personality, 123, 254

Products Marketing Corporation (design), 168
Program Surveys Division, 80, 91
Progressive Era, 16, 36
propaganda, 10, 34–35, 42, 74–75, 78–83, 91, 97, 128, 153, 167, 187
Psychological Corporation (market research), 48, 57–62, 76, 92, 103, 255
Psychologische Forschungen, 85
Publicis (advertising), 277
Public Opinion Quarterly, 60, 76, 80

Quelle GmbH (retail), 255
quota sampling, 111–12

radio, 4, 27, 34, 42, 52, 60–65, 76–80, 117, 135–36, 252, 269
Rainbelle (umbrella), 170–71
Rainier Brewing Company (food), 208
Rand, Paul (designer), 175
Rand McNally (publishing), 184
randomized sampling, 92, 102, 111–12, 252; area sampling, 92, 111–12
Rappaport, Maurice (psychologist), 209
Rapson, Ralph (designer), 229
Rasch (wallpaper), 238
RAVAG (media), 52, 61
Reckwitz, Andreas (sociologist), 12–13, 135, 149; paradox of innovation, 12–13, 275
Reemtsma (tobacco), 33, 251, 265
refugee, 2–3, 15–16. *See also* exile
Renault (automobiles), 255
Research Bureau for Retail Training, 58
Resor, Stanley (advertiser), 47, 49
retail, 1–3, 5, 12, 14, 21–24, 28, 30, 32, 34, 41, 57–59, 91, 103, 106, 112, 139–40, 146, 159–61, 170, 194, 196–97, 202–4, 210–15, 219, 223, 227, 250, 255, 265
Rhode Island School of Design, 137, 142, 173
Riesman, David (sociologist), 123, 273
Risom, Jens (designer), 161, 228–29, 237
Rochester Institute of Technology, 190
Rockefeller, Nelson (politician), 234, 236
Rockefeller Foundation, 16, 51, 55, 58, 60, 73, 76, 184
Rodes, Toby (manager), 238–39
Rodgers, Daniel (historian), 16
Rohde, Gilbert (designer), 137
Rolls Royce (automobiles), 239
Rome, 144, 208

INDEX

Roosevelt, Franklin D. (politician), 98
Roper, Elmo (pollster), 48, 76, 84, 109, 112, 118
Rosenthal (ceramics), 238, 260–61, 265
Rosenthal, Philip (entrepreneur), 262–63
Rowland, Allison (historian), 67
Rudofsky, Bernard (designer), 140–41, 159, 213, 273
Rudolf Mosse Annoncen-Expedition (advertising), 109
Russia, 28, 95, 149, 160, 173, 225

Saarinen, Eero (architect), 224, 229–31
Saarinen, Eliel (architect), 143, 229
Saarinen, Loja (designer), 144, 232, 236
Saatchi & Saatchi (advertising), 277
Sachplakat (object poster), 31
Safeway (retail), 208, 211, 215
Saint-Tropez, 150
Sakier, George (designer), 137–38, 140, 172
Saks Fifth Avenue (retail), 27, 146
S&W Foods (food), 208
San Francisco, 6, 182, 205–8, 231
San Francisco Museum of Art, 207
Saturday Evening Post, 23, 48
Scandinavia, 4, 223, 229, 232, 261, 277
Schaefer, Herwin (designer), 261–62, 268
Schawinsky, Xanti (designer), 170, 174, 180, 189
Schenectady, 58
Schumpeter, Joseph (economist), 7–8, 13, 37
scientific management, 8, 13, 24, 47
scientific marketing, 24, 35, 46, 47–49, 102, 108, 113, 125, 127, 249, 251, 273
Schlink, Frederick (consumer advocate), 36
Schlumbohm, Peter (designer), 143, 267
Schniewind, Carl (curator), 158
Schott (glass), 32, 179
Schulz, Richard (designer), 232
Schwartz, Frederic (historian), 173
Schwarzkopf, Stefan (historian), 126
Schwitters, Kurt (artist), 189
S. C. Johnson & Son (hygiene), 184
Scott, Walter D. (marketing scholar), 47, 49, 59
Sears Roebuck (retail), 23, 44, 146, 181, 184, 192
Seattle, 208
segmented marketing, 9–10, 33, 42, 45, 55, 102–3, 123, 132, 215, 219, 253, 276
Seidels Reklame, 30

Seitlin, Percy (advertiser), 155
Sert, Josep (designer), 172–73, 182
Seyffert, Rudolf (economist), 32
Shaw, Arch W. (marketing scholar), 47, 49
Sheldon, Roy (designer), 8–10, 20, 24–26, 35–37, 95, 100, 128, 145, 158
Shils, Edward (sociologist), 83, 123
shopping mall, 1, 271
Siemens (electric), 31, 251, 261
Simonson, Peter (historian), 63
Sinel, Joseph (designer), 28, 137
Smith, Zay (art director), 181
Smith College, 85
Social Democratic Party (SPD), Germany, 246
social engineering, 6, 18, 33–34, 36, 72, 74, 79, 95, 97, 165, 175–76, 189–92, 207–8, 219, 221, 273–74
social field, 43, 86, 89, 95, 68, 276
social reform, 4, 6, 18, 22, 33, 35–36, 42, 47, 55, 132, 164, 251, 279
Social Research Inc. (SRI) (market research), 100, 123, 252
Society for the Promotion of German-American Trade, 251
Society of Automotive Engineers, 148
Society of Industrial Designers (SID), 137–38, 147, 161–62, 190, 260
Sociometry, 120
Sommaripa, Alexis (manager), 44–45, 60
Sony (technology), 278
Sorenson, Abel (designer), 230
Southwick, Barbara (designer), 229
Spiegel, Bernt (market researcher), 252, 254, 276
Spiegel, Der, 239, 241
Spiegel Inc. (retail), 76, 180
Stahle, Norma (administrator), 177
Stam, Mart (architect), 166, 174
Standard Oil (energy), 76, 215
Stanford University, 85, 209
Stanton, Frank (manager), 45, 61, 117, 157, 234–36
Stanton-Lazarsfeld Program Analyzer, 61, 67
State University of Iowa, 84, 86, 88
Steinway (pianos), 136
Stimson, Henry (politician), 238
Stirling Getchell (advertising), 117
St. Louis, 188, 193
St. Louis Star Tribune, 193
Stockholm, 238, 240
Stolper, Gustav (publisher), 90

369

Stouffer, Samuel (communications scholar), 79
Strengell, Marinanne (designer), 144
Studebaker (automobiles), 148–49, 162
Stuttgart, 32, 54, 224–25, 228, 237, 239–40, 261, 263, 267–68
styling, 9, 11, 18, 21, 26–27, 48, 103, 131, 134–36, 139, 142, 146, 148, 187, 194, 196–97, 200, 215, 261–62, 272–73; streamlining, 18, 134–36, 142, 145, 149, 164, 172, 182, 193, 215
Suchard (food), 197
supermarket, 41, 131, 196, 205, 208–12, 250, 266, 268
survey research, 42, 60, 63, 67, 74, 79–80, 83, 93, 111–12, 198, 251; costs, 119; phone surveys, 61
Swainson, Anne (designer), 161
Swarthmore College, 85
Sweden, 35, 109, 140, 167
Swift, Harold (entrepreneur), 177
Swift & Co. (food), 103, 119, 200
Switzerland, 35, 144, 225, 229, 231, 247
Sylvania Electric Co. (electric), 184

tachistoscope (eye camera), 67, 201, 210
tastemakers, 11, 131, 161, 195, 223, 241, 274
Taut, Bruno (designer), 30
Teague, Walter (designer), 28, 136, 138–40, 144, 191, 204, 207
Technical University Munich, 165
Tecta (furniture), 239
Telesis Group, 207–8
thematic apperception test (TAT), 122
Theobald, Paul (publisher), 189
Thonet (furniture), 167
Tide—Newsmagazine for Advertising and Marketing, 56–57, 64, 111
Tiffany's (jewelry), 142, 241
Tigerman, Bobbye (historian), 233
Tillich, Paul (theologian), 22
Time, 64, 76, 191
Tobler (foods), 30
Tokyo, 208
Trans World Airlines (TWA) (transportation), 146, 214
Traveling Exhibition Service (TES), 249, 268
Troost (advertising), 255
Tupperware parties, 89
two-step flow of communication, 67, 81

Ullmann, Hedi (social scientist), 63
Ulm, 248, 263
UNESCO, 247
Union Carbide (chemicals), 268
United Airlines (transportation), 177, 181, 184, 192
United Service Organization (USO), 196, 228
United States: Armed Forces, 185; Bureau of Agricultural Economics (BAE), 91; Bureau of Foreign and Domestic Commerce, 59; Department of Agriculture (USDA), 80, 84, 155; Department of Commerce, 44, 94, 103, 199; Federal Reserve, 92–94, 234; Information Service (USIS), 228; Office of Civilian Defense, 143, 185; Office of Price Administration, 79, 92; Office of Strategic Services (OSS), 248; Office of War Information (OWI), 64, 79–80, 92; Postal Service, 214; Treasury Department, 92; War Department (Department of Defense), 64, 196
University of Berlin, 108, 142–43
University of Chicago, 67, 84–85, 91
University of Göttingen, 90
University of Michigan, 93, 274
University of Missouri, 252
University of North Carolina, 190
University of Oslo, 247
University of Pittsburgh, 58
University of Syracuse, 263
University of Vienna, 45, 51, 53
University of Wisconsin, 85
Urban, Joseph (designer), 136, 194
US Gypsum (gypsum), 186
US Package Designer Council, 264
US Steel (steel), 111, 215

Vallye, Anna (historian), 190
Van Doren, Harold (designer), 135, 137, 164, 198, 204
Verkaufspraxis, 33
Vershofen, Wilhelm (market researcher), 32
Vicary, James (psychologist), 127–28
Vienna, 2, 17, 29, 30, 32, 35, 45–47, 50–55, 59–63, 72, 95, 115, 137, 143–45, 225, 247, 255, 259, 266, 271; Arbeiterkammer, 51; Red Vienna, 35, 47, 51–52, 55, 66, 71
Vienna School (of market research), 42, 45–46, 55, 63–73, 74–76, 83, 92, 95, 113, 115

370

Vogt, Victor (publisher), 33
Vogue, 146, 152–53
Voice of America, 246
Volkswagen (automobiles), 167, 239, 257
von Allesch, Marianna (designer), 28
von Mises, Ludwig (economist), 52
von Nessen, Walter (designer), 28
von Saldern, Axel (designer), 263
von Zahn, Peter (journalist), 254

Wagenfeld, Wilhelm (designer), 32, 179, 238, 261
Walgreens (retail), 184
Wall Street Journal, 118, 122, 163
Walter Knoll Co. (furniture), 225, 237
Wanamaker's (retail), 154
Warburg, Anita (art patron), 207
Warner, Lloyd (sociologist), 123
War of the Worlds (H. G. Wells), 62
Watson, John (psychologist), 47, 57, 74, 77
Weaver, Henry (market researcher), 103, 197
Weber, Kem (designer), 26, 28, 137, 145, 205
Weimar, 32, 165, 173, 179, 190
Weintraub & Co. (advertising), 176
Wells, H. G. (author), 62
Werkbund, 28, 31–33, 132, 134, 143, 162, 164–67, 173, 177, 192, 205, 225, 228, 260–62, 283
Wertheimer, Max (psychologist), 84–85, 90–91, 96, 109, 125, 189
Western Electric (communications), 136
Westernization, 17, 220, 237, 245, 249

Westinghouse (electric), 148
Wiener Werkstätte, 32
Wiesen, Jonathan (historian), 34
Wildenhein, Frans (designer), 190
window dressing/shop windows
Wirtschaftspsychologische Forschungsstelle, 32, 45, 50–55, 57–58, 70, 83, 115
WMF (housewares), 238, 261
World Association of Public Opinion Research (WAPOR), 72
World's Fair, 1939–40, New York, 2, 4, 136, 154, 168, 195, 207, 231
World War I, 15, 21, 24, 26, 32, 34, 79, 81, 97, 146, 165, 173
World War II, 7, 10, 16, 42, 78–80, 102, 122, 188, 220, 226, 228, 243, 248; war bond drive, 80
Wrigley, William (entrepreneur), 177
W. S. Crawford (advertising), 206

Young & Rubicam (advertising), 56, 76, 251, 278
Yugoslavia, 247

Zeisel, Eva (designer), 69, 144, 159, 267, 274
Zeisel, Hans (market researcher), 45–46, 51, 59, 62–66, 69–70, 72, 106, 144
Zeisel, Ilse (social scientist), 67
Zepf, Toni (designer), 157
Ziegfeld Follies (entertainment), 136
Zündapp-Werke (motorcycles), 109
Zurich, 54, 143, 230, 238, 252, 255